Ferrante Neri, Carlos Cotta, and Pablo Moscato (Eds.)

Handbook of Memetic Algorithms

Studies in Computational Intelligence, Volume 379

Editor-in-Chief
Prof. Janusz Kacprzyk
Systems Research Institute
Polish Academy of Sciences
ul. Newelska 6
01-447 Warsaw
Poland
E-mail: kacprzyk@ibspan.waw.pl

Further volumes of this series can be found on our homepage: springer.com

Vol. 357. Nadia Nedjah, Leandro Santos Coelho,
Viviana Cocco Mariani, and Luiza de Macedo Mourelle (Eds.)
Innovative Computing Methods and their Applications to Engineering Problems, 2011
ISBN 978-3-642-20957-4

Vol. 358. Norbert Jankowski, Włodzisław Duch, and Krzysztof Grąbczewski (Eds.)
Meta-Learning in Computational Intelligence, 2011
ISBN 978-3-642-20979-6

Vol. 359. Xin-She Yang, and Slawomir Koziel (Eds.)
Computational Optimization and Applications in Engineering and Industry, 2011
ISBN 978-3-642-20985-7

Vol. 360. Mikhail Moshkov and Beata Zielosko
Combinatorial Machine Learning, 2011
ISBN 978-3-642-20994-9

Vol. 361. Vincenzo Pallotta, Alessandro Soro, and Eloisa Vargiu (Eds.)
Advances in Distributed Agent-Based Retrieval Tools, 2011
ISBN 978-3-642-21383-0

Vol. 362. Pascal Bouvry, Horacio González-Vélez, and Joanna Kolodziej (Eds.)
Intelligent Decision Systems in Large-Scale Distributed Environments, 2011
ISBN 978-3-642-21270-3

Vol. 363. Kishan G. Mehrotra, Chilukuri Mohan, Jae C. Oh, Pramod K. Varshney, and Moonis Ali (Eds.)
Developing Concepts in Applied Intelligence, 2011
ISBN 978-3-642-21331-1

Vol. 364. Roger Lee (Ed.)
Computer and Information Science, 2011
ISBN 978-3-642-21377-9

Vol. 365. Roger Lee (Ed.)
Computers, Networks, Systems, and Industrial Engineering 2011, 2011
ISBN 978-3-642-21374-8

Vol. 366. Mario Köppen, Gerald Schaefer, and Ajith Abraham (Eds.)
Intelligent Computational Optimization in Engineering, 2011
ISBN 978-3-642-21704-3

Vol. 367. Gabriel Luque and Enrique Alba
Parallel Genetic Algorithms, 2011
ISBN 978-3-642-22083-8

Vol. 368. Roger Lee (Ed.)
Software Engineering, Artificial Intelligence, Networking and Parallel/Distributed Computing 2011, 2011
ISBN 978-3-642-22287-0

Vol. 369. Dominik Ryżko, Piotr Gawrysiak, Henryk Rybinski, and Marzena Kryszkiewicz (Eds.)
Emerging Intelligent Technologies in Industry, 2011
ISBN 978-3-642-22731-8

Vol. 370. Alexander Mehler, Kai-Uwe Kühnberger, Henning Lobin, Harald Lüngen, Angelika Storrer, and Andreas Witt (Eds.)
Modeling, Learning, and Processing of Text Technological Data Structures, 2011
ISBN 978-3-642-22612-0

Vol. 371. Leonid Perlovsky, Ross Deming, and Roman Ilin (Eds.)
Emotional Cognitive Neural Algorithms with Engineering Applications, 2011
ISBN 978-3-642-22829-2

Vol. 372. António E. Ruano and Annamária R. Várkonyi-Kóczy (Eds.)
New Advances in Intelligent Signal Processing, 2011
ISBN 978-3-642-11738-1

Vol. 373. Oleg Okun, Giorgio Valentini, and Matteo Re (Eds.)
Ensembles in Machine Learning Applications, 2011
ISBN 978-3-642-22909-1

Vol. 374. Dimitri Plemenos and Georgios Miaoulis (Eds.)
Intelligent Computer Graphics 2011, 2011
ISBN 978-3-642-22906-0

Vol. 375. Marenglen Biba and Fatos Xhafa (Eds.)
Learning Structure and Schemas from Documents, 2011
ISBN 978-3-642-22912-1

Vol. 376. Toyohide Watanabe and Lakhmi C. Jain (Eds.)
Innovations in Intelligent Machines – 2, 2012
ISBN 978-3-642-23189-6

Vol. 377. Roger Lee (Ed.)
Software Engineering Research, Management and Applications 2011, 2011
ISBN 978-3-642-23201-5

Vol. 378. János Fodor, Ryszard Klempous, and Carmen Paz Suárez Araujo (Eds.)
Recent Advances in Intelligent Engineering Systems, 2011
ISBN 978-3-642-23228-2

Vol. 379. Ferrante Neri, Carlos Cotta, and Pablo Moscato (Eds.)
Handbook of Memetic Algorithms, 2012
ISBN 978-3-642-23246-6

Ferrante Neri, Carlos Cotta, and Pablo Moscato (Eds.)

Handbook of Memetic Algorithms

Editors

Dr. Ferrante Neri
University of Jyväskylä
Dept. of Mathematical Information
Technology
P.O. Box 35
FI-40014 Jyväskylä
Finland
E-mail: neferran@cc.jyu.fi

Dr. Pablo Moscato
University of Newcastle
School of Electrical Engineering &
Computer Science
University Drive
Callaghan NSW 2308
Australia
E-mail: moscato@cs.newcastle.edu.au

Dr. Carlos Cotta
Universidad Málaga
Escuela Técnica Superior de Ingeniería
Informática
Campus de Teatinos, s/n
29071 Málaga
Spain
E-mail: ccottap@lcc.uma.es

ISBN 978-3-642-26942-4

ISBN 978-3-642-23247-3 (eBook)

DOI 10.1007/978-3-642-23247-3

Studies in Computational Intelligence

ISSN 1860-949X

© 2012 Springer-Verlag Berlin Heidelberg

Softcover reprint of the hardcover 1st edition 2012

This work is subject to copyright. All rights are reserved, whether the whole or part of the material is concerned, specifically the rights of translation, reprinting, reuse of illustrations, recitation, broadcasting, reproduction on microfilm or in any other way, and storage in data banks. Duplication of this publication or parts thereof is permitted only under the provisions of the German Copyright Law of September 9, 1965, in its current version, and permission for use must always be obtained from Springer. Violations are liable to prosecution under the German Copyright Law.

The use of general descriptive names, registered names, trademarks, etc. in this publication does not imply, even in the absence of a specific statement, that such names are exempt from the relevant protective laws and regulations and therefore free for general use.

Typeset & Cover Design: Scientific Publishing Services Pvt. Ltd., Chennai, India.

Printed on acid-free paper

9 8 7 6 5 4 3 2 1

springer.com

If the world doesn't adapt itself to you, you have to adapt yourself to it. (Gil Grissom)

Imagination could conceive almost anything in connection with this place. (Howard Phillips Lovecraft)

We must be the change we want to see in the world. (Mahatma Gandhi)

To my friends in Jyväskylä and worldwide, to my parents in Bari, global thanks for patience and support (Ferrante Neri)

To Rocío(s), Carlos and Alicia, the local-optimizers of my life (Carlos Cotta)

To those that in elementary schools teach our children about the power of evolution, and to those that use this power to make the world a better place (Pablo Moscato)

Preface

Memetic Algorithms (MAs) are computational intelligence structures combining multiple and various operators in order to address optimization problems. The diversity in the operator selection is at the basis of MA success and their capability of facing complex problems. Besides the details correlated to specific implementations, the importance and need of MAs is in the fact that they opened a new scenario in front of the scientific community. More specifically, MAs suggested to the computer science community that optimization problems can be more efficiently tackled by hybridizing and combining existing algorithmic structures rather than using existing paradigms. A crucially important contribution of MAs has been to offer a new perspective in algorithmic design. Before MA diffusion, the various paradigms were considered as "separated islands" to be elected as a solver for a given problem. On the contrary, MAs assume that a paradigm should not be necessarily selected. A solver can be generated by combining the strong points of various paradigms and obtaining a solver which is capable to outperform each paradigm, separately. This approach is the basics of the problem oriented algorithmic design which is, on one hand, the natural consequence of the No Free Lunch theorems, on the other hand, the founding concept for the automatic and real time design of problem solvers. The latter will likely be the future of computational intelligence as machines, in the future, will need to analyse and "understand" the problems before automatically proposing a suitable solver.

This book organizes, in a structured way, all the the most important results in the field of MAs since their earliest definition until now. This is one of the few books explicitly addressing MAs, algorithmic aspects, and specific implementations and is the only book which offers a systematic set of "recipes" to tackle, by means of memetic approaches, a broad set of optimization problems. Optimization in the presence of both discrete and continuous representation is analysed as well as constrained and multi-objective problems in both stationary case and in the presence of uncertainties. Each chapter describes the algorithmic solutions for facing one of the above-mentioned problems. A big emphasis is also given to the automatic coordination of algorithmic components by means of self-adaptive, co-evolutionary, and diversity-adaptive schemes. In addition, this book attempts to be self-consistent as it gives a description of a set of possible modules composing a MA. In addition, a set of successful examples in real-world applications in engineering is also given.

The book is structured in four parts. In the first part, containing Chapters 1–4, the basic concepts and elements composing Memetic Algorithms (MAs) are introduced. Chapter 1 defines basic concepts and definitions about optimization, complexity and metaheuristics. Chapter 2 describes the general structure and issues about Evolutionary Algorithms (EAs). General issues concerning the operation of EAs are discussed and different EA variants presented. Then, the problem of choosing the right EA instance, that is, the problem of designing and tuning evolutionary algorithms is presented. Chapter 3 defines Local Search and gives some examples of Local Search Algorithms by distinguishing the main algorithmic structures in continuous and combinatorial spaces. Chapter 4 gives a definition of MA, analyzes the reason of its success and distinguishes between MAs and Memetic Computing.

In the second part, containing Chapters 5–14, methodological aspects about algorithmic design and how to handle problem difficulties are studied. For each class of problems a review on the subject is given and some study cases are displayed for clarity. Chapter 5 discusses parametrization problems and balance of global and local search within MA frameworks. MAs in discrete and combinatorial optimization problems are analyzed in Chapters 6 and 7, respectively. Chapter 6 focuses on the design of semantic combination operators, development of dedicated local search procedures and management of population diversity. Other important issues, such as design of rich evaluation functions and constraint handling techniques, are also discussed. Two case studies with the purpose of showing how these issues can be effectively implemented in practice are also included in Chapter 6. Combinatorial problems and MA performance is the main focus of Chapter 7 where the concept of fitness landscapes is introduced and advanced fitness landscape analysis techniques are presented. A comparative analysis of the performance is carried out for the Travelling Salesman Problem an the Binary Quadratic Programming Problem. MAs for continuous optimization are presented in Chapter 8 after an overview on popular global optimizers for continuous problems. Particularities of memetic approaches for continuous optimization are highlighted in a novel taxonomy. Constrained optimization problems are addressed in Chapter 9 where a review on MAs for constrained problems is given and and two algorithmic implementations are presented in greater details. In the subsequent two chapters, the automatic coordination of local search components within evolutionary frameworks is discussed. Chapter 10 discusses diversity-based adaptive systems and focuses on fitness diversity techniques for adaptive MAs. A comparative analysis of recently proposed diversity metrics is also given. Chapter 11 presents recent research results about self-adaptive evolution of the memes and co-evolutionary MAs. It is shown how adaptive schemes containing local search information encoded within solutions and evolving in parallel populations connected to the population of solutions can lead to the design of flexible memetic frameworks. The chapter describes a framework for this research and previous findings with self-adaptive methods concerning representation and scalability. It then goes on to consider in more depth issues relevant to co-evolutionary systems such as credit assignment and the ratio of population sizes which can be thought of as the memetic "load" that an evolving population can support. Chapter 12 discusses another trending topic in MAs, namely the combination of MAs with complete techniques (i.e., techniques capable of provably finding the global optimum, or guaranteeing approximation bounds), and with incomplete variants thereof. The book also focuses on MAs for specific classes of optimization problems. Chapter 13 presents MA implementations for multi-objective optimization problems. Chapter 14 shows recent MA implementations for optimization problems in the presence of uncertainties.

The third part contains Chapter 15 and 16 and gives some examples of domain specific MA implementations and applications in given fields. Chapter 15 presents relevant MA applications in engineering and design while Chapter 16 summarizes the most significative applications of MAs in Bioinformatics.

Finally the fourth part, containing the epilogue of this book, Chapter 17 within the context of biographical notes and anecdotes, present the ideas that guided MAs at their earliest definition stage and how many open problems posed before can guide the future development of the field.

We wish to express our sincere gratitude for all the external contributors of this book who allowed us to produce a solid and high quality work covering a large spectrum in the field of MAs. Last but not least, we thank our families and friends for the constant support during the production of this volume.

Jyväskylä, Finland, June 2011   Ferrante Neri
Málaga, Spain, June 2011   Carlos Cotta
Newcastle, Australia, June 2011   Pablo Moscato

Contents

Part I: Foundations

1 Basic Concepts 3
Ferrante Neri, Carlos Cotta
   1.1   What Is Optimization? 3
   1.2   Optimization Can Be Hard 5
   1.3   Using Metaheuristics 6

2 Evolutionary Algorithms 9
Ágoston E. Eiben, James E. Smith
   2.1   Motivation and Brief History 9
   2.2   What Is an Evolutionary Algorithm? 10
   2.3   Components of Evolutionary Algorithms 12
        2.3.1   Representation (Definition of Individuals) 13
        2.3.2   Evaluation Function (Fitness Function) 14
        2.3.3   Population 14
        2.3.4   Parent Selection Mechanism 15
        2.3.5   Variation Operators 15
        2.3.6   Survivor Selection Mechanism (Replacement) 17
        2.3.7   Initialisation 17
        2.3.8   Termination Condition 18
   2.4   The Operation of an Evolutionary Algorithm 18
   2.5   Evolutionary Algorithm Variants 21
   2.6   Designing and Tuning Evolutionary Algorithms 25
   2.7   Concluding Remarks 27

3 Local Search 29
Marco A. Montes de Oca, Carlos Cotta, Ferrante Neri
   3.1   Basic Concepts 29
   3.2   Neighborhoods and Local Optima 30
   3.3   Classifications of Local Search 31
   3.4   Local Search in Combinatorial Domains 33

		3.4.1	Hill Climbing	34
		3.4.2	Simulated Annealing	35
		3.4.3	Tabu Search	36
	3.5	Local Search in Continuous Domains		36
		3.5.1	Classification of Local Search Techniques for Continuous Domains	37
		3.5.2	Commonly Used Local Search Techniques in Memetic Algorithms for Continuous Domains	38

4 A Primer on Memetic Algorithms 43
Ferrante Neri, Carlos Cotta

4.1	Introduction	43
4.2	The Need for Memetic Algorithms	44
4.3	A Basic Memetic Algorithm Template	46
4.4	Design Issues	49
4.5	Conclusions and Outlook	50
4.6	Memetic Algorithms and Memetic Computing	51

Part II: Methodology

5 Parametrization and Balancing Local and Global Search 55
Dirk Sudholt

5.1	Introduction	55
5.2	Balancing Global and Local Search	56
	5.2.1 Early Works and the Effect of Local Search	56
	5.2.2 Aspects That Determine the Optimal Balance	58
	5.2.3 How to Find an Optimal Balance	60
5.3	Time Complexity of Local Search	61
	5.3.1 Polynomial and Exponential Times to Local Optimality	62
	5.3.2 Intractability of Local Search Problems	63
5.4	Functions with Superpolynomial Performance Gaps	66
	5.4.1 Functions Where the Local Search Depth Is Essential	67
	5.4.2 Functions Where the Local Search Frequency Is Essential	69
5.5	Conclusions	71

6 Memetic Algorithms in Discrete Optimization 73
Jin-Kao Hao

6.1	Introduction	73
6.2	Survey of Memetic Algorithms for Discrete Optimization	74
	6.2.1 Rationale	74
	6.2.2 Memetic Algorithms in Overview	75
	6.2.3 Performance of Memetic Algorithms for Discrete Optimization	77

	6.3	Special Design Considerations	77
		6.3.1 Design of Dedicated Local Search	77
		6.3.2 Design of Semantic Combination Operator	81
		6.3.3 Population Diversity Management	83
		6.3.4 Other Issues	85
	6.4	Case Studies	86
		6.4.1 Graph Coloring Problems	87
		6.4.2 Maximum Parsimony Phylogeny	89
	6.5	Conclusions	93
7	**Memetic Algorithms and Fitness Landscapes in Combinatorial Optimization**		**95**
	Peter Merz		
	7.1	Introduction	95
	7.2	MAs in Combinatorial Optimization	95
		7.2.1 Combinatorial Optimization	96
		7.2.2 MA Outline	96
		7.2.3 Related Meta-Heuristics	98
	7.3	Why and When MAs Work	98
		7.3.1 The Concept of Fitness Landscapes	99
		7.3.2 NK-Landscapes	99
		7.3.3 Analysis of Fitness Landscapes	100
	7.4	Case Study I: The TSP	105
		7.4.1 Fitness Landscape	106
		7.4.2 State-of-The-Art Meta-Heuristics for the TSP	108
	7.5	Case Study II: The BQP	111
		7.5.1 Fitness Landscape	112
		7.5.2 State-of-the-Art Meta-Heuristics for the BQP	115
		7.5.3 A Memetic Algorithm Using Innovative Recombination	116
	7.6	Conclusion	119
8	**Memetic Algorithms in Continuous Optimization**		**121**
	Carlos Cotta, Ferrante Neri		
	8.1	Introduction and Basic Concepts	121
	8.2	Global and Local Continuous Optimization	122
	8.3	Global Optimization Algorithms	123
		8.3.1 Stochastic Global Search, Brute Force and Random Walk	124
		8.3.2 Evolution Strategy and Real Coded Evolutionary Algorithms	124
		8.3.3 Particle Swarm Optimization	127
		8.3.4 Differential Evolution	129
	8.4	Particularities of Memetic Approaches for Continuous Optimization	131

9	**Memetic Algorithms in Constrained Optimization**	135
	Tapabrata Ray, Ruhul Sarker	
	9.1 Introduction	135
	9.2 Constrained Optimization	136
	9.3 Classification of MAs	138
	9.4 MAs with Conventional Representation	138
	9.5 MAs with Alternative Representations	140
	9.6 Numerical Case Studies	142
	9.6.1 Case Study 1: Infeasibility Empowered Memetic Algorithm for Constrained Optimization Problems: MA with Conventional Representation	142
	9.6.2 Case Study 2: MA with Alternative Representation	147
	9.7 Summary and Conclusions	151
10	**Diversity Management in Memetic Algorithms**	153
	Ferrante Neri	
	10.1 Introduction	153
	10.2 Handling the Diversity of Memetic Algorithms: A Short Survey	154
	10.3 Fitness Diversity Adaptation	157
	10.3.1 Fitness Diversity Metrics	158
	10.3.2 Coordination of the Search: The "Natura non Facit Saltus" Principle	162
	10.4 Conclusion	164
11	**Self-adaptative and Coevolving Memetic Algorithms**	167
	James E. Smith	
	11.1 Introduction	167
	11.2 Background	168
	11.2.1 MAs with Multiple LS Operators	168
	11.2.2 Self-adaptation in EAs	169
	11.2.3 Co-evolutionary Systems	169
	11.3 A Framework for Self-adaption and Co-evolution of Memes and Genes	170
	11.3.1 Specifying Local Search	171
	11.3.2 Adapting the Specification of Local Search	171
	11.4 Test Suit and Methodology	173
	11.4.1 The Test Suite	174
	11.4.2 Experimental Set-Up and Terminology	174
	11.5 Self-adaptation of Fixed and Varying Sized Rules	175
	11.5.1 Self-adapting the Choice from a Fixed Set of Memes . . .	175
	11.5.2 Self-adaptation of Meme Definitions	176
	11.5.3 Results on Trap Functions	176
	11.5.4 Analysis of Results and Evolution of Rule Base	177
	11.5.5 Benchmarking the Self-adaptive Systems	178
	11.5.6 Summary of Self-adaptive Results	181

	11.6	Extension to True Co-evolution: the Credit Assignment	
		Problem	181
		11.6.1 Results: Reliability	182
		11.6.2 Results: Efficiency	184
	11.7	Varying the Population Sizes	185
	11.8	Conclusions	188

12 Memetic Algorithms and Complete Techniques 189
Carlos Cotta, Antonio J. Fernández Leiva, José E. Gallardo

	12.1	Introduction	189
	12.2	Background	190
	12.3	Classification of Hybridization Approaches	191
	12.4	Integrative Combinations	192
	12.5	Collaborative Combinations	195
	12.6	Conclusions	199

13 Multiobjective Memetic Algorithms 201
Andrzej Jaszkiewicz, Hisao Ishibuchi, Qingfu Zhang

	13.1	Introduction	201
	13.2	Basic Definition and Concepts	202
		13.2.1 Basic Concepts	202
		13.2.2 Aggregation Functions	204
		13.2.3 Weighted Sum Approach	204
		13.2.4 Tchebycheff Approach	205
	13.3	Adaptation of Memetic Algorithms for Multiobjective	
		Optimization – Basic Concepts	205
		13.3.1 Dominance-Based Evaluation Mechanisms	206
		13.3.2 Aggregation Function-Based Evaluation	
		Mechanisms	207
		13.3.3 Problem Landscapes in Multiobjective Optimization ..	208
		13.3.4 Archive of Potentially Pareto-optimal Solutions	209
		13.3.5 Evaluation of Multiobjective Memetic Algorithms	209
	13.4	Examples of Multiobjective Memetic Algorithms	210
		13.4.1 MOGLS of Ishibuchi and Murata	210
		13.4.2 M-PAES	210
		13.4.3 NSGA-II with LS	211
		13.4.4 MOGLS of Jaszkiewicz	211
		13.4.5 RM-MEDA	212
		13.4.6 MOEA/D	213
		13.4.7 MGK Population Heuristic	213
		13.4.8 Memetic Approach by Chen and Chen	214
		13.4.9 SPEA2 with LS	214
		13.4.10 Interactive Memetic Algorithm by Dias et al.	214
		13.4.11 SMS-EMOA with Local Search	214
	13.5	Implementation of Multiobjective Memetic Algorithms	214
	13.6	Conclusions	217

14	Memetic Algorithms in the Presence of Uncertainties	219
	Yoel Tenne	
	14.1 Motivation	219
	14.2 Uncertainty Due to Approximation	220
	14.3 Uncertainty Due to Robustness	224
	14.4 Uncertainty Due to Noise	229
	14.5 Uncertainty Due to Time-Dependency	232
	14.6 Conclusion	237

Part III: Applications

15	Memetic Algorithms in Engineering and Design	241
	Andrea Caponio, Ferrante Neri	
	15.1 Introduction	241
	15.2 Applications of MAs in Engineering Problems	242
	15.2.1 Engineering Applications in Single-Objective Optimization	242
	15.2.2 Engineering Applications in Multi-Objective Optimization	252
	15.3 A Study Case: The Fast Adaptive Memetic Algorithm	255
	15.3.1 An Insight into the Problem	255
	15.3.2 Fast Adaptive Memetic Algorithm	257
	15.4 Conclusions	259

16	Memetic Algorithms in Bioinformatics	261
	Regina Berretta, Carlos Cotta, Pablo Moscato	
	16.1 Introduction	261
	16.2 Microarray Data Analysis	262
	16.2.1 Clustering	264
	16.2.2 Feature Selection	265
	16.3 Phylogenetics	266
	16.4 Protein Structure Analysis and Molecular Design	267
	16.5 Sequence Analysis	269
	16.6 Systems Biology	270

Part IV: Epilogue

17	Memetic Algorithms: The Untold Story	275
	Pablo Moscato	
	17.1 Motivation, or Something Like That	275
	17.2 In the Beginning, There Was no Evolutionary Computation	276
	17.3 Caltech and the Red Door Cafe	281
	17.4 Landscapes and the Correlation of Local Optima	285
	17.5 Hierarchical Objective Functions and Memetic Algorithms That Run on a "Segment"	289
	17.6 A Royal Visit to Argentina	292

17.7	To Brazil, without the Beaches	296
17.8	Fixed-Parameter Tractability, and the Complexity of Recombination	299
17.9	Newcastle, Australia, and Biomedical Research Closer to the Beach	303
17.10	Future Opportunities (if We Constrain the Beast)	305

References 311

Author Index 361

Subject Index 363

List of Contributors

Regina Berretta
Centre for Bioinformatics, Biomarker
Discovery and Information-Based Medicine,
The University of Newcastle, University
Drive, Callaghan, NSW, 2308, Australia
e-mail: Regina.Berretta@
   newcastle.edu.au

Andrea Caponio
Technical University of Bari, Via E. Orabona
5, 70121 Bari, Italy
e-mail:
caponio@deemail.poliba.it

Carlos Cotta
ETSI Informática, Universidad de Málaga,
Campus de Teatinos,
29071 Málaga, Spain
e-mail: ccottap@lcc.uma.es

Ágoston E. Eiben
Free University,
Amsterdam, The Netherlands
e-mail: gusz@cs.vu.nl

Antonio J. Fernández Leiva
ETSI Informática, Universidad de Málaga,
Campus de Teatinos,
29071 Málaga, Spain
e-mail: afdez@lcc.uma.es

José E. Gallardo
ETSI Informática, Universidad de Málaga,
Campus de Teatinos,
29071 Málaga, Spain
e-mail: pepeg@lcc.uma.es

Jin-Kao Hao
LERIA, Université d'Angers, 2 Boulevard
Lavoisier,
49045 Angers Cedex 01, France
e-mail:
jin-kao.hao@univ-angers.fr

Hisao Ishibuchi
Department of Computer Science and
Intelligent Systems, Osaka Prefecture
University, 1-1 Gakuen-cho, Nakaku, Sakai,
Osaka 599-8531, Japan
e-mail:
hisaoi@cs.osakafu-u.ac.jp

Andrzej Jaszkiewicz
Poznan University of Technology, Institute
of Computing Science, Piotrowo 2, 60-965
Poznan, Poland
e-mail:
jaszkiewicz@cs.put.poznan.pl

Peter Merz
University of Kaiserslautern, Department of
Computer Science,

67653 Kaiserslautern, Germany
e-mail: peter.merz@ieee.org
and
University of Applied Sciences and Arts Hannover, Department of Computer Science and Business Administration, 30459 Hannover, Germany
e-mail: peter.merz@fh-hannover.de

Marco Montes de Oca
IRIDIA, CoDE, Université Libre de Bruxelles, Brussels, Belgium
e-mail: mmontes@ulb.ac.be

Pablo Moscato
Centre for Bioinformatics, Biomarker Discovery and Information-based Medicine, The University of Newcastle, University Drive, Callaghan NSW 2308, Australia
e-mail: Pablo.Moscato@newcastle.edu.au

Ferrante Neri
Department of Mathematical Information Technology, P.O. Box 35 (Agora), 40014, University of Jyväskylä, Finland
e-mail: ferrante.neri@jyu.fi

Tapabrata Ray
School of Engineering and Information Technology, University of New South Wales at Australian Defence Force Academy, Canberra ACT 2600, Australia
e-mail: t.ray@adfa.edu.au

Ruhul Sarker
School of Engineering and Information Technology, University of New South Wales at Australian Defence Force Academy, Canberra ACT 2600, Australia
e-mail: r.sarker@adfa.edu.au

James E. Smith
Department of Computer Science, University of the West of England, Bristol, BS16 1QY, UK
e-mail: james.smith@uwe.ac.uk

Dirk Sudholt
School of Computer Science, The University of Birmingham Edgbaston, Birmingham B15 2TT, UK
e-mail: d.sudholtcs.bham.ac.uk

Yoel Tenne
Department of Mechanical Engineering and Science-Faculty of Engineering, Kyoto University, Yoshida-honmachi, Sakyo-ku, Kyoto 606-8501, Japan
e-mail: yoel.tenne@
    ky3.ecs.kyoto-u.ac.jp

Qingfu Zhang
The School of Computer Science & Electronic Engineering University of Essex, Colchester, CO4 3SQ, UK
e-mail: qzhang@essex.ac.uk

Acronyms

ACA	Adaptive Checkers Algorithm
ACO	Ant Colony Optimization
ADM	Adaptive Dual Mapping
AES	Average Evaluation of Success
AGLMA	Adaptive Global-Local Memetic Algorithm
AnDE	Annealing Differential Evolution
ANN	Artificial Neural Network
B&B	Branch and Bound
BE	Bucket Elimination
BLX	Blended Crossover
BP	Backpropagation algorithm
BQP	Binary quadratic Programming
BS	Beam Search
CA	Checkers Algorithms
CDMOMA	Cross-Dominance Multi-Objective Memetic Algorithm
CE	Controlled Evaluation
CHC	Cross generational elitist selection, Heterogeneous recombination, and Cataclysmic mutation
CI	Computational Intelligence
CMA-ES	Covariance Matrix Adaptation Evolution Strategy
COMA	Co-evolutionary Memetic Algorithm
COP	Combinatorial Optimization Problem
ConOP	Constrained Optimization Problem
CVRP	Capacitated Vehicle Routing Problem
CCVRP	Cumulative Vehicle Routing Problems
DE	Differential Evolution
DFSS	Design For Six Sigma
DiBIP	Diversity-Based Information Preservation Crossover
DOR	Dynastically Optimal Recombination

DPX	Distance Preserving Crossover
DT	Discrete Tomography
EA	Evolutionary Algorithm
EC	Evolutionary Computation
EDA	Estimation of Distribution Algorithm
EIT	Electrical Impedance Tomography
EMDE	Enhanced Memetic Differential Evolution
EO	Extremal Optimization
EP	Evolutionary Programming
ET	Exact Technique
FAMA	Fast Adaptive Memetic Algorithm
FDA	Fitness Diversity Adaptation
FDC	Fitness Distance Correlation
FET	Full Employment Theorem
FPGA	Field Programmable Gate Array
FPT	Fixed-Parameter Tractable
GA	Genetic Algorithm
GARSS	Genetic Algorithm with Robust Selection Scheme
GCHC	Greedy Crossover Hill Climbing
GP	Genetic Programming
GPX	Greedy Partition Crossover
HC	Hill Climbing
H-IFF	Hierarchical-if-and-only-if
HJA	Hooke-Jeeves Algorithm
HK	Held-Karp bound
IDEA	Infeasibility Driven Evolutionary Algorithm
IEMA	Infeasibility Empowered Memetic Algorithm
ILK	Iterated Lin-Kernighan heuristic
IPE	Inexact Pre-Evaluation
k-COLOR	graph k-Coloring Problem
LKH	Lin-Kernighan Heuristic
LP	Linear Programming
LS	Local Search
LSD	Least Significant Difference
LTFE	Life Time Fitness Evaluation
MA	Memetic Algorithm
MaxCMO	Maximum Contact Map Overlap
MAX-SAT	Maximum Satisfiability
MC	Memetic Computing
MC-VRP	Multi-Compartment Vehicle Routing Problem
MDE	Memetic Differential Evolution
MOEA/D	Multi-Objective Evolutionary Algorithm based on Decomposition
MOGLS	Multi-Objective Genetic Local Search

MOMA	Multi-Objective Memetic Algorithm
MOO	Multi-Objective Optimization
MOP	Multi-objective Optimization Problem
MORA	Multi-Objective Rosenbrock Algorithm
MSE	Mean-Squared Error
MST	Minimum Spanning Tree
NEWUOA	NEW Unconstrained Optimization Algorithm
NFLT	No Free Lunch Theorem
NSGA	Non-dominated Sorting Genetic Algorithm
PALS	Problem-Aware Local Search
PBIL	Population-Based Incremental Learning
PCR	Polymerase Chain Reaction
PD	Proportional Derivative
PDMOSA	Pareto Dominance Multi-Objective Simulated Annealing
PF	Pareto Front
PI	Proportional Integral
PID	Proportional Integral Derivative
PLS	Polynomial Local Search
PMSM	Permanent Magnet Synchronous Motor
PNS	Progressive Neighborhood Search
PTAS	Polynomial Time Approximation Scheme
PSO	Particle Swarm Optimization
QAP	Quadratic Assignment Problem
QB	Queen-Bee algorithm
RBF	Radial Basis Function
REVAC	Relevance Estimation and VAlue Calibration
RM-MEDA	Regularity Model-Based Multi-objective Estimation of Distribution Algorithm
SA	Simulated Annealing
SIA	Swarm Intelligence Algorithm
SLS	Stochastic Local Search
SMHC	Steepest Mutation hill Climbing
SNP	Single Nucleotide Polymorphism
SNR	Signal-to-Noise Ratio
SOM	Self-Organizing Map
SPEA2	modified Strength Pareto Evolutionary Algorithm
SPMDE	Super Fit Memetic Differential Evolution
SPO	Sequential Parameter Optimization
SPOT	Sequential Parameter Optimization Toolbox
SQP	Sequential Quadratic Programming
SR	Success Rate
SS	Scatter Search
TR	Trust Region

TRI	Triggered Random Immigrants
TS	Tabu Search
TSP	Traveling Salesman Problem
UPGMA	Unweighted Pair Group Method with Arithmetic Mean
VEGA	Vector Evaluated Genetic Algorithm
VLS	Variable Local Search
VLSI	Very-Large Scale Integration
VRP	Vehicle Routing Problem
WCSP	Weighted constraint Satisfaction Problem

Part I
Foundations

Chapter 1
Basic Concepts

Ferrante Neri and Carlos Cotta

1.1 What Is Optimization?

In every day life, we always have to make decisions, e.g. the path to choose in order to go back home from work, the brand of milk in a supermarket, whether to watch football or a movie on TV, etc. Some of these choices appear to us obvious while some other choices require some thinking. Regardless of the context, decisions are usually made in order to reach a certain goal or satisfy a given necessity. For example, in the case of going back home from work, a reasonable goal would be to choose a path which leads us back home in the shortest possible time. Let us assume that the path should be performed by walking. In this case, the solution for the problem is likely to be the shortest path. This would be a simple optimization problem. If the goal would be to be at home at the earliest time after having bought something in the city center, e.g. a visit a shop, we have to exclude some of the possible paths. More specifically, we have to take into account only the paths which pass through the shop. The path having the latter features are said to be feasible while all the others are infeasible. The newly stated problem is a constrained optimization problem. If an additional goal, beside being back at home in the shortest possible time, is to take the opportunity for having some physical activities by means of a long walk, two conflicting objectives must be taken into account and a compromise must be accepted (e.g. a path that is not too long as to get home reasonably early but also not too short as to have at least some physical activity). Due to the presence of two simultaneous and conflicting goals, the latter is a multi-objective optimization problem.

Ferrante Neri
Department of Mathematical Information Technology, P.O. Box 35 (Agora), 40014,
University of Jyväskylä, Finland
e-mail: `ferrante.neri@jyu.fi`

Carlos Cotta
Dept. de Lenguajes y Ciencias de la Computación. Universidad de Málaga,
Campus de Teatinos, 29071 Málaga, Spain
e-mail: `ccottap@lcc.uma.es`

More mathematically, let us consider a solution x, i.e. a vector of n design variables $(x_1, x_2, \ldots, x_i, \ldots, x_n)$. Each of the design variable x_i can take values from a domain $\mathscr{D}_i$ (e.g., an interval $[x_i^L, x_i^U]$ if variables are continuous, or a certain discrete collection of values otherwise). The Cartesian product of these domains for each design variable is called the decision space $\mathscr{D}$. Let us consider a set of functions $f_1, f_2, \ldots, f_m$ defined in $\mathscr{D}$ and returning real values. Under these conditions, the most general statement of an optimization problem is given by the following formulas:

$$
\begin{aligned}
& Maximize/Minimize && f_m && m = 1, 2, \ldots, M \\
& subject-to && g_j(x) \leqslant 0 && j = 1, 2, \ldots, J \\
& && h_k(x) = 0 && k = 1, 2, \ldots, K \\
& && x_i^L \leqslant x_i \leqslant x_i^U && i = 1, 2, \ldots, n
\end{aligned}
\tag{1.1}
$$

where g_j and h_k are inequality and equality constraints, respectively.

From the definition above, we can easily see that if $m = 1$ the problem is single-objective, while for $m > 1$ the problem is multi-objective. The presence/absence of the functions g_j and h_k make the problem more or less severely constrained. Finally, the continuous or combinatorial nature of the problem is given by the fact that $\mathscr{D}$ is a discrete or dense set. In other words, all the problems considered in this book can be considered as specific cases of the general definition in equations (1.1).

In the continuous case, for each m the detection of a maximum or minimum point requires the detection of those points characterized by a null gradient, i.e.:

$$
\nabla f = \begin{bmatrix} \frac{\partial}{\partial x_1} \\ \frac{\partial}{\partial x_2} \\ \ldots \\ \frac{\partial}{\partial x_n} \end{bmatrix} = \bar{0}
\tag{1.2}
$$

In general, in a multidimensional continuous decision space $\mathscr{D}$, there are several points satisfying the condition in eq. (1.2). Some of these points are minima, some are maxima and some are saddle points. While solving an optimization problem, e.g., a minimization, it is fundamental to distinguish the three kinds of point. In order to distinguish them, the determinant of the Hessian matrix should be discussed. More specifically, the Hessian matrix is:

$$
H(x) = \begin{bmatrix} \frac{\partial^2 f}{\partial x_1^2} & \frac{\partial}{\partial x_1} \frac{\partial}{\partial x_2} f & \cdots & \frac{\partial}{\partial x_1} \frac{\partial}{\partial x_n} f \\ \frac{\partial}{\partial x_2} \frac{\partial}{\partial x_1} f & \frac{\partial^2 f}{\partial x_2^2} & \cdots & \cdots \\ \cdots & \cdots & \cdots & \cdots \\ \frac{\partial}{\partial x_n} \frac{\partial}{\partial x_1} f & \cdots & \cdots & \frac{\partial^2 f}{\partial x_n^2} \end{bmatrix}
\tag{1.3}
$$

In order to check whether a point x_0 is a minimum, a maximum, or a saddle point, the determinant Δ of the Hessian matrix must be checked. If

1 Basic Concepts   5

$$\Delta > 0 \text{ and } \frac{\partial^2 f}{\partial x_0^2} > 0, \tag{1.4}$$

x_0 is a local minimum; if

$$\Delta > 0 \text{ and } \frac{\partial^2 f}{\partial x_0^2} < 0, \tag{1.5}$$

x_0 is a local maximum; if $\Delta < 0$, x_0 is a saddle point.

The situation is much more subtle in the case of combinatorial domains, in which the notion of locality for optima is associated to a particular definition of neighborhood among the discrete elements in $\mathscr{D}$.

1.2 Optimization Can Be Hard

In real-world applications, it is usually not so important to detect local optima. The global optimum is usually of interest for engineers and practitioners. Thus, in principle, all the null gradient points should be detected and analyzed before selecting the global optimum. In practical problems, this set of operations is not always possible as often the objective function is not differentiable within the entire decision space, or is not even available in an explicit analytical form (being e.g. a procedure, a simulation, or an experiment measurement). In addition, it must be remarked that from an engineering/application viewpoint it is fundamental to detect a solution which displays a high performance and it is usually irrelevant whether or not this solution corresponds to a null gradient.

When regarded from a computational perspective, the above ideas can be characterized in terms of computational complexity. Assuming a certain computational framework (e.g., Turing machines), it is possible to measure the amount of resources (time or space to give two distinguished examples) that a certain algorithm requires in order to fulfill its objective, e.g., finding the global optimum for a certain optimization problem. By analyzing the growth of such resource consumption in terms of the size of the problem instance considered it is possible to define complexity classes of problems. More precisely, we can denote as **REC**$(f(n))$ the class of problems for which there exists an algorithm (not necessarily the same algorithm for all problems in the class) that solves any instance of size n using at most $f(n)$ units of resource REC. It is customary –yet sometimes unrealistic– to consider that a problem is tractable if it can be solved in polynomial time, i.e., if it belongs to class **TIME**(n^k) for some fixed k. In case of decision problems (those for which a yes/no response is sought), this definition amounts to the well-known class P.

Using the notion of reduction (an *efficient*[1] mechanism for transforming an instance of problem A into an instance of problem A'), we can define a problem A as **C**-hard if any problem in class **C** can be reduced to A (hence A is at least as hard to solve as any problem in **C**). If a problem is **C**-hard and also belongs to class **C**, it is

---

[1] The notion of efficiency here refers to the particular complexity class under consideration, e.g., polynomial time when studying classes in the polynomial hierarchy [701].

termed C-complete. Problems complete for a class are useful in characterizing the actual complexity of the class.

It turns out that many interesting problems are NP-complete when their decision version is considered, that is, they can be solved in polynomial time by a non-deterministic Turing machine (or alternatively, a yes-solution can be verified in polynomial time). Clearly, class P is a subset of class NP and, although yet unproven, it is widely believed that P is a proper subset of NP, i.e., P$\neq$NP. This means that no efficient –polynomial-time– algorithm is known to solve the problem to optimality. Furthermore, many real-world problems can also be shown to be hard to approximate, i.e., there exist no efficient algorithm capable of providing solutions whose quality is guaranteed to be within a certain distance of the optimum (several complexity classes can be defined in terms of the approximation ratios attainable [907]).

This complexity barrier can be dealt with using two different (and complementary) approaches. The first one is the use of parameterized complexity techniques. These techniques try to factor out some part of the problem input as a parameter k, and provide **TIME**$(f(k)n^c)$ (where c is a constant that does not depend on the parameter k and $f(\cdot)$ is an arbitrary function of k) algorithms for these problems. Assuming realistic instances of the problems would just exhibit low parameter values k, these algorithms turn out to provide efficient solutions to the problems under consideration (which are thus termed fixed-parameter tractable). The second potential approach is the use of metaheuristics, as discussed next.

1.3 Using Metaheuristics

When hypotheses on the optimization problem cannot be made, a general purpose optimization algorithm/procedure must be implemented for solving the problem or at least detecting some solutions with a high performance. General purpose algorithms are usually referred as metaheuristics from the ancient Greek words μετα and ευρισκω, i.e., literally "*I search beyond*" or more generally "*beyond the search*", in the sense that the search can be done at an abstract level to the result of another search procedure.

Metaheuristics have been developed during the last decades jointly with the progress of computational hardware and, nowadays, there exists a huge variety of general purpose optimization algorithms. Some of them get their inspiration from the nature, e.g. evolutionary principles, physical phenomena, animal behaviour, etc., in order to tackle the problem. These nature inspired methods are also known as Computational Intelligence Optimization algorithms since they use Computational Intelligence (CI) to face optimization problems. Traditionally, CI was identified as subject including Fuzzy Systems, Neural Networks and Evolutionary Computation. This definition appears today too restrictive and outdated, since other recently defined algorithmic structures, such as Swarm Intelligence, can also fit within CI. Amongst these emergent metaheuristics or, if we prefer, CI optimization algorithms, Memetic Algorithms (MAs) represent a successful story which developed during the last two decades and are year after year becoming an important CI paradigm which allows the solution of complex optimization problems.

This book attempts to explain in depth the algorithmic and implementation aspects of this paradigm, its variations in optimization problems under specific circumstances, some implementation in specific application domains, and finally the historical context where the terms have been coined and the early implementations have been performed. More specifically, this book is divided into four parts, the first about basic concepts and algorithmic components, the second is about specific MA implementations and problems, and the third part is about MA applications. Finally, the last part gives some historical background and biographical notes regarding the earliest definition of MAs.

Acknowledgements. F. Neri is supported by the Academy of Finland, Akatemiatutkija 130600, Algorithmic Design Issues in Memetic Computing. C. Cotta is partially supported by Spanish MICINN under project NEMESIS (TIN2008-05941) and by Junta de Andalucía under project TIC-6083.

Chapter 2
Evolutionary Algorithms

Ágoston E. Eiben and James E. Smith

2.1 Motivation and Brief History

Developing automated problem solvers (that is, algorithms) is one of the central themes of mathematics and computer science. Similarly to engineering, where looking at Nature's solutions has always been a source of inspiration, copying 'natural problem solvers' is a stream within these disciplines. When looking for the most powerful problem solver of the universe, two candidates are rather straightforward:

- the human brain, and
- the evolutionary process that created the human brain.

Trying to design problem solvers based on these answers leads to the fields of neurocomputing and evolutionary computing respectively. The fundamental metaphor of evolutionary computing (EC) relates natural evolution to problem solving in a trial-and-error (a.k.a. generate-and-test) fashion.

In natural evolution, a given environment is filled with a population of individuals that strive for survival and reproduction. Their fitness – determined by the environment – tells how well they succeed in achieving these goals, i.e., it represents their chances to live and multiply. In the context of a stochastic generate-and-test style problem solving process we have a collection of candidate solutions. Their quality – determined by the given problem – determines the chance that they will be kept and used as seeds for constructing further candidate solutions.

Surprisingly enough, this idea of applying Darwinian principles to automated problem solving dates back to the forties, long before the breakthrough of computers [270]. As early as in 1948 Turing proposed "genetical or evolutionary search"

Ágoston E. Eiben
Free University, Amsterdam, The Netherlands
e-mail: gusz@cs.vu.nl

James E. Smith
UWE, Bristol, UK
e-mail: James.Smith@uwe.ac.uk

Table 2.1. The basic evolutionary computing metaphor linking natural evolution to problem solving

EVOLUTION		PROBLEM SOLVING
environment	⟵⟶	problem
individual	⟵⟶	candidate solution
fitness	⟵⟶	quality

and already in 1962 Bremermann actually executed computer experiments on "optimization through evolution and recombination". During the sixties three different implementations of the basic idea have been developed at three different places. In the USA Fogel introduced evolutionary programming, [269, 271], while Holland called his method a genetic algorithm [325, 389, 645]. In Germany Rechenberg and Schwefel invented evolution strategies [761, 801]. For about 15 years these areas developed separately; it is since the early nineties that they are envisioned as different representatives ("dialects") of one technology that was termed evolutionary computing [32, 36, 37, 235, 596]. It was also in the early nineties that a fourth stream following the general ideas has emerged: Koza's genetic programming [41, 483]. The contemporary terminology denotes the whole field by evolutionary computing, or evolutionary algorithms, and considers evolutionary programming, evolution strategies, genetic algorithms, and genetic programming as sub-areas.

2.2 What Is an Evolutionary Algorithm?

As the history of the field suggests, there are many different variants of evolutionary algorithms. The common underlying idea behind all these techniques is the same: given a population of individuals within some environment that has limited resources, competition for those resources causes natural selection (survival of the fittest). This in turn causes a rise in the fitness of the population. Given a quality function to be maximised, we can randomly create a set of candidate solutions, i.e., elements of the function's domain, commonly called individuals. We then apply the quality function to these as an abstract fitness measure – the higher the better. On the basis of these fitness values some of the better individuals are chosen to seed the next generation. This is done by applying recombination and/or mutation to them. Recombination is an operator that is applied to two or more selected individuals (the so-called parents) producing one or more new candidates (the children). Mutation is applied to one individual and results in one new individual. Therefore executing the operations of recombination and mutation on the parents leads to the creation of a set of new individuals (the offspring). These have their fitness evaluated and then compete – based on their fitness (and possibly age)– with the old ones for a place in the next generation. This process can be iterated until an individuals with sufficient quality (a solution) is found or a previously set computational limit is reached.

There are two fundamental forces that form the basis of evolutionary systems:

- Variation operators (recombination and mutation) create the necessary diversity within the population, and thereby facilitate novelty.
- Selection acts as a force increasing the mean quality of solutions in the population. As opposed to variation operators, selection reduces diversity.

The combined application of variation and selection generally leads to improving fitness values in consecutive populations. It is easy (although somewhat misleading) to view this process as if evolution is optimising (or at least "approximising") the fitness function, by approaching the optimal values closer and closer over time. An alternative view is that evolution may be seen as a process of adaptation. From this perspective, the fitness is not seen as an objective function to be optimised, but as an expression of environmental requirements. Matching these requirements more closely implies an increased viability, which is reflected in a higher number of offspring. The evolutionary process results in a population which is increasingly better adapted to the environment.

It should be noted that many components of such an evolutionary process are stochastic. Thus, although during selection fitter individuals have a higher chance of being selected than less fit ones, typically even the weak individuals have a chance of becoming a parent or of surviving. During the recombination process, the choice of which pieces from the parents will be recombined is made at random. Similarly for mutation, the choice of which pieces will be changed within a candidate solution, and of the new pieces to replace them, is made randomly. The general scheme of an evolutionary algorithm is given in Fig. 1 in a pseudocode fashion.

Algorithm 1. The general scheme of an evolutionary algorithm in pseudocode

1	*INITIALISE population* with random individuals;
2	*EVALUATE* each individual;
3	**repeat**
4	*SELECT* parents;
5	*RECOMBINE* pairs of parents;
6	*MUTATE* the resulting offspring;
7	*EVALUATE* new individuals;
8	*SELECT* individuals for the next generation;
9	**until** *TERMINATION CONDITION is satisfied* ;

It is easy to see that this scheme falls into the category of generate-and-test algorithms. The evaluation (fitness) function represents a heuristic estimation of solution quality, and the search process is driven by the variation and selection operators. Evolutionary algorithms possess a number of features that can help to position them within the family of generate-and-test methods:

- EAs are population based, i.e., they process a whole collection of candidate solutions simultaneously.

- EAs mostly use recombination, mixing information from two or more candidate solutions to create a new one.
- EAs are stochastic.

The various dialects of evolutionary computing that we have mentioned previously all follow these general outlines, differing only in technical details as shown in the overview table (2.2) later on in this chapter. In particular, the representation of a candidate solution is often used to characterise different streams. Typically the representation (i.e., the data structure encoding an individual) has the form of; strings over a finite alphabet in genetic algorithms (GAs), real-valued vectors in evolution strategies (ESs), finite state machines in classical evolutionary programming (EP), and trees in genetic programming (GP). The origin of these differences is mainly historical. Technically, one representation might be preferable to others if it matches the given problem better; that is, it makes the encoding of candidate solutions easier or more natural. For instance, when solving a satisfiability problem with n logical variables, the straightforward choice is to use bit-strings of length n, hence the appropriate EA would be a genetic algorithm. To evolve a computer program that can play checkers, trees are well-suited (namely, the parse trees of the syntactic expressions forming the programs), thus a GP approach is likely. It is important to note that the recombination and mutation operators working on candidates must match the given representation. Thus, for instance, in GP the recombination operator works on trees, while in GAs it operates on strings. In contrast to variation operators, the selection process only takes fitness information into account, and so it works independently from the choice of representation. Therefore differences between the selection mechanisms commonly applied in each stream are a matter of tradition rather than of technical necessity.

2.3 Components of Evolutionary Algorithms

In this section we discuss evolutionary algorithms in detail. There are a number of components, procedures, or operators that must be specified in order to define a particular EA. The most important components, indicated by italics in Fig. 1, are:

- Representation (definition of individuals)
- Evaluation function (or fitness function)
- Population
- Parent selection mechanism
- Variation operators, recombination and mutation
- Survivor selection mechanism (replacement)

To create a complete, run-able, algorithm, it is necessary to specify each of these components and to define the initialisation procedure and a termination condition.

2.3.1 Representation (Definition of Individuals)

The first step in defining an EA is to link the "real world" to the "EA world", that is, to set up a bridge between the original problem context and the problem-solving space where evolution takes place. Objects forming possible solutions within the original problem context are referred to as **phenotypes**, while their encoding, that is, the individuals within the EA, are called **genotypes**. This first design step is commonly called **representation**, as it amounts to specifying a mapping from the phenotypes onto a set of genotypes that are said to represent them. For instance, given an optimisation problem where the possible solutions are integers, the given set of integers would form the set of phenotypes. In this case one could decide to represent them by their binary code, so for example the value 18 would be seen as a phenotype, and 10010 as a genotype representing it. It is important to understand that the phenotype space can be very different from the genotype space, and that the whole evolutionary search takes place in the genotype space. A solution – a good phenotype – is obtained by decoding the best genotype after termination. Therefore it is desirable that the (optimal) solution to the problem at hand – a phenotype – is represented in the given genotype space.

Within the Evolutionary Computation literature many synonyms can be found for naming the elements of these two spaces.

- On the side of the original problem context the terms **candidate solution**, phenotype, and **individual** are all used to denote points in the space of possible solutions. This space itself is commonly called the **phenotype space**.
- On the side of the EA, the terms genotype, **chromosome**, and again individual are used to denote points in the space where the evolutionary search actually takes place. This space is often termed the **genotype space**.
- There are also many synonymous terms for the elements of individuals. A placeholder is commonly called a variable, a **locus** (plural: loci), a position, or – in a biology-oriented terminology – a **gene**. An object in such a place can be called a value or an **allele**.

It should be noted that the word "representation" is used in two slightly different ways. Sometimes it stands for the mapping from the phenotype to the genotype space. In this sense it is synonymous with **encoding**, e.g., one could mention binary representation or binary encoding of candidate solutions. The inverse mapping from genotypes to phenotypes is usually called **decoding**, and it is necessary that the representation should be invertible so that for each genotype there is at most one corresponding phenotype. The word representation can also be used in a slightly different sense, where the emphasis is not on the mapping itself, but on the "data structure" of the genotype space. This interpretation is the one we use when, for example, we speak about mutation operators for binary representation.

2.3.2 Evaluation Function (Fitness Function)

The role of the **evaluation function** is to represent the requirements the population should adapt to. It forms the basis for selection, and so it facilitates improvements. More accurately, it defines what "improvement" means. From the problem-solving perspective, it represents the task to be solved in the evolutionary context. Technically, it is a function or procedure that assigns a quality measure to genotypes. Typically, this function is composed from a quality measure in the phenotype space and the inverse representation. To stick with the example above, if the task is to find an integer x that maximises x^2, the fitness of the genotype 10010 could be defined as the square of its corresponding phenotype: $18^2 = 324$.

The evaluation function is commonly called the **fitness function** in EC. This might cause a counterintuitive terminology if the original problem requires minimisation, because the term fitness is usually associated with maximisation. Mathematically, however, it is trivial to change minimisation into maximisation, and vice versa.

Quite often, the original problem to be solved by an EA is an optimisation problem. In this case the name **objective function** is often used in the original problem context, and the evaluation (fitness) function can be identical to, or a simple transformation of, the given objective function.

2.3.3 Population

The role of the **population** is to hold (the representation of) possible solutions. A population is a multiset[1] of genotypes. The population forms the unit of evolution. Individuals are static objects that do not change or adapt; it is the population that does. Given a representation, defining a population may be as simple as specifying how many individuals are in it, that is, setting the population size. Alternatively, in some sophisticated EAs a population has an additional spatial structure, defined via a distance measure or a neighbourhood relation. This may be thought of as akin to the way that "real" populations evolve within the context of a spatial structure dictated by the individuals' geographical position on earth. In such cases the additional structure must also be defined in order to fully specify a population. In contrast to variation operators, that act on one or more parent individuals, the selection operators (parent selection and survivor selection) work at the population level. In general, they take the whole current population into account, and choices are always made relative to what is currently present. For instance, the best individual *of a given population* is chosen to seed the next generation, or the worst individual *of the given population* is chosen to be replaced by a new one. In almost all EA applications the population size is constant and does not change during the evolutionary search.

The **diversity** of a population is a measure of the number of *different* solutions present. No single measure for diversity exists. Typically people might refer to the number of different fitness values present, the number of different phenotypes

---

[1] A multiset is a set where multiple copies of an element are possible.

present, or the number of different genotypes. Other statistical measures such as entropy are also used. Note that the presence of only one fitness value in a population does not necessarily imply that only one phenotype is present, since many phenotypes may have the same fitness. Equally, the presence of only one phenotype does not necessarily imply only one genotype. The converse is, however, not true: if only one genotype is present then this implies only one phenotype and fitness value are.

2.3.4 Parent Selection Mechanism

The role of **parent selection** or **mating selection** is to distinguish among individuals based on their quality, and in particular, to allow the better individuals to become parents of the next generation. An individual is a **parent** if it has been selected to undergo variation in order to create offspring. Together with the survivor selection mechanism, parent selection is responsible for pushing quality improvements. In EC, parent selection is typically probabilistic. Thus, high-quality individuals have more chance of becoming parents than those with low quality. Nevertheless, low-quality individuals are often given a small, but positive chance; otherwise the whole search could become too greedy and get stuck in a local optimum.

2.3.5 Variation Operators

The role of **variation operators** is to create new individuals from old ones. In the corresponding phenotype space this amounts to generating new candidate solutions. From the generate-and-test search perspective, variation operators perform the "generate" step. In principle, there is no restriction on how such variation operators work. The variation operators in the traditional EA dialects are usually divided into two types based on their **arity**, distinguishing unary and n-ary ($n > 1$) operators. Such a division can also be made for the newest members of the EA family, such as differential evolution [733] or particle swarm optimisation methods [457].

2.3.5.1 Mutation

A unary variation operator is commonly called **mutation**. It is applied to one genotype and delivers a (slightly) modified mutant, the **child** or **offspring**. A mutation operator is always stochastic: its output – the child – depends on the outcomes of a series of random choices. It should be noted that an arbitrary unary operator is not necessarily seen as mutation. For example, it might be tempting to use the term mutation to describe a problem-specific heuristic operator which acts on one individual[2]. However, in general mutation is supposed to cause a random, unbiased change. For this reason it might be more appropriate not to call heuristic unary operators mutation. The role of mutation has historically been different in various EC dialects. Thus, in genetic programming for instance, it is often not used at all, whereas in genetic algorithms it has traditionally been seen as a background

---

[2] Such operators are used frequently in memetic algorithms.

operator, used to fill the gene pool with "fresh blood", and in evolutionary programming it is the sole variation operator responsible for the whole search work.

It is worth noting that variation operators form the evolutionary implementation of elementary steps within the search space. Generating a child amounts to stepping to a new point in this space. From this perspective, mutation has a theoretical role as well: it can guarantee that the space is connected. There are theorems which state that an EA will (given sufficient time) discover the global optimum of a given problem. These often rely on this "connectedness" property that each genotype representing a possible solution can be reached by the variation operators [236]. The simplest way to satisfy this condition is to allow the mutation operator to "jump" everywhere: for example, by allowing that any allele can be mutated into any other with a nonzero probability. However, it should also be noted that many researchers feel these proofs have limited practical importance, and many implementations of EAs do not in fact possess this property.

2.3.5.2 Recombination

A binary variation operator is called **recombination** or **crossover**. As the names indicate, such an operator merges information from two parent genotypes into one or two offspring genotypes. Like mutation, recombination is a stochastic operator: the choices of what parts of each parent are combined, and how this is done, depend on random drawings. Again, the role of recombination differs between EC dialects: in genetic programming it is often the only variation operator, and in genetic algorithms it is seen as the main search operator, whereas in evolutionary programming it is never used. Recombination operators with a higher arity (using more than two parents) are mathematically possible and easy to implement, but have no biological equivalent. Perhaps this is why they are not commonly used, although several studies indicate that they have positive effects on the evolution [234].

The principle behind recombination is simple – by mating two individuals with different but desirable features, we can produce an offspring that combines both of those features. This principle has a strong supporting case – for millennia it has been successfully applied by plant and livestock breeders to produce species that give higher yields or have other desirable features. Evolutionary algorithms create a number of offspring by random recombination, and we hope that while some will have undesirable combinations of traits, and most may be no better or worse than their parents, some will have improved characteristics. The biology of the planet Earth, where with a *very* few exceptions lower organisms reproduce asexually, and higher organisms always reproduce sexually [569, 570], suggests that recombination is the superior form of reproduction. However recombination operators in EAs are usually applied probabilistically, that is, with a non-zero chance of not being performed.

It is important to remember that variation operators are representation dependent. Thus for different representations different variation operators have to be defined. For example, if genotypes are bit-strings, then inverting a 0 to a 1 (1 to a 0) can be used as a mutation operator. However, if we represent possible solutions by tree-like structures another mutation operator is required.

2.3.6 Survivor Selection Mechanism (Replacement)

The role of **survivor selection** or **environmental selection** is to distinguish among individuals based on their quality. In that, it is similar to parent selection, but it is used in a different stage of the evolutionary cycle. The survivor selection mechanism is called after the creation of the offspring from the selected parents. As mentioned in Sect. 2.3.3, in EC the population size is almost always constant, which means that a choice has to be made about which individuals will be allowed in to the next generation. This decision is often based on their fitness values, favouring those with higher quality, although the concept of age is also frequently used. In contrast to parent selection, which is typically stochastic, survivor selection is often deterministic. Thus, for example, two common methods are the fitness-based method of ranking the unified multiset of parents and offspring and selecting the top segment, or the age-biased approach of selecting only from the offspring.

Survivor selection is also often called **replacement** or the replacement strategy. In many cases the two terms can be used interchangeably, and so the choice of which to use is often arbitrary. A good reason to use the name survivor selection is to keep terminology consistent: steps 1 and 5 in Fig. 1 are both named selection, distinguished by an adjective. A preference for using replacement can be motivated if there is a large difference between the number of individuals in the population and the number of newly-created children. In particular, if the number of children is very small with respect to the population size, e.g., 2 children and a population of 100. In this case, the survivor selection step is as simple as choosing the two old individuals that are to be deleted to make places for the new ones. In other words, it is more efficient to declare that everybody survives unless deleted and to choose whom to replace. If the proportion is not skewed like this, e.g., 500 children made from a population of 100, then this is not an option, so using the term survivor selection is appropriate.

2.3.7 Initialisation

Initialisation is kept simple in most EA applications, the first population is seeded by randomly generated individuals. In principle, problem-specific heuristics can be used in this step, to create an initial population with higher fitness. Whether this is worth the extra computational effort, or not, very much depends on the application at hand. There are, however, some general observations concerning this issue based on the so-called anytime behaviour of EAs. These are discussed in Sect. 2.4.

2.3.8 Termination Condition

We can distinguish two cases of a suitable **termination condition**. If the problem has a known optimal fitness level, probably coming from a known optimum of the given objective function, then in an ideal world our stopping condition would be the discovery of a solution with this fitness, albeit perhaps only within a given precision $\varepsilon > 0$. However, EAs are stochastic and mostly there are no guarantees of reaching such an optimum, so this condition might never get satisfied, and the algorithm may never stop. Therefore we must extend this condition with one that certainly stops the algorithm. The following options are commonly used for this purpose:

1. The maximally allowed CPU time elapses.
2. The total number of fitness evaluations reaches a given limit.
3. The fitness improvement remains under a threshold value for a given period of time (i.e., for a number of generations or fitness evaluations).
4. The population diversity drops under a given threshold.

Technically, the actual termination criterion in such cases is a disjunction: optimum value hit *or* condition x satisfied. If the problem does not have a known optimum, then we need no disjunction. We simply need a condition from the above list, or a similar one that is guaranteed to stop the algorithm.

2.4 The Operation of an Evolutionary Algorithm

Evolutionary algorithms have some rather general properties concerning how they work. To illustrate how an EA typically works, we will assume a one-dimensional objective function to be maximised. Fig. 2.1 shows three stages of the evolutionary search, showing how the individuals might typically be distributed in the beginning, somewhere halfway, and at the end of the evolution. In the first phase, directly after initialisation, the individuals are randomly spread over the whole search space (Fig. 2.1, left). After only a few generations this distribution changes: because of selection and variation operators the population abandons low-fitness regions and starts to "climb" the hills (Fig. 2.1, middle). Yet later (close to the end of the search, if the termination condition is set appropriately), the whole population is concentrated around a few peaks, some of which may be suboptimal. In principle it is possible that the population might climb the "wrong" hill, leaving all of the individuals positioned around a local but not global optimum. Although there is no universally accepted definition of what the terms mean, these distinct phases of the search process are often categorised in terms of **exploration** (the generation of new individuals in as yet untested regions of the search space), and **exploitation** (the concentration of the search in the vicinity of known good solutions). Evolutionary search processes are often referred to in terms of a trade-off between exploration and exploitation. Too much of the former can lead to inefficient search, and too much of the latter can lead to a propensity to focus the search too quickly. **Premature convergence** is the well-known effect of losing population diversity too quickly, and

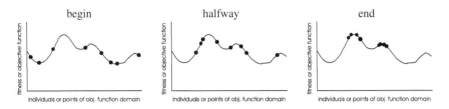

Fig. 2.1. Typical progress of an EA illustrated in terms of population distribution

getting trapped in a local optimum. This danger is generally present in evolutionary algorithms.

The other effect we want to illustrate is the **anytime behaviour** of EAs. We show this by plotting the development of the population's best fitness (objective function) value over time (Fig. 2.2). This curve is characteristic for evolutionary algorithms, showing rapid progress in the beginning and flattening out later on. This is typical for many algorithms that work by iterative improvements to the initial solution(s). The name "anytime" comes from the property that the search can be stopped at any time, and the algorithm will have some solution, even if it is suboptimal.

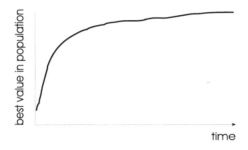

Fig. 2.2. Typical progress of an EA illustrated in terms of development over time of the highest fitness in the population

Based on this anytime curve we can make some general observations concerning initialisation and the termination condition for EAs. In Section 2.3.7 we questioned whether it is worth putting extra computational effort into applying intelligent heuristics to seed the initial population with better-than-random individuals. In general, it could be said that the typical progress curve of an evolutionary process makes it unnecessary. This is illustrated in Fig. 2.3. As the figure indicates, using heuristic initialisation can start the evolutionary search with a better population. However, typically a few (k in the figure) generations are enough to reach this level, making the worth of extra effort questionable in general.

The anytime behaviour also gives some general indications regarding the choice of termination conditions for EAs. In Fig. 2.4 we divide the run into two equally long sections. As the figure indicates, the progress in terms of fitness increase in

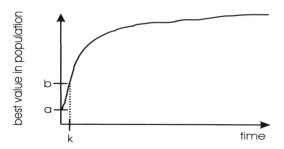

Fig. 2.3. Illustration of why heuristic initialisation might not be worth additional effort. Level a shows the best fitness in a randomly initialised population, level b belongs to heuristic initialisation

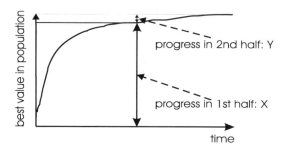

Fig. 2.4. Illustration of why long runs might not be worth performing. X shows the progress in terms of fitness increase in the first half of the run, while Y belongs to the second half

the first half of the run (X) is significantly greater than in the second half (Y). This provides a general suggestion that it might not be worth allowing very long runs. In other words, because of frequently observed anytime behaviour of EAs, we might surmise that effort spent after a certain time (number of fitness evaluations) are unlikely to result in better solution quality.

We close this review of EA behaviour by looking at EA performance from a global perspective. That is, rather than observing one run of the algorithm, we consider the performance of EAs for a wide range of problems. Fig. 2.5 shows the 1980s view after Goldberg [325]. What the figure indicates is that robust problem solvers –as EAs are claimed to be– show a roughly evenly good performance over a wide range of problems. This performance pattern can be compared to random search and to algorithms tailored to a specific problem type. EAs clearly outperform random search. In contrast, a problem-tailored algorithm performs much better than an EA, but only on the type of problem for which it was designed. As we move away from this problem type to different problems, the problem-specific algorithm quickly loses performance. In this sense, EAs and problem-specific algorithms form two opposing extremes. This perception played an important role in positioning EAs and stressing the difference between evolutionary and random search, but it gradually changed in the 1990s based on new insights from practise as well as from

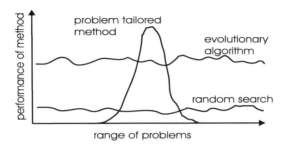

Fig. 2.5. 1980s view of EA performance after Goldberg [325]

theory. The contemporary view acknowledges the possibility of combining the two extremes into a hybrid algorithm. This insight is the main premise behind memetic algorithms that form the subject matter of the present book.

2.5 Evolutionary Algorithm Variants

Throughout this chapter we present evolutionary computing as *one* problem-solving paradigm, mentioning four historical types of EAs as "dialects". These dialects have emerged independently to some extent (except GP that grew out of GAs) and developed their own terminology, research focus, and technical solutions to realise particular evolutionary algorithm features. The differences between them, however, are not crisp – there are many examples of EAs that are hard to place into one of the historical categories. It is one of our main messages that such a division is not highly relevant, even though it may be helpful in some cases. Existing literature however, often uses the names of these dialects to position a particular method and we feel that a good introduction should also include some information about them. To this end, we provide a simple summary in Table 2.2.

It is worth to note that the borders between the four main EC streams have diminishined over the last decade. Approaching EAs from a "unionist" perspective it is better not to distinguish different EAs by the traditional stream they belong to, but by their main algorithmic components: representation, recombination operator, mutation operator, parent selection operator, and survivor selection operator. Reviewing the details of the commonly used operators and related parameters exceeds the scope of this chapter. Hence, we are forced to use (the names of) them without further explanation here and refer to a modern text book, such as [239] or [193], for those details. Table 2.3 provides an illustration showing how particular choices can lead to a typical genetic algorithm or evolution strategy, thus linking the two perspectives.

Considering Table 2.3, one may notice that it does not provide all details needed for a complete specification of an evolutionary algorithm. For instance, the population size is not specified. This observation raises the issue of algorithm parameters and, one step further, the issue of algorithm design.

Table 2.2. Overview of the main EA dialects

| Component | EA Dialect | | | |
or feature	GA	ES	EP	GP
Typical problems	Combinatorial optimisation	Continuous optimisation	Optimisation	Modelling
Typical representation	Strings over a finite alphabet	Vectors of real numbers	Appl. specific often as in ES	Trees
Role of recombination	Primary variation operator	Important, but secondary	Never applied	Primary/only variation operator
Role of mutation	Secondary variation operator	Important, sometimes the only operator	The only variation operator	Secondary, sometimes not used at all
Parent selection	Random, biased by fitness	Random, uniform	Each individual creates one child	Random, biased by fitness
Survivor selection	Random, biased by fitness	Deterministic, biased by fitness	Random, biased by fitness	Random, biased by fitness

Table 2.3. A typical GA and ES as an instantiation of the generic EA scheme

	GA	ES
Representation	bit-strings	real-valued vectors
Recombination	1-point crossover	intermediary
Mutation	bit-flip	Gaussian noise by $N(0, \sigma)$
Parent selection	2-tournament	uniform random
Survivor selection	generational	(μ, λ)
Extra	none	self-adaptation of σ

In the broad sense, algorithm design includes all decisions needed to specify an algorithm for solving a given (type of) problem. A decision to use evolutionary algorithms implies a general algorithmic framework – the one described in the beginning of this chapter. Using such an algorithmic framework implies that the algorithm designer adopts many design decisions (that led to the framework) and only needs to specify a "few" details. The principal challenge for algorithm designers is caused by the fact that the design details largely influence the performance of the algorithm. A well designed EA can perform orders of magnitude better than one based on poor choices. Hence, algorithm design in general, and EA design in particular, is an optimization problem itself, where the objective to be optimised is the performance of the EA.

As stated above, designing an EA for solving a given problem requires filling in the details of the generic EA framework appropriately. To denote these details one can use the term *EA parameters*. Using this terminology, designing an EA for a given application amounts to selecting good values for the parameters. For instance, the definition of an EA might include setting the parameter `crossoveroperator` to `onepoint`, the parameter `crossoverrate` to 0.5, and the parameter `populationsize` to 100. In principle, this is a sound naming convention, but intuitively, there is a difference between choosing a good crossover operator from a given list, e.g., {`onepoint, uniform, averaging`}, and choosing a good value for the related crossover rate $p_c \in [0, 1]$. This difference can be formalised if we distinguish parameters by their domains. The parameter `crossoveroperator` has a finite domain with no sensible distance metric or ordering, whereas the domain of the parameter p_c is a subset of IR with the natural structure for real numbers. This difference is essential for searchability of the design space. For parameters with a domain that has a distance metric, or is at least partially ordered, one can use heuristic search and optimization methods to find optimal values. For the first type of parameters this is not possible because the domain has no exploitable structure. The only option in this case is sampling.

The difference between these two types of parameters has already been noted in evolutionary computing, but various authors use various naming conventions. For instance, [47] uses the names *qualitative* and *quantitative* parameters respectively, [951] distinguishes between *symbolic* and *numeric* parameters, while [67] calls them *categorical* and *numerical*. Furthermore, [819] calls unstructured parameters *components* and the elements of their domains operators and in the corresponding terminology a parameter is instantiated by a value, while a component is instantiated by allocating an operator to it. In the context of statistics and data mining one distinguishes two types of variables (rather than parameters) depending on the presence of an ordered structure, but a universal terminology is lacking here too. Commonly used names are *nominal* vs. *ordinal* and *categorical* vs. *ordered* variables. Looking at it from a technical perspective, the very essence of the matter is the presence/absence of a (partial) ordering which is pivotal to searchability. This aspect is best captured through the names *ordered* and *unordered* parameters.

Table 2.4. Possible pairs of terms to distinguish the two types of EA parameters

Type I	Type II
qualitative parameter	quantitative parameter
symbolic parameter	numeric parameter
categorical parameter	numerical parameter
component	parameter
nominal variable	ordinal variable
categorical variable	ordered variable
unordered parameter	ordered parameter

For a clear distinction between these cases we propose to use the terms *qualitative parameter* and *quantitative parameter* and to call the elements of the parameter's domain *parameter values*.[3] In practice, quantitative parameters are mostly numerical values, e.g., the parameter crossover rate uses values from the interval $[0, 1]$, and qualitative parameters are often symbolic, e.g., `crossoveroperator`. However, in general, quantitative parameters and numerical parameters are not the same, because it is possible to have an ordering on a set of symbolic values - for example colours may be ordered by how they appear in the rainbow. Note that the terminology we propose here does not refer to the presence/absence of the (partial) ordering. In this respect, ordered vs. unordered could have been be better, but we prefer quantitative and qualitative for non-technical reasons, feeling that their use is more natural.

It is important to note that the number of parameters of EAs is not specified in general. Depending on particular design choices one might obtain different numbers of parameters. For instance, instantiating the qualitative parameter `parentselection` by `tournament` implies a new quantitative parameter `tournamentsize`. However, choosing for `roulettewheel` does not add any parameters. This example also shows that there can be a hierarchy among parameters. Namely, qualitative parameters may have quantitative parameters "under them". If an unambiguous treatment requires we can call such parameters *sub-parameters*, always belonging to a qualitative parameter.

Distinguishing qualitative and quantitative parameters naturally leads to distinguishing two levels in designing a specific EA for a given problem. In the resulting terminology we say that the high-level qualitative parameters define the EA, while the low-level quantitative parameters define a variant of this EA. Table 2.5 illustrates this matter.

Adopting this naming convention we can give a detailed answer to the question that forms the title of this chapter: What are Evolutionary Algorithms? An evolutionary algorithm is a partial instantiation of the generic EA framework where the values to instantiate qualitative parameters are defined, but the quantitative parameters are not. After specifying all details, including the values for all parameters, we obtain *an EA instance*. This terminology enables precise formulations, meanwhile it enforces care with phrasing. Clearly, this distinction between EAs and EA instances is similar to distinguishing problems and problem instances. For example, "TSP" represents the set of all possible problem configurations of the travelling salesman problem, whereas an instance is one specific problem, e.g., the 10 cities TSP with a given distance matrix D and Euclidean metric. If rigorous terminology is required then the right phrasing is "to apply an EA instance to a problem instance".

---

[3] Parameter values belonging to qualitative parameters, e.g., one-point-crossover, uniform-crossover, or tournament-selection, ranked-biased-selection, are usually called operators. This is fully consistent with our proposal here and can be seen as a matter of an additional naming convention.

2 Evolutionary Algorithms

Table 2.5. Three EA instances specified by the qualitative parameters: Representation, recombination, mutation, parent selection, survivor selection, and the quantitative parameters : mutation rate (p_m), mutation step size (σ), crossover rate (p_c), population size (μ), offspring size (λ), and tournament size. The EA instances in columns EA_1 and EA_2 are just variants of the same EA. The EA instance in column EA_3 belongs to a different EA.

	EA_1	EA_2	EA_3
Representation	bitstring	bitstring	real-valued
Recombination	1-point	1-point	averaging
Mutation	bit-flip	bit-flip	Gaussian $N(0,\sigma)$
Parent selection	tournament	tournament	uniform random
Survivor selection	generational	generational	(μ,λ)
p_m	0.01	0.1	0.05
σ	n.a.	n.a	0.1
p_c	0.5	0.7	0.7
μ	100	100	10
λ	n.a.	n.a.	70
tournament size	2	4	n.a.

2.6 Designing and Tuning Evolutionary Algorithms

As mentioned above, designing an good EA is in fact an optimisation problem. This problem is far from trivial, because there is very little known in general about the influence of EA parameters on EA performance. Most researchers and practitioners agree that the parameters of EAs interact with each other in a complex, non-linear way and even after 30 years of research there are only vague heuristics for designing a good EA instance for a given problem. In practice, the EA is often chosen intuitively or driven by habits, e.g., one may have a personal preference for GAs, while others' default could be ES. After that, parameter values are mostly selected by conventions (mutation rate should be low), ad hoc choices (why not use uniform crossover), and experimental comparisons on a limited scale (testing combinations of three different crossover rates and three different mutation rates).

Figure 2.6 shows the general scheme of the EA design process attempting to optimise algorithm performance on a given problem.[4] The designer is testing different parameter values, whose utility is determined by the performance of the corresponding EA instance on the given problem instance. Formally, such a design session is a trial-and-error (a.k.a. generate-and-test) procedure, resulting in specific values for the parameters of the EA in question. Given that all parameters of an EA must be specified before it can be applied, finding good parameter values is an absolutely necessary condition for any application, hence an immediate need for all researchers

---
[4] To be very precise: optimise the performance of an EA instance on a given problem instance.

Fig. 2.6. Illustration of 3-tier hierarchy behind EA design showing the control flow (left), and the information flow (right).

and practitioners. In this light, it is odd that the evolutionary computing community has not adopted algorithmic optimisers to solve the EA parameter tuning problem. Ironically, it has been noted long ago that the EA tuning problem falls in the problem class where EAs are claimed to be competitive solvers:

- the given problem has many parameters leading to a large search space,
- the problem has parameters of different types (e.g., reals, integers, symbolic values),
- there are complex non-linear interactions between the parameters leading to a complex non-linear objective function,
- the objective function has many local optima,
- there is noise in the data hindering exact calculations.

This insight has motivated the so-called meta-EAs, whose first representatives, meta-GAs, have been developed already in the late eighties to tune GA parameters [335]. However, meta-GAs or meta-ES [381] have never been used on a large scale. This really suboptimal situation is slowly changing over the last couple of years. The new development takes place along different research lines. First, meta-EAs are being "unearthed", enriched with additional features and tested for their ability to find good EA parameters, see, for instance, [951]. Another line of research concerns generic parameter tuners, developed and used to optimise EA parameters. The *Sequential Parameter Optimization Toolbox* (SPOT), [47, 48, 49] uses an iterative procedure, repeatedly testing parameter vectors and using the results to fit a model to predict the utility of other parameter vectors. Over the course of a run, SPOT simultaneously improves the prediction model and the parameter vectors. The *Relevance Estimation and VAlue Calibration* method (REVAC) implicitly creates probability distributions regarding the parameters (one probability distribution per parameter) in such a way that parameter values that proved to be good in former trials have a higher probability then poor ones. Initially, all distributions represent a uniform random variable and after each new test they are updated based on the new information. After terminating the tuning process, i.e., stopping REVAC, these distributions can be retrieved and analysed, showing not only the range of promising parameter values, but also disclosing information about the relevance of each parameter,

[650, 651, 819]. As for the (near) future, it seems safe to predict that the increasing maturity of such parameter tuners will lead to their adoption in the EC community. This, in turn, can increase the performance of EAs on a large scale and deliver novel insights and knowledge about the relationships between EA parameters and EA performance.

2.7 Concluding Remarks

We have described the basic evolutionary paradigm and how it encompasses a wide range of iterative population-based global search methods. Representatives from this class of methods have now been successfully applied to a huge range of different application domains as can be witnessed by the ever increasing volume of papers, conferences and journals. The prime difference between evolutionary and memetic algorithms (MAs) is that, as we have described them, EAs do not consider a step of self-improvement within the cycle - they just work on the outcome of randomised variation. In contrast Memetic Algorithms introduce a stage of individual (rather than population) learning, so that a solution (or its genotype) is (often systematically) perturbed and replaced by the new solution (or possibly its genotype) if that has higher fitness, *independently* of the rest of the population.

Chapter 3
Local Search

Marco A. Montes de Oca, Carlos Cotta, and Ferrante Neri

3.1 Basic Concepts

At an abstract level, memetic algorithms can be seen as a broad class of population-based stochastic local search (SLS) methods, where a main theme is "exploiting *all available knowledge* about a problem," see also Moscato and Cotta [618], page 105. The most wide-spread implementation of this theme is probably that of improving some or all individuals in the population by some local search method. This combination of a population-based, global search and a single-solution local search is a very appealing one. The global search capacity of the evolutionary part of a memetic algorithm takes care of exploration, trying to identify the most promising search space regions; the local search part scrutinizes the surroundings of some initial solution, exploiting it in this way. This idea is not only an appealing one, it is also practically a very successful one. In fact, for a vast majority of combinatorial optimization problems and, as it is also becoming more clear in recent research, also for many continuous optimization problems this combination leads to some of best performing heuristic optimization algorithms.

The role of the local search is fundamental and the selection of its search rule and its harmonization within the global search schemes make the global algorithmic success of memetic frameworks. The local search can be integrated within the

---

Marco A. Montes de Oca
IRIDIA, CoDE, Université Libre de Bruxelles, Brussels, Belgium
e-mail: mmontes@ulb.ac.be

Carlos Cotta
Department of Lenguajes y Ciencias de la Computación, University of Malaga, Spain
e-mail: ccottap@lcc.uma.es

Ferrante Neri
Department of Mathematical Information Technology, P.O. Box 35 (Agora), 40014, University of Jyväskylä, Finland
e-mail: ferrante.neri@jyu.fi

evolutionary cycle mainly in two ways. The first is the so called "life-time learning", i.e., the application of the local search to a candidate solution. In this case the metaphor is the cultural development of the individuals which is then transmitted to the other solutions over the subsequent generations. The second way is the application of the local search during the solution generation, i.e., the generation of a perfect child. This class of memetic implementations aims at selecting the most convenient offspring amongst the potential offspring solutions. This aim can be achieved, for example, by applying a local search to select the most convenient cutting point in a genetic algorithm crossover, or by using some kind of complete technique for this purpose [164, 165] – see Chapter 12.

In this chapter, we focus on the local search techniques and give a concise overview of local improvement methods both for combinatorial and for continuous optimization problems. Rather than giving a detailed account of all intricacies of local search algorithms, we focus on the main concepts and review some of the main local search variants. We start by first introducing some basic notions about neighborhoods and local optima. Next, we discuss the characterization of the local search and illustrate in details, some examples of local search algorithms in combinatorial and continuous domains. Without loss of generality, we assume here that we deal with minimization problems. Obviously, everything discussed here can easily be adapted to the maximization case.

3.2 Neighborhoods and Local Optima

The notion of neighborhood is essential to understand local search. Intuitively, a solution s' is termed a *neighbor* of s if the former can be reached from the latter in a single step (by the application of a so-called move operator). The neighborhood $\mathcal{N}(s)$ of a solution s is the set of all its neighbors. Notice firstly that neighborhood relationships are often –but not always– symmetrical. Secondly, the existence of a move operator allows not having to define neighborhoods explicitly (by enumerating the neighbors) but just implicitly (by referring to the potential transitions attainable upon application of the move operator). Moves can be typically regarded as modifications of some parts of a solution. Under an appropriate distance measure between solutions, these moves can thus be seen as "local", hence the name of the search paradigm. This said, the notion of closeness between neighboring solutions is not straightforward; as a matter of fact, neighborhood relationships can be quite complex – see [624].

Local search algorithms in combinatorial and continuous spaces have some intrinsic differences due to the differences in the types of the underlying search spaces. Let us denote the search space by S in the following. While combinatorial search spaces are finite for finite size problems, continuous search spaces are infinite and not enumerable. From these differences some other differences also result in the notions of local optima and the way how one is searching for improved candidate solutions. Intuitively, a local optimum is a solution for which in its local neighborhood no better solution exists.

3 Local Search

In combinatorial problems, the number of candidate solutions in the neighborhood of a current candidate solution s is enumerable and a local optimum can be defined as a a candidate solution s^l for which it holds that $\forall s \in \mathcal{N}(s^l)$ we have $f(s) \leq f(s^l)$, where $f : s \mapsto R$ is the objective function. It is also easy to identify whether a candidate solution is a local optimum, since one simply needs to enumerate all neighboring candidate solutions and check whether they are better or not than the current candidate solution. If none such exists, the solution is a local optimum. If the number of neighbors is polynomial in the instance size and the objective function is computable in polynomial time, which is the typical case for many neighborhood definitions and optimization problems, then this check can be also done in polynomial time.

For continuous optimization problems, the decision space is in principle a dense set and is thus composed of an infinite amount of points. This makes the enumeration impossible for the search of the optimum. It must be remarked that in the representation within a (finite) machine, dense sets cannot be represented with an infinite set and the sets are technically finite and the distance between each pairs of points is at least equal to the machine precision. An extensive description of this topic is given in Chapter 8, where continuous optimization problems are discussed. We can formally define a *local optimum* in a continuous space as a point $s° \in S$, such that

$$f(s°) \leq f(s), \tag{3.1}$$

for all points $s \in S$ that satisfy the relation $0 \leq ||s° - s|| \leq \varepsilon$. The set of points encircled in the region limited by the magnitude of ε is the *neighborhood* of the local optimum $s°$. Note that any global optimum is also a local optimum; however, the opposite is not necessarily true.

According to an alternative representation, in continuous optimization, a *global optimum* is a solution $s^\star \in S$, such that

$$s^\star = \underset{s \in S}{\operatorname{argmin}} f(s), \tag{3.2}$$

where $S \subseteq \mathbb{R}^n$ is the feasible search space, and $f : S \mapsto \mathbb{R}$ is the cost function to minimize. When $S = \mathbb{R}^n$ the problem of finding the optimum of f is called *unconstrained*, otherwise it is said to be *constrained*.

3.3 Classifications of Local Search

At a general level, for both combinatorial and continuous domains, local search algorithms can be classified from various perspectives:

1. According to the nature of the search logic:

    - **Stochastic:** the generation of the trial solution occurs in a randomized way
    - **Deterministic:** the generation of the trial solution is deterministic

2. According to the amount of solutions involved
   - **Single-solution**: the algorithm processes and perturbs only one solution
   - **Multiple-solution**: the algorithm processes more that one solution which are usually employed for interacting and jointly generate trial solutions
3. According to the pivot rule:
   - **Steepest Descent**: the algorithm generates a set of solutions and selects the most promising only after having explored all the possibilities
   - **Greedy**: the algorithm performs the replacement as soon as detects a solution outperforming the current best and starts over the exploration

On the basis of these classifications two important considerations must be carried out. First, every local search algorithm and, more generally, every optimization algorithm, can be seen as a logical procedure composed of two sets of operations: 1) generation of trial solutions 2) selection of trial solutions. In this light the first two classifications above characterize some crucial aspects of the trial solution generation while the latter regards the selection phase. Second, the classifications should not be considered in a binary way but more as properties of the procedure phases. For example, an algorithm is not either fully stochastic or fully deterministic but is likely to have a certain degree of stochastic logic and determinism. For two given algorithms a meaningful statement is to establish which among those two is "more" deterministic. To give a more concrete example, the simplex proposed in [653] cannot be considered a stochastic algorithm but due to the randomized re-sampling of the solutions is more stochastic than the Powell algorithm proposed in [728]. In a similar way, the classification according to the pivot rule must be interpreted as a property of the selection phase. An algorithm which explores/enumerates the entire neighborhood of a trial solution is possible as well as an algorithm which centers the search in the new current best solution. On the other hand, several intermediate possibilities can also be implemented and lead to efficient algorithms.

In this light, also the concepts of local and global searches should be seen as a progressive property of the algorithm. Intuitively, a local search is thought of as an algorithmic structure converging to the closest local optimum while the global search should have the potential of detecting the global optimum. De facto this definition is not rigorous and does not correspond to the actual behavior of the algorithms. For example, a hill-descender with a fully deterministic candidate solution generation can potentially converge to the global optimum of a highly multi-modal fitness landscape if a proper initial radius is set. Dually, an evolutionary algorithm with a population composed of a few individuals can converge to a suboptimal solution in the neighborhood of the best individual of the initial population. In this sense, an algorithm is not simply either global or local but can be characterized by a certain degree of local/global search features.

3.4 Local Search in Combinatorial Domains

Combinatorial spaces constitute a formidable and challenging application domain for local search techniques. Unlike continuous spaces in which neighborhood relationships are naturally defined in terms of Euclidean distance or any other suitable metric on $\mathbb{R}^n$, the definition of neighborhood in discrete domains is a major step in the resolution of the problem under consideration. Following the terminology used in forma analysis [750], we can consider situations in which the representation of solutions is orthogonal –i.e., solutions are represented as a collection of variables v_i, each of them taking values from a domain $\mathscr{D}_i$ and any combination of values for different variables being feasible– and situations in which the representation is non-orthogonal. In the former case the (feasible) search space S is

$$S = \prod_{i=1}^{n} \mathscr{D}_i, \tag{3.3}$$

i.e., the Cartesian product of variable domains. It is then possible to define neighborhood relationships on the basis of modifications of single variables. An illustrative example is that of binary strings, the typical representation in traditional genetic algorithms. Solutions thus belong to $\mathbb{B}^n$, and we can define a collection of nested neighborhoods $\mathscr{N}_i(s) = \{s' \mid H(s,s') = i\}$, where $H(s,s')$ is the Hamming distance between s and s'. This example is easily generalizable to integer representations, and can be further extended by considering the L_1 norm (Manhattan distance) or alike in case there was some locality between numerically close variable values.

The case of non-orthogonal representations is more complex since not every combination of variable values is feasible in them. This means that moves must be done solution-wise rather than variable-wise. A popular example is that of permutations: solutions are in this case arrangements of a collection of n objects (the integers $1, \cdots, n$ typically), and no single variable can be modified in isolation. There are numerous possibilities for perturbing permutations though. For example:

- *swap*: two elements are selected and interchanged.
- *insert*: an element is selected, removed from its location and inserted at another place.
- *invert*: a subsequence of adjacent elements is selected and their ordering is reversed.

just to name a few (please check chapter 7 of [764] for an overview of perturbation strategies for permutations in the context of the Traveling Salesman Problem). The situation can obviously be more complex in cases where other more sophisticated structures are used as representation of solutions, e.g., trees [19, 152, 723].

The neighborhoods sketched above are central in the functioning of single solution metaheuristics. These algorithms process one solution and after subsequent modifications returns a solution which is supposed to be similar to the starting solution but characterized by a higher performance. Within the context of MAs, these local searches process one solution within the evolutionary framework and improve during their "life-time".

Algorithm 2. A Local Search Algorithm

```
 1  Procedure Local Search (s);
 2  begin
 3      INITIALIZEMEMORY(M);
 4      repeat
 5          N ← PICKNEIGHBORHOODSTRUCTURE(M);
 6          s' ← PICKNEIGHBOR(N, s);
 7          SELECT(s, s', M);
 8          UPDATEMEMORY(s, M);
 9      until TERMINATIONCRITERION(M);
10      return s;
11  end
```

Algorithm 2 provides a rather general outline of a single-solution metaheuristic. This algorithm receives an initial solution and iteratively picks a neighbor and decides whether to accept this neighbor as the new current solution or not. This process can be modulated by a memory structure that the algorithm may use in order to decide which neighborhood should be used to select the neighbor, whether to accept the latter as new current solution or not, and even to support some high-level strategy for intensifying or diversifying the search.

One of the most distinctive features of local search strategies in combinatorial domains is the possibility of performing some kind of incremental evaluation of neighbors, i.e., computing $f(s')$ as $f(s') = f(s) + \Delta f(s, s')$, where $\Delta f(s, s')$ is a term that depends on the perturbation exerted on s to obtain s' and can be typically computed in a simple and efficient way. This often means that the cost of exploring the neighborhood of a solution is not much higher than a few full evaluations, thus allowing the practical use of intensive local search procedures. The following subsections give some examples of single-solution local search algorithms.

3.4.1 Hill Climbing

The simplest way to perform the local search in a combinatorial space is obtained by perturbing a trial solution and replacing it with the perturbed solution when the newly generate candidate solution outperforms the solution before perturbation. This procedure is termed Hill Climbing (HC) –or hill descending in a minimization context– and can be characterized in terms of the pseudocode depicted in Algorithm 2 as a memoryless, single-neighborhood local search procedure.

There can be several variants of the algorithm depending on, for instance, the pivot rule as mentioned in Sect. 3.3. Thus, we have steepest-ascent HC (an algorithm that explores the full neighborhood $N(s)$ of the current solution s, picks the best neighbor, accepts it if it is better than the incumbent) and random HC (an algorithm that picks a single random neighbor $s' \in N(x)$ and accepts it if is better than s). In the first case, the algorithm terminates upon finding a local optimum (i.e., a solution s which is better than any other $s' \in N(x)$) or exhausting its computational budget; in the second case only the computational limit applies, unless the algorithm keeps track of the neighbors generated and is capable of avoiding duplicates and/or detecting when the neighborhood is fully explored.

In some cases the neighborhood is very large and a full exploration is not possible (at least using the simple schema of HC – local branching procedures for exploring very large neighborhoods are possible though [263]). Such situations are usually dealt with by considering a random sampling of a certain size of the neighborhood, or by resorting to a simpler random HC. Another important issue is the presence of plateaus, i.e., when the best neighbor is neither better nor worse than the current solution. In that case the algorithm may opt for terminating, or may try to navigate through the plateau by accepting this best neighbor even if it is not strictly better than the incumbent. This is done for example in GSAT [804], a powerful solver for the MAX-SAT problem. Some strategy for avoiding cycling (i.e., oscillating search between a couple of solutions) may be required in this case. these are typically used in tabu search – see Sect. 3.4.3.

3.4.2 Simulated Annealing

While simple, the selection strategy depicted before for HC is unable to cope with rugged search spaces in which local optima are manifold, at least as an independent search technique. The search will terminate in a local optimum and it will have to be restarted from a different initial solution in order to locate other local optima (and eventually the global optimum). Such a restarting strategy is a simple way of endowing HC with global optimization capabilities, yet in some sense can be seen as a brute-force approach. More sophisticated strategies are however possible in order to escape from local optima. In particular, uphill moves (i.e., moves to worse solutions) must be at some point accepted. This is precisely the case of Simulated Annealing (SA) [122, 468]. SA is a single solution metaheuristic which performs the search of new solutions according to the logic described for HC but, in addition to performing the replacement when $f(s') \leqslant f(s)$, a worsening in the performance is accepted with a certain probability. This probability depends exponentially on the run-time, i.e., it is high at the beginning of the (local) optimization process and low at its end. The acceptance of only improvements may result into the fact that the algorithm gets stuck on a suboptimal solution. This condition is clearly undesired because if the algorithm does not succeed at improving its candidate solution for a large number of function calls, the algorithmic budget is wasted and the final solution is likely to have a poor performance. Thus, the main idea behind SA is to avoid such situation by refreshing the solution. The acceptance of a slightly worse solution should prevent the algorithm from getting stuck in some suboptimal solutions. To be precise, the neighboring configuration is accepted with probability P given by

$$P = \begin{cases} 1, & \text{if } \Delta f > 0 \\ e^{-\frac{\Delta f}{T}}, & \text{otherwise} \end{cases} \quad (3.4)$$

where T is a time-varying parameter termed *temperature*. This parameter modulates the acceptance probability, since the larger the temperature, the worse an acceptable neighbor can be. This parameter is decreased from its initial value T_0 to a final value $T_k < T_0$ via a process termed cooling schedule. There exist many cooling schedules in the literature, such as arithmetic, geometric or hyperbolic cooling [867]. In addition, there exist adaptive cooling strategies that take into account the evolution of the search, performing cooling and reheating as required, e.g., [243].

3.4.3 Tabu Search

Tabu Search (TS) is memory-based local search metaheuristic [309, 310, 317] with global optimization capabilities. It can be regarded a sophisticated extension of basic HC in which the best neighboring solution is chosen as the next configuration, even if it is worse than the current one. Notice that in case the neighborhood relationship is symmetric, it might happen that the best neighbor of solution s was s' and vice versa. In that case a memory-less search would simply cycle between these two solutions. TS avoids this by keeping a so-called *tabu list* of movements: a neighboring solution is accepted only if the corresponding move is not tabu. The actual meaning of a move being *tabu* may vary depending on the problem and designer's choice. Thus, it may be possible that it is tabu to restore a modified variable to a previous value, or it may be tabu to modify that variable at all (an analogous reasoning can be done when the moves are done solution-wise –e.g., as in permutations– rather than variable-wise). The tabu status of a move is not permanent: it only lasts for a number of search steps, whose value is termed *tabu tenure*. The tabu tenure can be fixed, or may vary during the search (even randomly within certain limits). The latter is useful to hinder long cycles in the search. In addition, it is common to consider an aspiration criterion that allows overriding the tabu status of a move, e.g., a move is accepted if it improves the best known solution even if it is tabu.

Algorithm 3 depicts the basic structure of a TS algorithm. It must be noted that a full-fledged TS algorithm is often endowed with additional strategies for intensifying and diversifying the search (e.g., frequency-based strategies that promote or penalize attribute values that occur frequently), and may also incorporate multiple neighborhoods among which the algorithm oscillates strategically.

3.5 Local Search in Continuous Domains

In this section, we turn our attention to optimization in continuous domains. In contrast to combinatorial optimization, where the search space is finite, in continuous optimization the search space is, in theory, infinite. Of course, in practice, the search space is also discrete as one is constrained by the precision of the representation of floating point numbers in the host computer. Nevertheless, its cardinality is so huge that for practical purposes it can be regarded as continuous.

Finding an extremum in a continuous domains is a different story. It is a very important task but it's difficult in general. The main difference with respect to the local

3 Local Search

Algorithm 3. Tabu Search

```
1  Procedure Tabu Search (s, 𝒩);
2  begin
3      INITIALIZETABULIST(M);
4      s* ← s;                                  // best known solution
5      repeat
6          L ←GENERATENEIGHBORLIST(𝒩,s);
7          s' ← PICKBEST(L);
8          if ISTABU(s',M) and not ASPIRATIONCRITERIA(s',s*) then
9              s' ← PICKBESTNONTABU(L)
10         endif
11         UPDATETABULIST(s,s',M);
12         s ← s';
13         if f(s*) > f(s) then
14             s* ← s
15         endif
16     until TERMINATIONCRITERION(M);
17     return s*;
18 end
```

search in discrete domain is the concept of gradient. More specifically, in continuous optimization, a local search can make use of gradient information in order to quickly descend the basin of attraction and thus support a global optimization framework which performs the selection of the search directions by means of the only fitness comparisons. For example, a trivial MA composed of an evolutionary framework and a hill-climber combines two alternative search logics: the fitness comparison of distant solutions generated within the decision space offered by the evolutionary framework and the gradient based search logic offered by the hill-climber. A proper combination of global and local search guarantees the success of a MA.

This section focuses on local search for continuous optimization problems. After some basic concepts and a classification of LS techniques for continuous domains, four popular algorithms are briefly described and a short survey on other techniques is given.

3.5.1 Classification of Local Search Techniques for Continuous Domains

The main criterion to classify local search methods for continuous domains is based on the order of the derivatives used for exploring the search space. Based on this criterion, optimization methods can be classified as follows:

Zeroth-order methods. Methods that belong to this class are called *direct search* methods. According to Trosset's definition [892], direct search methods are those that work with ordinal relations between objective function values. They do not use the actual values to model, directly or indireclty, higher order properties of the objective function.

First-order methods. This class of methods rely on direct access to the objective function and to the gradient vector at any point. Methods referred to as "derivative-free", belong to this category if the zeroth-order information is used to approximate higher order properties of the objective function.

Second-order methods. These methods use objective function values, (numerical approximations of) gradient vectors, and (numerical approximations of) Hessian matrices.

A second classification criterion is whether a method is stochastic or deterministic. Stochastic methods make randomized choices during their execution, while deterministic methods do not. These randomized choices encompass solution generation and/or solution selection [393]. In memetic algorithms, both types of techniques are used; however, deterministic methods are more commonly used. A brief description of some of these techniques is presented in the following section.

3.5.2 Commonly Used Local Search Techniques in Memetic Algorithms for Continuous Domains

In this section, we describe some of the most commonly used local search techniques in the literature of memetic algorithms for continuous optimization. Additionally, each of the algorithms that are described in detail belong to different strategies for local optimization, namely, downhill, gradient, quasi-Newton, and trust-region strategies.

3.5.2.1 Downhill Strategy: Simplex Method

This method, which was proposed by Nelder and Mead [653], is used for minimizing an n-dimensional objective function f. It is based on a *simplex*, which is composed of a set of $n+1$ points $\mathscr{P} = \{P_0, P_1, \ldots, P_n\}$ in the search space. At each iteration of the algorithm, the simplex can modified by at least one of three operations: *reflection*, *contraction*, and *expansion*. We denote P_h and P_l as the points with the highest and lowest objective function values, respectively. $\bar{P}$ is the centroid of the points P_i such that $f(P_i) < f(P_h)\ \forall P_i \in \mathscr{P}$. The reflection of P_h, denoted by $P^\star$, is defined as

$$P^\star = (1+\alpha)\bar{P} - \alpha P_h, \tag{3.5}$$

where $\alpha > 0$ is a parameter called *reflection coefficient*. If $f(P_l) < f(P^\star) < f(P_h)$, then P_h is replaced by $P^\star$, and the algorithm starts a new iteration. If $f(P^\star) < f(P_l)$, then $P^\star$ is expaned to $P^{\star\star}$ as follows

$$P^{\star\star} = \gamma P^\star + (1-\gamma)\bar{P}, \tag{3.6}$$

where $\gamma > 1$ is a parameter called *expansion coefficient*. If $f(P^{\star\star}) < f(P_l)$, then P_h is replaced by $P^{\star\star}$, and the algorithm starts a new iteration. However, if $f(P^{\star\star}) >$

$f(P_l)$, then it is said that the expansion operation failed and P_h is replaced by P^* before continuing to the next iteration.

If $f(P^*) > f(P_i) \; \forall P_i \in \mathscr{P} \setminus \{P_h\}$, then P_h is replaced by P^* only if $f(P^*) < f(P_h)$, otherwise P_h remains unchanged. After this operation, a new point P^{**} is generated as follows

$$P^{**} = \beta P_h + (1-\beta)\bar{P}, \tag{3.7}$$

where $0 < \beta < 1$ is a parameter called *contraction coefficient*. P_h is replaced by P^{**} unless $f(P^{**}) > min\{f(P_h), f(P^*)\}$, in which case all points P_i are replaced by $(P_i + P_l)/2$ before the next iteration begins.

The termination criterion can be either a minimum displacement in the search space, a minimum decrease in the objective function value, or a maximum number of iterations. This method has been used many times as a local search component of other algorithms (e.g., [123, 250, 383, 540, 607, 680, 719]).

3.5.2.2 Gradient Strategy: Powell's Direction Set Method

This method was proposed by M. J. D. Powell [728]. It tries to minimize an objective function $f : \mathbb{R}^n \to \mathbb{R}$ by constructing a set of *conjugate directions* through a series of line searches. Directions $v_i \; i \in \{1:n\}$ are said to be conjugate with respect to an $n \times n$ positive definite matrix A, if

$$v_i^T A v_j = 0, \quad \forall i, j \in \{1:n\}, \; i \neq j. \tag{3.8}$$

Furthermore, to be conjugate, directions $v_i \; i \in \{1:n\}$ must be linearly independent.

Conjugate search directions are attractive because if A is the Hessian matrix of the objective function, it can be minimized in exactly n line searches [731]. Although Powell's method does not need information about the objective function's derivatives, it can be considered a gradient strategy because it uses (implicitly) second order properties of the objective function [801].

The basic procedure of this method is the following: First, from an initial point $P_0 \in \mathbb{R}^n$, it performs n line searches using the unit vectors e_i as initial search directions u_i. At each step, the new initial point from which the next line search is carried out is the point where the previous line search found a relative minimum. A point P_n denotes the minimum discovered after all n line searches. Second, the method eliminates the first search direction by doing $u_i = u_{i+1} \; \forall i \in \{1:n-1\}$, and replacing the last direction u_n for $P_n - P_0$. Next, a move to the minimum along the direction u_n is performed.

Performing n iterations of the procedure described above, would minimize a quadratic objective function using a total of $n(n+1)$ line searches. However, under some circumstances, it is possible that the set of constructed directions become linearly dependent, which would make the algorithm fail. To prevent this from happening, after n or $n+1$ iterations, the set of search directions can be reset to either the original unit vectors, or to the columns of an orthogonal matrix (e.g., obtained by computing the principal components of the old direction set). Another approach is to substitute search directions in such a way as to maximize the determinant of the normalized direction vectors.

The method terminates when the magnitude of change of all variables is smaller than a predefined threshold. Powell's method has been used in numerous hybrid algorithms (e.g., [609, 680, 683, 766, 806]).

3.5.2.3 Quasi-Newton Strategy: Davidon-Fletcher-Powell Method

A Newton strategy uses the objective function's first and second order derivatives to rapidly optimize quadratic forms. In most practical cases, however, a quasi-Newton strategy is preferred. In a quasi-Newton strategy, the inverse of the objective function's Hessian is not computed directly, it is approximated from the objective function's gradient.

The Davidon-Fletcher-Powell method, also known as the variable metric strategy, was proposed by Fletcher and Powell [266], who based their proposal on Davidon's work [182]. It is an iterative procedure that works as follows. First, the user needs to specify an initial guess P_0 for the solution. An $n \times n$ matrix H_0 must also be initialized. Normally, $H_0 = I$. For any iteration k, we have

$$P_{k+1} = P_k - s_k H_k^T \nabla f(P_k), \tag{3.9}$$

where the step length s_k is computed by line search and it is the value that minimizes the objective function along the direction $-H_k^T \nabla f(P_k)$ from the current solution P_k.

The matrix H_k is updated as follows

$$H_{k+1} = H_k + \frac{y_k y_k^T}{y_k^T z_k} - \frac{H_k z_k (H_k z_k)^T}{z_k^T H_k z_k}, \tag{3.10}$$

where $y_k = P_{k+1} - P_k = -s_k H_k^T \nabla f(P_k)$, and $z_k = \nabla f(P_{k+1}) - \nabla f(P_k)$.

When the derivatives of the objective function are not available, the modifications introduced by Stewart [849] can be used. The Davidon-Fletcher-Powell method has been used as a local search component in hybrid local-global optimization algorithms, for instance in [29, 336, 510, 515].

3.5.2.4 Trust-Region Strategy: NEWUOA

In this category of methods, the objective function is approximated by a local model defined over a neighborhood of the current best solution. The quality of this model is trusted only in this neighborhood, therefore the name *trust region* [142].

A trust-region method works as follows: At any iteration, a model of the objective function is defined over a trust region of a certain radius. The trust region is centered on the current best-so-far solution P_k. A trial point s_k is then generated such that a new point $P_k + s_k$ sufficiently reduces the model, and it is still within the trust region. An evaluation of the objective function at $P_k + s_k$ is performed, and the value returned is compared to the prediction of the model. If the prediction is sufficiently accurate, the new point is accepted as the new best-so-far solution, and the next iteration is executed. In the next iteration, the trust-region radius can be larger or equal to the one used in the previous iteration. However, if the prediction is not sufficiently accurate, the new point is rejected and the next iteration is executed with a reduced trust-region radius.

Existing trust-region methods differ in the way they define the model of the objective function, the criteria used to accept a new solution, and the way they update both the model and the trust-region radius. Here, we briefly describe a state-of-the-art trust-region method for unconstrained continuous optimization. This method is called NEWUOA (NEW Unconstrained Optimization Algorithm) [730]. For approximating an n-dimensional objective function, NEWUOA uses a quadratic interpolation of $\mathcal{O}(n)$ points within the trust region (a common value being $2n+1$ points). It is possible to use a linear number of interpolation points if the Hessian of the model changes as little as possible from one iteration to the next. Once the model is computed, it is minimized using a truncated conjugate gradient method [952]. The point that minimizes the model is accepted if some conditions on the accuracy of the interpolation and the trust region size are met. Besides the number of interpolation points, the initial and final trust region radii are parameters of the method.

Trust-region methods have been applied in the context of global optimization in [682, 687, 878, 956]

3.5.2.5 Other Methods

The family of methods described in the previous section have been the most common choice for performing local search in memetic and other hybrid local-global search continuous optimization algorithms. However, in principle, any method that explores a solution's neighborhood can be used as a local search mechanism. In fact, there are several hybrid algorithms proposed in the literature that use algorithms that can be used as global optimizers. Interestingly, the main distinctive feature of these mechanisms is that they belong to the class of stochastic local search methods (cf. Section 3.5.1).

Examples of this approach are Solis and Wets' minimization technique [834], which is used, for example, in [383, 528, 607]. The covariance matrix adaptation evolution strategy [361] (a state-of-the-art continuous optimization technique at the time of writing) has been used as a local search method in [607, 608, 643]. A particle swarm optimization algorithm [457] has been used with genetic algorithms [326] to refine elite solutions [436]. Simulated annealing [468] has been also used as a local search method in [658]. As a final example, we want to mention the simultaneous perturbation stochastic approximation method [838], which has also been used as a gradient-estimation local search method in [515].

Acknowledgements. The authors thank Thomas Stützle for his help in early drafts of this chapter. This work was supported by the META-X project, an *Action de Recherche Concertée* funded by the Scientific Research Directorate of the French Community of Belgium, by Spanish MICINN under project NEMESIS (TIN2008-05941) and Junta de Andalucía under project TIC-6083, and by the Academy of Finland, Akatemiatutkija 130600, Algorithmic Design Issues in Memetic Computing.

Chapter 4
A Primer on Memetic Algorithms

Ferrante Neri and Carlos Cotta

4.1 Introduction

Memetic Algorithms (MAs) are population-based metaheuristics composed of an evolutionary framework and a set of local search algorithms which are activated within the generation cycle of the external framework, see [376]. The earliest MA implementation has been given in [621] in the context of the Travelling Salesman Problem (TSP) while an early systematic definition has been presented in [615]. The concept of meme is borrowed from philosophy and is intended as the unit of cultural transmission. In other words, complex ideas can be decomposed into memes which propagate and mutate within a population. Culture, in this way, constantly undergoes evolution and tends towards progressive improvements. Strong ideas tend to resist and be propagated within a community while weak ideas are not selected and tend to disappear. In the metaphor, the ideas are the search operators: the fittest tend to be employed while the inadequate ones are likely to disappear.

This chapter gives an initial description of MA frameworks explaining the literature context of their generation and success as well as their general structures. More specifically, Section 4.2 analyzes the context where MAs have been introduced and puts into relationship the algorithmic flexibility of the memetic paradigm with the the No Free Lunch Theorem. Section 4.3 shows the outline of a general MA implementation. Section 4.5 gives a quick overview on the MA application and employment in literature. Finally, Section 4.6 explains the difference between MAs and the general emerging trend of Memetic Computing.

Ferrante Neri
Department of Mathematical Information Technology, P.O. Box 35 (Agora), 40014, University of Jyväskylä, Finland
e-mail: `ferrante.neri@jyu.fi`

Carlos Cotta
Departamento de Lenguajes y Ciencias de la Computación, Escuela Técnica Superior de Ingeniería Informática, Universidad de Málaga, Campus de Teatinos, 29071 Málaga, Spain
e-mail: `ccottap@lcc.uma.es`

4.2 The Need for Memetic Algorithms

In order to understand in depth the role and need of MAs, it is fundamental to consider the historical context within which MAs have been defined. In 1988, when the first MAs were defined, Genetic Algorithms (GAs) were extremely popular among computer scientists and their related research was oriented towards the design of algorithms having a superior performance with respect to all the other algorithms present in literature. This approach is visible in many famous texts published in those years, e.g. [325]. Unlike all the algorithms proposed at that time, a MA was not a specific algorithm but was something much more general than an optimization algorithm: since MAs consist of the concept of combining global and local search algorithms, they represented a broad and flexible class of algorithms which somehow contained the previous work on Evolutionary Algorithms (EAs) and thus, constituted a new philosophy in optimization. Probably due to their excessively innovative contents, MAs had to face for about one decade, the skepticism of the scientific community which repeatedly rejected the memetic approach as a valuable possibility in optimization.

Since 1997, researchers in optimization had to dramatically change their view about the subject. More specifically, in the light of increasing interest in general purpose optimization algorithms, it has become important, in the end of 90's to understand the relationship between how well an algorithm a performs on a given optimization problem f on which it is run on the the basis of the features of the problem f. A slightly counter intuitive result has been derived by Wolpert and Macready in [940] which states that for a given pair of algorithms A and B:

$$\sum_f P(x_m|f,A) = \sum_f P(x_m|f,B) \tag{4.1}$$

where $P(x_m|f,A)$ is the probability that algorithm A detects the optimal solution for a generic objective function f and $P(x_m|f,B)$ is the analogue probability for algorithm B. In [940] the statement eq. 4.1 is proved for both static and time-dependent case and are named "No Free Lunch Theorems" (NFLT). In other words, in 1997 it was mathematically proved that the average performance of any pair of algorithms across all possible problems is identical. Thus, if an algorithm performs well on a certain class of problems then it necessarily pays for that with degraded performance on the set of all remaining problems as this is the only way that all algorithms can have the same performance averaged over all functions [940]. Strictly speaking, the proof of NFLT is made under the hypothesis that both the algorithms A and B are non-revisiting, i.e. the algorithms do not perform the fitness evaluation of the same candidate solution more often than once during the optimization run. Although this hypothesis is de facto not respected for most of the computational intelligence optimization algorithms, the concept that there is no universal optimizer had a significant impact on the scientific community.

It should be highlighted that a class of problems on which an algorithm performs well is not defined by the nature of the application but rather by the features of the fitness function within the search space. For example an optimization problem is characterized by:

- the shape and properties of a corresponding fitness landscape (see definitions below),
- multi-modality,
- separability of the problem,
- absence or presence of a noise in the values of the objective function (optionally, the type of noise),
- time dependency of the objective function (dynamic problems)
- shape and connectivity of the search domain

In evolutionary biology, the idea of studying evolution by visualizing the distribution of fitness values as a kind of landscape was first introduced by Wright [941].

More formally, the fitness landscape (S, f, d) of a problem instance for a given problem consists of a set of points S, a fitness function f which assigns values (fitness) to solutions from S, and a distance measure $d : S \times S \to \mathbb{R}$ which defines the spacial structure of the landscape. This rather abstract concept has proven to be useful for understanding the functionality of various optimization methods, see [581] and [583].

One of the most important properties of the fitness landscape is epistasis whose concept has been borrowed from biology where it refers to the degree to which the genes are correlated. As it is well known, a function is separable if it can be rewritten as a sum of functions of just one variable. The separability is closely related to the concept of epistasis. In the field of evolutionary computation, the epistasis measures how much the contribution of a gene to the fitness of the individual depends on the values of other genes. Nonseparable functions are more difficult to optimize as the accurate search direction depends on two or more genes. On the other hand, separable functions can be optimized for each variable in turn. However, epistasis does not provide any piece of information on how the fitness values are topologically related to each other. By knowing the epistasis of an optimization problem, it cannot be established whether the fitness values form a smooth progression resulting in a solitary optimum or whether they form a spiky pattern of many isolated optima [438].

The impossibility of understanding each detail of the fitness landscape depends not only on the fitness function but also on the search algorithm [438] since an observed landscape appears to be an artefact of the algorithm used or, more specifically, of the neighborhood structure induced by the operators used by the algorithm [433]. The neighborhood structure is defined as a set of points that can be reached by a single move of a search algorithm [375]. Closely related to the concept of the neighborhood structure is the notion of a basin of attraction induced by this structure. More specifically, a basin of attraction of a local optimum x is the set of points X of the search space such that a search algorithm starting from any point from X ends in the local optimum x. A special note should be made regarding the landscapes with plateaus, i.e. regions in search domain where the function has constant or nearly constant values. If a search method is trapped on such region it cannot get any information regarding the gradient or even its estimates. Generally speaking, this situation is rather complicated and special algorithmic components should be used in this case. Finally, an important feature of a fitness landscape is the presence or absence of symmetry. Special components can be included in the algorithms for symmetrical problems.

In addition, two features can be mentioned which appear to be semi-defining when distinguishing the classes of problems on which an algorithm performs well. The first one is dimensionality of the problem. Two problems with high dimensionality of the search domain can be put into the same class, however an algorithm that performs well for one of them might not necessarily work well for the other one. At the same time, two specialized algorithms for these two problems will have some common features intended to overcome difficulties arising from high dimensionality. The second semi-defining feature is computational cost of a single evaluation of the objective function. Clearly, two problems with computationally expensive objective functions can have different features mentioned above that will put them into different classes. However, these problems are unsolvable (in practice) if treated as computationally cheap functions, therefore algorithms for such problems should have common type components which allow proper handling of the computational cost.

There is generally a performance advantage in incorporating prior knowledge into the algorithm, however the results of NFLT do not deem the use of unspecialized algorithms futile. It is impossible to determine the fraction of practical problems for which an algorithm yields good results rapidly, therefore a practical free lunch is possible. NFLT constitute, in a certain sense, the "Full Employment Theorem" (FET) for optimization professionals. In computer science and mathematics, the term FET is used to refer to a theorem that shows that no algorithm can optimally perform a particular task done by some class of professionals. In this sense, as no efficient general purpose solver exists, there is always scope for improving algorithms for better performance on particular problems. Since MAs, as mentioned above, represent a broad class of algorithms which combine various algorithmic components, a suitable combination is necessary for a given problem. Since, during the last decade, computer scientists had to observe the features of their optimization problem in order to propose an ad-hoc optimization algorithm, the approach of combining various search operators within the algorithmic design became a common practice. In this sense, the development of NFLT implicitly encouraged the use and development of MAs, which became extremely popular and often necessary, in computer science at first, and in engineering and applied science more recently, thus constituting the FET for MAs.

4.3 A Basic Memetic Algorithm Template

As mentioned in previous sections, MAs blend together ideas from different search methodologies, and most prominently ideas from local search techniques and population-based search. Indeed, from a very general point of view a basic MA can be regarded as one (or several) local search procedure(s) acting on a set *pop* of $|pop| \geqslant 2$ solutions which engage in periodical episodes of cooperation via recombination procedures. This is shown in Algorithm 4.

Algorithm 4. A Basic Memetic Algorithm

1 **function** BasicMA (**in** *P*: Problem, **in** *par*: Parameters): Solution;
2 **begin**
3   *pop* ← Initialize(*par*, *P*);
4   **repeat**
5   *newpop$_1$* ← Cooperate(*pop*, *par*, *P*);
6   *newpop$_2$* ← Improve(*newpop$_1$*, *par*, *P*);
7   *pop* ← Compete (*pop*, *newpop$_2$*);
8   **if** Converged(*pop*) **then**
9   *pop* ← Restart(*pop*, *par*);
10   **endif**
11   **until** *TerminationCriterion(par)* ;
12   **return** *GetNthBest(pop, 1)*;
13 **end**

This template requires some explanation. First of all, the Initialize procedure is responsible for creating the initial set of $|pop|$ solutions. While traditional evolutionary algorithms usually resorted to simply generating $|pop|$ solutions at random (in some cases following a systematic procedure to ensure a good coverage of the search space), MAs typically attempt to use high-quality solutions as starting point. This can be done either using a more sophisticated mechanism (for instance, some constructive heuristic) to inject good solutions in the initial population [861], or by using a local-search procedure to improve random solutions (see Algorithm 5).

Algorithm 5. Injecting high-quality solutions in the initial population.

1 **function** Initialize(**in** *par*: Parameters, **in** *P*: Problem): Bag{Solution};
2 **begin**
3   *pop* ← ∅;
4   **for** *j* ← *1* **to** *par.popsize* **do**
5   *i* ← RandomSolution(*P*);
6   *i* ← LocalSearch (*i*, *par*, *P*);
7   *pop* ← *pop* ∪ {*i*};
8   **endfor**
9   **return** *pop*;
10 **end**

As for the TerminationCriterion function, it typically amounts to checking a limit on the total number of iterations, reaching a maximum number of iterations without improvement, or having performed a certain number of population restarts.

Algorithm 6. The pipelined Cooperate procedure.

```
1  function Cooperate (in pop: Bag{Solution}, in par: Parameters, in P: Problem):
   Bag{Solution};
2  begin
3      lastpop ← pop;
4      for j ← 1 to par.numop do
5          newpop ← ∅;
6          for k ← 1 to par.numapps^j do
7              parents ← Select (lastpop, par.arity^j);
8              newpop ← newpop ∪ ApplyOperator (par.op^j, parents, P);
9          endfor
10         lastpop ← newpop;
11     endfor
12     return newpop;
13 end
```

The procedures Cooperate and Improve constitute the core of the MA. Starting with the former, its most typical incarnation is based on two operators for selecting solutions from the population and recombining them. Of course, this procedure can be readily extended to use a collection of variation operators applied in a pipeline fashion. As shown in Algorithm 6, this procedure comprises *numop* stages, each one corresponding to the iterated application of a particular operator op^j that takes $arityin^j$ solutions from the previous stage, generating $arityout^j$ new solutions.

As to the Improve procedure, it embodies the application of a local search procedure to solutions in the population. Notice that in an abstract sense a local search method can be modeled as a unary operator, and hence it could have been included within the Cooperate procedure above. However, local search plays such an important role in MAs that it deserves separate treatment. Indeed, there are several important design decisions involved in the application of local search to solutions, i.e., to which solutions should it be applied, how often, for how long, etc. See also next section.

Next, the Compete procedure is used to reconstruct the current population using the old population *pop* and the newly generated population $newpop_2$. Borrowing the terminology from the evolution strategy [761, 800] community, there exist two main possibilities to carry on this reconstruction: the *plus* strategy and the *comma* strategy. The latter is usually regarded as less prone to stagnation [32], with the ratio $|newpop|/|pop| \simeq 6$ being a common choice [34]. Since this option can be somewhat computationally expensive if the fitness function is complex and time-consuming, a popular alternative is using a plus strategy with a low value of $|newpop|$, analogous to the so-called *steady-state* replacement strategy in GAs [930]. This option usually provides a faster convergence to high-quality solutions, although care has to be taken with premature convergence to suboptimal regions of the search space. This leads to the last component of the template shown in Algorithm 4, the restarting procedure.

Algorithm 7. The Restart procedure.

```
1  function Restart (in pop: Bag{Solution}, in par: Parameters, in P: Problem):
   Bag{Solution};
2  begin
3      newpop ← ∅;
4      for j ← 1 to par.preserved do
5          i ← GetNthBest(pop, j);
6          newpop ← {i};
7      endfor
8      for j ← par.preserved + 1 to par.popsize do
9          i ← RandomSolution(P);
10         i ← LocalSearch (i, par, P);
11         newpop ← {i};
12     endfor
13     return newpop;
14 end
```

First of all, it must be decided whether the population has degraded or has not, using some measure of information diversity in the population such as Shannon's entropy [184]. Once the population is considered to be at a degenerate state, the restart procedure is invoked. Again, this can be implemented in a number of ways. A very typical strategy is to keep a fraction of the current population, generating new (random or heuristic) solutions to complete the population, as shown in Algorithm 7. The procedure shown therein is also known as the *random-immigrant* strategy [130]. Another possibility is to activate a *strong* or *heavy* mutation operator in order to drive the population away from its current location in the search space.

4.4 Design Issues

The general template of MAs depicted in the previous section must be instantiated with precise components in order to be used for solving a specific problem. MAs are commonly implemented as EAs endowed with a local search component, and therefore the theoretical corpus available for the former can be used to guide some aspects of the design process, e.g., the representation of solutions in terms of meaningful information units [183, 751].

The most MA-specific design decisions are those related to the local search component, not just from the point of view of parameterization (see below) but also with the actual inner working of the component and its interplay with the remaining operators. This latter issue is well exemplified in the work of Merz and Freisleben on the TSP [285]. They consider the use of the Lin-Kernighan heuristic [524], a highly intensive local search procedure, and note that the average distance between local optima is similar to the average distance between a local optimum and the global optimum. For this reason, they introduce a distance-preserving crossover (DPX) operator that generate offspring whose distance from the parents is the same as the distance between the parents themselves. Such an operator is likely to be less effective if a less powerful local improvement method, e.g., 2-opt, was used, inducing a different distribution of local optima.

Once a local search procedure is selected, an adequate parameterization must be determined, i.e., how often it must be applied, how to select the solutions that will undergo local improvement, and how long must improvement epochs last. These are delicate issues since there exists theoretical evidence [494, 857] that an inadequate parameter setting can turn the algorithmic solution from easily solvable to non-polynomially solvable. Regarding the probability of application of local search, its precise values largely depend on the problem under consideration [411], and its determination is in many cases an art. For this reason, adaptive and self-adaptive mechanisms have been defined in order to let the algorithm learn what the most appropriate setting is. The term partial lamarckianism [151, 396, 717] is used to denote these strategies where not every individual is subject to local search.

As to the selection of individuals that will undergo local search, most common options are random-selection, and fitness-based selection, where only the best individuals are subject to local improvement. For example, Nguyen *et al.* [665] consider an approach in which the population is sorted and divided into n levels (n being the number of local search applications), and one individual per level is randomly selected. Note that such a strategy can be readily deployed on a structured MA as defined by Moscato *et al.* [62, 94, 282, 576, 578], in which fitness-based layers are explicitly available. See also [80, 736, 737, 836] for other population management strategies.

4.5 Conclusions and Outlook

Memetic algorithms are a pragmatic, cross-disciplinary optimization paradigm that has emerged in the last quarter of a century to become nowadays one of the most widely used solving approaches. This is supported by a plethora of applications in disparate fields ranging from machine learning and knowledge discovery to planning, scheduling and timetabling, from bioinformatics to electronics, engineering, and telecommunications, or from economics to physics, just to mention a few. The reader may check [154, 375, 618, 619, 620, 626, 632], for a survey of these applications and pointers to the literature.

Throughout this chapter we have provided a brief introduction to the main issues regarding the definition and design of a basic memetic algorithm. However, it must be emphasized that the MA paradigm is very rich and has given rise to an ample set of variations and more sophisticated MA models. Among these, we can firstly cite multiobjective MAs (MOMAs). MOMAs are applied to problems which exhibit multiple, partially-conflicting objectives, and in which the notion of Pareto-dominance is therefore essential. Actually, MOMA approaches can be roughly classified into two major classes: scalarizing approaches [408, 409, 419, 421] (based on the use of some aggregation mechanism to combine the multiple objectives into a single scalar value), and Pareto-based approaches [471, 472] (considering the notion of Pareto-dominance for deciding transitions among neighboring solutions). MOMAs will be dealt in more detail in chapter 13 in this volume.

Adaptive MAs also deserve special attention. As mentioned in Section 4.4, decisions related to parameterization are essential in order to achieve an effective MA. It is therefore not surprising that attempts have been made to let the algorithm find by itself adequate values for these parameters [40, 536, 605, 606]. Furthermore, the term "meta-lamarckian learning" [680] has been coined to denote strategies in which the algorithm learns to select appropriate local search operators from a certain available collection (note the relationship with hyperheuristics [169]). A further step is taken in the so-called multi-memetic algorithms, in which each solution carries a gene that indicates which local search has to be applied on it (either indicating which one from a pre-existing collection, by parameterizing a general local search template, or by using a grammar to define new operators) [488, 490, 496]. At an even higher level, solutions and local-search operators can coevolve [830, 831]. Adaptive MAs will be dealt in more detail in chapter 11 in this volume.

Last but not least, there exist nowadays a growing trend in combining MAs with complete techniques such as branch-and-bound or branch-and-cut among others. There are many ways in which such a combination can be done. For example, an exact technique can be used as an internal operator of the MA [295, 742], as a post-processing technique [469], run in parallel with the MA [294, 297, 740], and even combine several of the previous approaches [299]. The combination of MAs with exact techniques will be dealt in more detail in chapter 12 in this volume.

4.6 Memetic Algorithms and Memetic Computing

It is fundamental to clarify the difference between MAs and Memetic Computing (MC) . As stated above, MAs are population-based evolutionary algorithms composed of an evolutionary framework and a list of local search algorithms activated within the generation cycle of the evolutionary framework, see [376]. While this book refers to MAs, it is worthy to take into account that recently the term MC became widely used amongst computer scientists. An early definition has been given in [689], where MC is defined as "...a paradigm that uses the notion of meme(s) as units of information encoded in computational representations for the purpose of problem solving". In other words, part of the scientific community tried to extend the concept of meme for problem solving, see [655], to something broader and more innovative. The fact that ad-hoc optimization algorithms can efficiently solve given problems is a well-known result from literature. On the other hand, the ultimate goal in artificial intelligence is the generation of autonomous and intelligent structures. In computational intelligence optimization, the goal is the automatic detection of the optimal optimization algorithm for each fitness landscape, or, in other terms, the on-line (i.e. during run-time) automatic design of optimization algorithms. MC can be seen then as a subject which studies complex structures composed of simple modules (memes) which interact and evolve adapting to the problem in order to solve it. This view of the subject leads to a more modern definition of MC.

Definition 4.1. Memetic Computing is a broad subject which studies complex and dynamic computing structures composed of interacting modules (memes) whose evolution dynamics is inspired by the diffusion of ideas. Memes are simple strategies whose harmonic coordination allows the solution of various problems.

In this light, MAs should be seen as a cornerstone and founding subset of MC.

Acknowledgements. C. Cotta is supported by Spanish MICINN under project NEMESIS (TIN2008-05941) and by Junta de Andalucía under project TIC-6083. F. Neri is supported by the Academy of Finland, Akatemiatutkija 130600, Algorithmic Design Issues in Memetic Computing.

Part II
Methodology

Chapter 5
Parametrization and Balancing Local and Global Search

Dirk Sudholt

5.1 Introduction

This chapter is devoted to the parametrization of memetic algorithms and how to find a good balance between global and local search. This is one of the most pressing questions when designing a hybrid algorithm. The idea of hybridization is to combine the advantages of different components. But if one components dominates another one, hybridization may become more hindering than useful and computational effort may be wasted. For the case of memetic algorithms, if the effect of local search is too strong, the algorithm may quickly get stuck in local optima of bad quality. Moreover, the algorithm is likely to rediscover the same local optimum over and over again. Lastly, an excessive local search quickly leads to a loss of diversity within the population.

The importance of the parametrization of memetic algorithms has already been recognized by Hart [366] in 1994. He posed the following questions, many of which have been reproduced in similar ways in later articles:

- How often should local search be applied?
- On which solutions should local search be used?
- How long should the local search be run?
- How efficient does a local search need to be?

We will mostly deal with the first and the third question in the sequel. In concrete implementations of memetic algorithms different parameters occur. Related to the first question is a strategy to call local search with a fixed frequency, the *local search frequency*. A similar strategy is to call local search probabilistically, with a fixed *local search probability*. With regard to the third question, often the running time of one local search is capped to a value called *local search depth*. Other mechanisms

---

Dirk Sudholt
School of Computer Science, The University of Birmingham Edgbaston,
Birmingham B15 2TT, UK
e-mail: d.sudholtcs.bham.ac.uk

can have a comparable effect. [411] restricted the neighborhood used for one iteration of local search to some fixed parameter k. The size of the neighborhood is also a crucial parameter in variable-neighborhood search algorithms [604]. Paenke, Jin, and Branke [697] used the lifetime of an individual to balance the effect of global- and individual-level adaptation in stochastic environments.

This list of mechanisms for balancing global and local search is by far not complete. While some considerations described in this chapter hold for a large variety of balancing techniques, we will consider the local search frequency and the local search depth as the most typical mechanisms.

We describe the outline of this chapter. In Section 5.2 we will survey applications and theoretical studies dealing with the parametrization. The effect of local search is discussed and aspects are described that have a strong impact on the optimal balance between global and local search. We also review approaches how to find such an optimal balance. Section 5.3 deals with the complexity of local search. We will ask how powerful local search is on its own and in which settings a local optimum can be found in polynomial time. For many practically important problems we cannot guarantee that local search always finds a local optimum in polynomial time. Even stronger, there is strong evidence that no algorithm can perform this task in polynomial time. Implications for memetic algorithm design are discussed. Finally, we will present artificial functions in Section 5.4 and running time analyses demonstrating that the parametrization of memetic algorithms can be extremely hard. This also strengthens the fact that there is no a priori optimal parametrization that works well for every problem. The chapter ends with conclusions in Section 5.5.

5.2 Balancing Global and Local Search

5.2.1 Early Works and the Effect of Local Search

The early work by Hart [366] and a subsequent extension to combinatorial optimization by Land [503] lead to many conclusions for the design of memetic algorithms. Hart investigated the impact of the local search frequency for the optimization of common test functions in continuous spaces like the Rastrigin function, the Griewank function, and modifications thereof. His experimental results suggest that genetic algorithms (GAs) with large populations are most effective when local search is used infrequently. He also claims that a large local search frequency is needed if the algorithm is not able to identify regions that are likely to contain global optima. As the introduction of elitism increases the degree of exploitation, compared to exploration, less local search is needed when using elitism. Hence, also the type of GA used with local search has a strong impact on performance. Hart also remarks that the use of local search has restricted many applications to use small population sizes because of the increased computational effort. This holds in particular when local search is applied to every individual in the population.

Regarding the selection of individuals for which local search is to be performed, Hart [366] proposes to decrease the local search frequency for each individual by the

5 Parametrization and Balancing Local and Global Search   57

number of duplicates contained in the population. This works around the problem of having redundant local searches on the same solutions. He generalizes this approach towards reducing the local search frequency with respect to the degree of similarity to other solutions in the population. To this end, a distance metric in genotype space is used. This closely resembles the well-known fitness sharing mechanism for preserving diversity [547]. Section III.C.3 Land [503] proposes several extensions and similar approaches. One is to choose a subset of the population such that the minimum distance between any two selected individuals is maximized. A second strategy is to ensure that every individual in the population is close to an individual that is selected for local search. This way, if the population consists of several clusters, we can hope that all clusters benefit from local search.

Hart [366] also investigated biasing the selection of individuals for local search towards fitter individuals. However, as argued by Section III.C.3 Land [503], this reinforces the dominance of the already fit individuals and hence leads to a rapid loss of diversity. In addition, good solutions are likely to be close to local optima, hence they will have the least benefit of applying more local search to them. Also, for solutions that are close to local optima improvements may be hard to find, which renders the local search less efficient.

Related to the last remark is the question how easy improvements can be achieved for specific solutions. Section III.C.3 Land [503] introduced the notion of a "local search potential" as a measure for the expected gain in fitness in relation to the computational effort. The local search potential can be estimated by performing few steps of local search, a so-called "local search sniff" and recording both the gain and the effort throughout the sniffing period. The average gain per unit of effort is then used as an estimation for its future effectiveness. The drawbacks of this approach is that these sniffs might use a fair amount of computational effort to yield reliable estimations. Moreover, there are no guarantees that the progress in early steps of local search will be an accurate prediction of future progress.

The use of local search is not restricted to evolutionary algorithms. Memetic approaches have also been used for various other paradigms such as estimation-of-distribution algorithms [6] or Ant Colony Optimization [215, 514]. The effect of local search can be quite different in other paradigms. In a recent study Neumann, Sudholt, and Witt [661] argued that the use of local search in ant colony optimization (ACO) can change the behavior of the algorithm drastically. Without local search, the sampling distribution for new solutions given by artificial pheromones usually follows the best-so-far solution. This enables the algorithm to follow paths and ridges in the search space. When introducing local search with a large local search depth, however, a newly discovered local optimum might be far away from the "center of gravity" of the sampling distribution. In ACO algorithms using the best-so-far rule (i.e. always rewarding the current best solution found so far), the pheromones are then directly adapted towards the new local optimum. Instead of following the path taken by local search to arrive at this local optimum, the direct adaptation of pheromones can make the algorithm sample solutions from a totally different area of the search space. Neumann et al [661] demonstrated for a constructed function where this effect may mislead the search and turn a polynomial optimization time into an exponential one, with high probability. However, they also proved for a slightly different function that this behavior can also prevent the algorithm from getting stuck in a local optima. Local search can then also help to reduce an exponential optimization time to a polynomial one.

5.2.2 Aspects That Determine the Optimal Balance

The optimal balance between global and local search clearly depends on the optimization problem at hand and the memetic algorithm applied to it. The latter not only includes the choice of the operators employed and issues of representation, but also various other parameters of the algorithm such as the population size, selection pressure, and the mutation rate. Even among the mentioned aspects and for plain evolutionary algorithms there is strong evidence that the precise choice of parameters can have a tremendous effect on performance. Theoretical studies have been performed, e. g., by Storch [852] and Witt [939] for the choice of the parent population size, Jansen, De Jong, and Wegener [417] for the choice of the offspring population size, Jansen and Wegener [416] for the choice of the mutation rate, and Lehre and Yao [511] for the ratio of the selection pressure in ranking selection and the mutation rate.

We therefore cannot expect to obtain design guidelines that do not depend on all the mentioned aspects and nevertheless always lead to good results. The existence of such guidelines is excluded by the well-known *no free lunch theorems* [401, 940]. These results state that when averaging over a class of problems that is closed under permutation, all algorithms (this includes all parametrizations for one specific algorithm) have equal average performance. It is, however, also clear that the setting of the no free lunch theorems is much too general to be of any relevance. The vast majority of functions considered are of no interest for optimization as they have exponential-size representations [228]. In Section 5.4 we will present much stronger results for one particular memetic algorithm. The considered functions do have polynomial-size representations and exhibit superpolynomial or exponential performance gaps for even small changes of the parametrization. This shows that for the considered algorithm there is no polynomial relation between optimal and non-optimal parameter values.

So, the parameters and design aspects of a memetic algorithm should not be viewed in isolation. The strongest dependency is probably the one between the local search depth and the local search frequency. Choosing one parameter value with disregard to the other one often does not make much sense. For instance, [411] discovered that in applications to the multi-objective permutation flowshop scheduling problem the optimal number k of neighbors visited in one iteration of local search was strongly negatively correlated with the local search frequency p_{LS}. The best performance was obtained when the product $k \cdot p_{LS}$ was within a range of 1 to 10.

Also, the balance of exploration and exploitation is important. In iterated local search algorithms [533] local search is typically used in every iteration and performed until a local optimum is found. So, local search is used to its utmost extend.

5 Parametrization and Balancing Local and Global Search 59

On the other hand, iterated local search algorithms tend to use strong perturbations, i. e., large mutations before applying local search. In this setting, a powerful explorative operator balances out a powerful exploitative operator. When the underlying evolutionary component of a memetic algorithm is more similar to a classical genetic algorithm, that is, if more emphasis is put on exploration by populations and the use of recombination and mutation, less local search should be used in order not to disrupt exploration.

The optimal balance between global and local search also depends on design and implementation issues. In some applications, local search is computationally expensive. This holds, for example, in the case of large or computationally expensive neighborhoods like the Lin-Kernighan neighborhood or pivoting rules such as steepest descent/ascent, where the whole neighborhood must be searched. Using pivoting rules such as first improvement or neighborhood reduction techniques can speed up the local search significantly and thus shift the "optimal" amount of local search.

In several applications it is possible to perform incremental fitness evaluations during local search. If the fitness can be efficiently updated in cases where only few components (bits, objects, edges, ...) are modified in an iteration of local search, local search tends to be much faster than the genetic component of the algorithm. One example is the TSP where the cost of a 2-Exchange operation can be computed by only looking at the 4 edges involved, see, e. g. [411, 583]. In fact, Jaszkiewicz [419] reported in a study on a multi-objective TSP problem that local search was able to perform 300 times more function evaluations per second than a multi-objective genetic algorithm. Also neighborhood reduction techniques turned out to be very useful for speeding up local search [583].

On the other hand, [411] argued that for flowshop scheduling recomputing the fitness after a local change of a schedule cannot be done much faster than computing the fitness from scratch. This is because even local changes may imply that the completion times for almost all jobs have to be recalculated. The execution time for one iteration of local search is thus a very important issue.

When considering multi-objective problems, it is important to maintain diversity in the population. Sindhya, Deb, and Miettinen [815] used a local search that optimizes an achievement scalarizing function. The local search helps with the convergence to the Pareto front, but it is also likely to create extreme points on the Pareto front. To this end, the authors used a dynamic schedule for choosing the local search probability. The local search probability linearly increases from 0 to the inverse population size and then drops to 0 again. The number of generations for one such cycle is proportional to the population size.

Concluding, there are many aspects that determine the optimal balance between global and local search. Many different parameter settings have been proposed, some of which are due to dynamic or adaptive schedules. Table 5.1 summarizes the above-mentioned aspects. In the following, we will describe approaches how such an optimal balance can be found.

Table 5.1. Overview on aspects that affect the optimal amount of local search.

	less local search	more local search
exploration by GA	weak exploration	strong exploration
mutation strength	small mutations	large perturbations
pivoting rule	steepest ascent/descent	first improvement
neighborhood size	large neighborhood	small neighborhood/reduction techniques
implementation of LS	expensive recalculations	incremental fitness evaluations
objectives	multi-objective problem	single-objective problem

5.2.3 How to Find an Optimal Balance

Several approaches have been proposed how to find a good parametrization for memetic algorithms. There are general approaches for finding good parameter settings that are not tailored towards memetic algorithms and hence are somewhat beyond the scope of this chapter. We briefly mention one such approach called sequential parameter optimization (SPO) introduced by Bartz-Beielstein, Lasarczyk, and Preuß [49]. SPO aims at finding the best parametrization by combining classical and modern statistical techniques. It can be seen as a search heuristic trying to optimize the performance of non-deterministic algorithms. SPO iteratively applies the following three steps. First, an experimental analysis of an algorithm with a given parametrization is performed. Then, the performance of the algorithm (including its parametrization) is estimated by means of a stochastic process model. In a third step, additional parameter settings in the parameter space are determined in a systematic way. For further details, we refer to Bartz-Beielstein [47], Bartz-Beielstein, Lasarczyk, and Preuß [49].

Goldberg and Voessner [329] and Sinha, Chen, and Goldberg [818] presented a system-level theoretical framework for optimizing global-local hybrids. Two different optimization goals are considered: maximizing the probability of reaching a solution within a given accuracy and minimizing the time needed to do so. The authors considered the impact of the local search depth for a hybrid that uses random search as a global component. They presented formulas for determining the optimal local search depth for the mentioned optimization goals. The formulas, however, are based on some simplifying assumptions and they do require knowledge on the structure of the problem that is usually not available in practice. The probabilities of reaching specific basins of attraction in one step of the global searcher have to be known as well as the average time local search takes to local optimality for each basin.

Another well-studied approach is to include domain knowledge into the design of memetic algorithms [583, 923]. This knowledge can be gained by analyzing the fitness landscape of the problem (instance) at hand. One useful measure for the ruggedness of a fitness landscape is the *correlation length*. It is, in turn, based on the *random walk correlation function* $r(s)$, also known as *autocorrelation*. The function $r(s)$ specifies the correlation between two points of a random walk that are s

time steps away. The random walk chooses the next point uniformly from a fixed neighborhood. Different neighborhoods may thus lead to different correlations. If the correlation is high, the correlation length is large and the fitness landscape is smooth. If the correlation is low, the correlation length is small and the fitness landscape is rugged. It has been observed that large correlation lengths lead to a large number of iterations until local search finds a local optimum. On the other hand, a small correlation length often means that local search may quickly get stuck in bad local optima [583]. Fitness landscape analysis can help to choose the right neighborhood and a suitable parametrization for the local search.

Last but not least, adaptive techniques may help to find a good parametrization. Memetic algorithms using many different local searchers are known as *multimeme algorithms* [658]; each local search operator is called a "meme." The choice of memes can be made adaptively or even self-adaptively, see the survey by Ong, Lim, Zhu, and Wong [683]. Also coevolutionary systems have been developed that coevolve a local searcher alongside the evolution of solutions [490, 830].

5.3 Time Complexity of Local Search

In order to fully understand the capabilities of local search, it is indispensable to know its limitations. In this section we describe theoretical results on the time complexity of local search and discuss implications on memetic algorithm design. We will look at local search in isolation and ask how long it takes until one call of local search finds a local optimum. From the perspective of memetic algorithms, we ask how efficient the local search component is in computing a local optimum from its basin of attraction. If local search cannot find local optima efficiently, a memetic algorithm will most likely show poor performance, even if the global component can locate the basin of attraction of the global optimum efficiently. We will also review a theory of intractability that applies to many important problems and memetic algorithms used in practice. It can be proven that under certain complexity theory assumptions and in the worst case local optima cannot be computed in polynomial time by any means, even for more sophisticated algorithms than local search. It is not the case that local search is too simple to locate local optima efficiently. Instead, the mentioned problems are so difficult that computing local optima is hard for any (arbitrarily sophisticated) search strategy.

The following presentation is based in parts on Michiels, Aarts, and Korst [598], Chapter 6. Define a *local search problem* as a combination of a combinatorial optimization problem, a neighborhood function mapping a solution to a subset of the search space, and an indication whether the problem is a maximization or a minimization problem. The goal of a local search problem is to compute a local optimum with respect to the goal of the optimization. Note that the neighborhood is an integral part of the problem. Using a different neighborhood function leads to a different local search problem.

The main question is how many iterations local search will need in order to find a local optimum. It is helpful to use the following perspective. Define the state graph of a problem as a directed graph where the set of vertices corresponds to the search space. The state graph includes an edge (x,y) if and only if y is a neighbor of x and y is strictly better than x. A local optimum thus corresponds to a sink, i.e., a vertex with no outgoing edges. The number of iterations needed to find a local optimum corresponds to the length of the path from the starting point to a sink. The precise choice of an outgoing edge is determined by the pivoting rule.

5.3.1 Polynomial and Exponential Times to Local Optimality

In many applications, local search finds an optimum in polynomial time. Assume the neighborhood is searchable in polynomial time and the number of function values is polynomially bounded. Then clearly all paths in the state graph only have polynomial length and local search will finish in polynomial time. Problems with only a polynomial number of function values include the NP-hard Minimum Graph Coloring problem if the number of colors used is taken as fitness function and the NP-hard MAXSAT problem, when one uses the number of satisfied clauses as objective function. Another NP-hard problem with this property is the graph partitioning problem. The fitness corresponds to the number of cut edges, which ranges from 0 to $n^2/4$, n being the number of vertices. Also weighted problems might show this property, for instance in special cases where the weights are integral, positive, and polynomially bounded. Land [503], Section III.A.1 gives a formal proof for a class of weighted graph partitioning problems and a weighted TSP.

Lin-Kernighan-type or variable-depth-type of local searches perform a chained sequence of local moves and fix solution components (edges, bits, vertices, ...) that have been changed until the end of local search. Hence, these local searches also trivially stop after polynomially many steps (see [859] for an analysis of memetic algorithms with variable-depth search). The effect is similar as for local searches with a maximum local search depth; local search stops in polynomial time without guarantee of having found a local optimum.

When the number of function values is superpolynomial, it might still be that all paths in the state graph have only polynomial length. But for some problems one can actually prove that in settings with exponentially many function values exponentially long paths exist. Englert, Röglin, and Vöcking [245] constructed an instance for the Euclidean TSP where the state graph for the 2-Opt algorithm has exponential length. Hence, in the worst case—with respect to the choice of the starting point and the pivoting rule—local search takes exponential time.

Similar results also hold for pseudo-Boolean optimization. Horn, Goldberg, and Deb [395] presented so-called *long path problems* which contain a fitness-increasing path in the state graph under the Hamming neighborhood (two solutions are neighbored if they only differ in exactly one bit). The length of the path is of order $\Theta(2^{n/2})$ if n is the number of bits. In addition, for every point x on the path every Hamming neighbor y of x has strictly lower fitness than x, unless y is itself a point on the path. In other words, the next successor on the path is the only neighbor with a better fitness. This property ensures that a local search using the Hamming neighborhood cannot leave the path and thus is forced to climb to its very end. This holds regardless of the pivoting rule as the pivoting rule cannot make any choices. All points not belonging to the path give hints to reach the start of the path, hence also on average over all starting points local search needs exponential time.

Note, however, that flipping 2 bits at a time or using a stochastic neighborhood such as standard bit mutations suffices to reach the end of the path efficiently by taking shortcuts. Rudolph [780] proved an upper bound of $O(n^3)$ for the expected optimization time of the simple algorithm (1+1) EA whose mutation operator flips each bit independently with probability $1/n$. He also formally defined a more robust generalization to long k-paths where at least k bits have to flip in order to take a shortcut. The parameter k can be chosen such that the length of the path is still exponential (say, of order $2^{\sqrt{n}}$) and the probability of taking a shortcut by standard bit mutations is still exponentially small. This yields an example where also using larger neighborhoods that can flip up to $k-1$ bits at a time need exponential time for suitable initializations. Also the stochastic neighborhood used by the (1+1) EA does not avoid exponential expected optimization times for suitable values of k, as proven by Droste, Jansen, and Wegener [227].

5.3.2 Intractability of Local Search Problems

NP-completeness theory is a well-known and powerful tool to prove that many important optimization problems are intractable, in a sense that no polynomial-time algorithm for the problem can exist, assuming P $\neq$ NP. There is a similar theory for local search problems that can be used to characterize local search problems where under reasonable assumptions no polynomial-time algorithm exists for finding local optima. This includes arbitrary algorithms that need not have much in common with local search algorithms. The foundation for this theory was laid by Johnson, Papadimitriou, and Yannakakis [429]. We give an informal introduction into this theory and refer the reader to Yannakakis [948] and Michiels et al [598], Chapter 6 for complete formal definitions. For this subsection we assume that the reader has basic knowledge on NP-completeness and refer to classical text books for further reading [303, 701, 925]. A brief treatment of NP-completeness is also given in Michiels et al [598], Appendix B.

The complexity class we will focus on is called PLS for "polynomial-time searchable." A local search problem Π is in PLS if there exist two polynomial-time algorithms with the following properties. One algorithm can be seen as an initialization operator. It simply computes some initial solution for Π in polynomial time. The second polynomial-time algorithm, given a solution s, either computes a better neighbor of s or reports that s is a local optimum. If a problem is in PLS, this means that there is a local search algorithm such that the initialization and each iteration of local search can be executed in polynomial time. This is not to be confused with the question how many iterations are needed in order to find a local optimum.

Similar to reductions in NP-completeness theory, there is the concept of a reduction between PLS-problems: we can relate the difficulties of two problems Π_1, Π_2

in PLS as follows. Denote a PLS-reduction from Π_1 to Π_2 by $\Pi_1 \leq_{\text{PLS}} \Pi_2$. A PLS-reduction demands a polynomial-time algorithm that maps a problem instance of Π_1 to an instance of Π_2 and a polynomial-time algorithm that maps a solution for Π_2 back to a solution for Π_1. In the latter mapping, we require that if the solution s_2 for Π_2 is a local optimum for Π_2 and s_2 is mapped to a solution s_1 for Π_1, then s_1 must be a local optimum for Π_1. Hence, if we want to solve Π_1, we can use the first algorithm to transform the instance for problem Π_1 into an instance of Π_2, then solve problem Π_2 to local optimality, and finally map the local optimum back to a local optimum for Π_1 using the second algorithm.

If $\Pi_1 \leq_{\text{PLS}} \Pi_2$ then we can conclude that Π_2 is "at least as hard" as Π_1. This means that if Π_1 cannot be solved in polynomial time, then Π_2 cannot be solved in polynomial time either. But if Π_2 is polynomial-time solvable, then Π_1 also is. This concept leads to the notion of PLS-completeness: a problem Π is PLS-complete if *every* problem in PLS can be PLS-reduced to it; in other words, Π is PLS-complete if it is at least as hard as every other problem in PLS. PLS-complete problems thus constitute the hardest problems in PLS. If it could be shown for one PLS-complete problem that a local optimum can always be found within polynomial time, then all problems in PLS would be solvable in polynomial time. Speaking in terms of complexity classes, we would then have P = PLS. However, as no polynomial-time algorithm has been found for *any* PLS-complete problem, it is widely believed that P $\neq$ PLS.

Theorem 5.1. *If P $\neq$ PLS, there exists no algorithm that always computes a local optimum for a PLS-complete local search problem in polynomial time.*

This result not only states that local search probably cannot find local optima for PLS-complete problems. It also says that no other, arbitrarily sophisticated algorithm can do better.

The theory of PLS-completeness has concrete implications as many well-known local search problems have been proven to be PLS-complete. We list some examples and refer to Michiels et al [598], Appendix C for a more detailed list.

Theorem 5.2. *The following local search problems are PLS-complete.*

- *Pseudo-Boolean optimization: maximize or minimize a function $\{0,1\}^n \to \mathbb{R}$ using the Hamming neighborhood*
- *MAX-2-SAT for the Hamming neighborhood as well as the Kernighan-Lin neighborhood*
- *MAXCUT for the Hamming neighborhood as well as the Kernighan-Lin neighborhood*
- *Metric TSP for the k-Exchange neighborhood as well as (a slightly modified variant of) the Lin-Kernighan neighborhood.*

So, there are PLS-completeness results for neighborhoods used by common local search algorithms. Memetic algorithms usually combine different neighborhoods for genetic operators and local searchers. Multimeme algorithms or variable-neighborhood search even use several neighborhoods for local search. Does PLS-completeness also hold in these settings?

5 Parametrization and Balancing Local and Global Search 65

The answer is yes. Recall that a PLS-reduction $\Pi_1 \leqslant_{\text{PLS}} \Pi_2$ demands that all local optima in Π_2 must be mapped to local optima in Π_1. In order to prove that a problem Π_2 is PLS-complete, it suffices to show that $\Pi_1 \leqslant_{\text{PLS}} \Pi_2$ for a PLS-complete local search problem Π_1. Assume that Π_1 is PLS-complete and consider the situation where Π_1 and Π_2 are based on the same combinatorial problem. Further assume that Π_2 uses a "larger" neighborhood in the following sense: if x and y are neighbored in Π_1 then they are also neighbored in Π_2. For example, Π_1 might be the TSP with a 2-Exchange neighborhood and Π_2 might be the TSP with a neighborhood of all 2-Exchange and 3-Exchange moves. Now, if x is a local optimum in Π_2 then it is also a local optimum in Π_1 (it might even have less neighbors to compete with). Hence, using the identity function for mapping local optima in Π_2 back to Π_1 establishes a PLS-reduction $\Pi_1 \leqslant_{\text{PLS}} \Pi_2$ and proves PLS-completeness for Π_2. Note that the term "neighborhood" can be used in a broad sense. In the above example, Π_2 might use different kinds of operators. For instance, instead of containing 2-Exchange and 3-Exchange moves, the neighborhood of Π_2 could contain 2-Exchange moves and Lin-Kernighan moves. Note, however, that the enlarged neighborhood must still be searchable in polynomial time as otherwise Π_2 would not be contained in PLS.

It is also possible to incorporate populations as described by Krasnogor and Smith [494]. A local search problem Π_1 whose state space reflects a single solution can be mapped to a local search problem Π_2 whose state space reflects all possible populations. The function value for Π_2 can be defined as the Π_1-value for the best individual in the population. The neighborhood function for Π_2 would contain all possible transitions to other populations using the neighborhood function of Π_1. As long as this neighborhood is searchable in polynomial time, a PLS-reduction $\Pi_1 \leqslant_{\text{PLS}} \Pi_2$ can simply map the best individual from the population of the problem Π_2 to Π_1. If the population cannot be improved by any operation in Π_2, then the best individual cannot be improved in Π_1. Hence, a locally optimal population for Π_2 implies a locally optimal individual for Π_1. With this PLS-reduction, we have shown that the population-enhanced problem Π_2 is PLS-complete as well.

The conclusion from these observations is the following: if we know that a local search problem Π_1 is PLS-complete, then all algorithms that result from Π_1 by extending the algorithm to populations, enlarging neighborhoods, or adding new operators are, in turn, PLS-complete. This holds under the condition that all considered neighborhoods are searchable in polynomial time. Krasnogor and Smith [494] formalize PLS-completeness results for memetic algorithms on the TSP that use the 2-Opt operator. Quoting from their work, "the addition of a population to the evolutionary heuristic does not improve the worst-case behavior beyond that of local search."

For the sake of completeness, we also mention that there is a stronger notion of PLS-completeness, called *tight PLS-completeness*. For tightly PLS-complete problems there can exist paths in the state graph of exponential length. This implies that local search needs exponential time in the worst case. This holds even regardless of the pivoting rule. Actually, all problems mentioned in Theorem 5.2 are tightly PLS-complete. To prove tight PLS-completeness, so called *tight PLS-reductions* are needed that additionally preserve the length of paths in the state graph, up to polynomial factors. Tight PLS-completeness is, however, not robust with respect to extensions of the neighborhood as larger neighborhoods might add shortcuts in the state graph.

How can we deal with PLS-complete problems? Recall that PLS-completeness only focusses on the worst-case behavior. Even if the worst case is hard, the average-case performance or the performance when starting with "typical" starting points generated by the global component might be much better. In fact, problem instances constructed to reveal exponential-length paths in the state space are mostly contrived and very dissimilar to problem instances encountered in practice. Furthermore, even if there is an intractability result for a general problem, it might be that one is actually solving an easier special case of the general problem. While the TSP using common neighborhoods is PLS-complete for general edge weights, local search trivially succeeds in polynomial time if the edge weights are positive and polynomially bounded integers. Though the general problem is (tightly) PLS-complete, the weight-restricted TSP is not.

5.4 Functions with Superpolynomial Performance Gaps

From general hardness results that hold for classes of algorithms under certain assumptions, we now move on the more concrete results for specific memetic algorithms. We will present results that prove the non-existence of a priori guidelines for the parametrization of the investigated memetic algorithms. For both the local search depth and the local search frequency there are functions where only specific parameter values can guarantee an effective running time behavior. With only small variations of the parameters, the typical running time experiences a phase transition from polynomial to superpolynomial or even exponential running times. The "optimal" parameter values for these functions can be chosen almost arbitrarily. This implies that for almost each fixed parametrization (whose value may depend on the problem size) there is a function for which this parameter is far from being optimal. This section is based on Sudholt [858]. Preliminary results were published in Sudholt [855, 856].

The non-existence of an all-purpose optimal parameter value is not surprising in the light of the no free lunch theorems [401, 940], but our statements are much stronger. For instance, they prove that the running times of "good" and "bad" parameter values are not polynomially related. Also, the no free lunch theorems only yield a mere existence proof and do not give any hints how separating functions might look like.

The downside of this approach is that these strong statements can only be obtained by fixing a memetic algorithm that is simple enough to be handled analytically. In particular, the algorithm does not use crossover. The algorithm is called $(\mu+\lambda)$ EA. It uses a fixed maximum local search depth denoted by δ and calls local search with a fixed frequency, every τ iterations. The local search used iteratively searches for neighbors with strictly larger fitness and stops if no such point exists or the maximum local search depth of δ iterations has been hit. It may be implemented using an arbitrary pivoting rule. The $(\mu+\lambda)$ MA operates with a population

Algorithm 8. Local search(y)

1 **for** δ *iterations* **do**
2   **if** *there is a* $z \in \mathcal{N}(y)$ *with* $f(z) > f(y)$ **then**
3          $y \leftarrow z$;
4   **else**
5   stop and **return** y;
6   **endif**
7 **endfor**
8 **return** y;

Algorithm 9. (μ+λ) Memetic Algorithm

1 Let $t \leftarrow 0$;
2 Initialize P_0 with μ individuals chosen uniformly at random;
3 **repeat**
4      $P'_t \leftarrow \emptyset$;
5   **for** $i \leftarrow 1$ **to** λ **do**
6   Choose $x \in P_t$ uniformly at random;
7   Create y by flipping each bit in x independently with prob. p_m;
8   **if** t mod $\tau = 0$ **then**
9              $y \leftarrow$ local search(y)
10   **endif**
11          $P'_t \leftarrow P'_t \cup \{y\}$;
12   **endfor**
13   Create P_{t+1} by selecting the best μ individuals from $P_t \cup P'_t$;   // Break ties in favor of P'_t
14      $t \leftarrow t + 1$;
15 **until** termination ;

of size μ and creates λ offspring in each generation. This is done by choosing randomly a parent, then mutating it, and, every τ generations, additionally applying local search to the result of the mutation. The population for the next generation is selected among the best parents and offspring.

5.4.1 Functions Where the Local Search Depth Is Essential

Now we describe how to construct a function f_D parametrized by an "ideal" value D for the local search depth, such that the following holds. Formal definitions can be found in Sudholt [858]. If the local search depth is chosen as $\delta = D$, then the (μ+λ) EA optimizes f_D efficiently. However, if the local search depth is only a little bit away from this ideal value, formally $|\delta - D| \geq \log^3 n$, then the ($\mu$+$\lambda$) EA needs superpolynomial time, with high probability. The precise result reads as follows.

Theorem 5.3. *Let* $D \geq 2\log^3 n$, $\lambda = O(\mu)$, *and* $\mu, \delta, \tau \in \text{poly}(n)$. *Initialize the* ($\mu$+$\lambda$) *MA with* μ *copies of the first point on the path, then the following holds with high probability:*

- if $\delta = D$, the $(\mu+\lambda)$ MA *optimizes* f_D *in polynomial time*
- if $|\delta - D| \geq \log^3 n$, the $(\mu+\lambda)$ MA *needs superpolynomial time on* f_D.

We only remark without giving a formal proof that the function can be adapted such that in the second case the stronger assumption $|\delta - D| \geq n^\varepsilon$ for some constant $\varepsilon > 0$ leads to exponential optimization times.

In the following, we describe the construction of the function f_D and the main proof ideas. The construction is based on the long k-paths already mentioned in Section 5.3. On this path it is very unlikely that mutation can find a shortcut as at least $k = \Omega(\sqrt{n})$ bits would have to flip simultaneously in one mutation. For simplicity, we assume that the algorithm starts with the whole population at the start of the path and only mention that the construction can be adapted for random initialization. All points that are neither on the path nor global optima are assigned a very low fitness, so that the algorithm only searches on the path. In fact, the mentioned points all receive the same low fitness value, so that local search stops immediately if called from a point that is surrounded by low-fitness individuals. In some sense, we have thus transformed an n-dimensional problem into a one-dimensional problem. The path points are assigned fitness values in the following way. The basic idea is that a global optimum can only be found with good probability if local search stops at specific points on the long k-path.

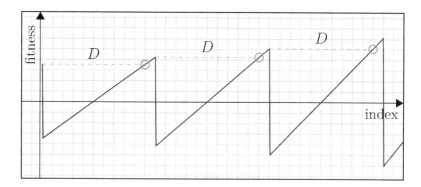

Fig. 5.1. Sketch of the function f_D. The x-axis shows the index on the long k-path. The y-axis shows the fitness. The thick solid line shows the fitness of the points on the long k-path. Encircled path points are close to a target region with respect to Hamming distance. The long k-path can be separated into n subsequent sections with increasing fitness, each one ending with a local optimum. For the sake of clarity, only the first three out of n sections are shown.

The path can be divided into sections on which the fitness is strictly increasing on the path. Each section ends with a local optimum. A sketch of the function is given in Figure 5.1, reproduced from Sudholt [858]. The absolute fitness values at the start of these sections is set so low that a section can only be climbed by local search, given that a preceding mutation creates a suitable starting point. For each section, a set of global optima is placed close to the path in a way that local search cannot locate a global optimum when climbing the section. (This is where we have to think of an n-dimensional problem again as this cannot be properly drawn in a one-dimensional picture.) However, if the local search depth is set in such a way that local search stops close to the global optima, then there is a good chance of jumping to a global optimum the next time this individual is selected for mutation.

Now, the main ideas of the proof are as follows. If the local search depth is smaller than $D - \log^3 n$, local search typically stops with a search point that is inferior to all points in the population. The new offspring is then immediately rejected by selection. The only way to avoid this is to make a large jump by mutation that flips at least $\log^3 n$ bits simultaneously. The probability for this event is superpolynomially small and the expected waiting time until this happens is superpolynomial. This establishes a superpolynomial lower bound in the case $\delta \leqslant D - \log^3 n$.

In case the local search depth attains the "ideal" value $\delta = D$, there is a constant probability that local search stops close to a set of global optima and a mutation flipping two bits creates a global optimum next time this individual is selected for mutation. Note that the algorithm only needs to be successful on one section in order to find a global optimum. With high probability this happens at least once within n trials and the algorithm succeeds in polynomial time.

If the local search depth is too high, i.e., $\delta \geqslant D + \log^3 n$, every time local search climbs a section it runs past the set of global optima and ends with the next local optimum. This holds since each section has length $D + \log^3 n$. From there, a global optimum can only be reached by a large mutation or if the population is able to approach the target set by moving downhill on the section from the local optimum. Note that a new offspring might survive even if it is worse than its parent in case the population contains individuals that are still worse than the offspring. However, using family-tree techniques [938], one can prove that with high probability the population is quickly taken over by the best individuals in the population before getting downhill.

5.4.2 Functions Where the Local Search Frequency Is Essential

Also the choice of the local search frequency can have a tremendous impact on the performance of the $(\mu+\lambda)$ MA. As the analysis presented in Sudholt [858] is quite involved, the results are limited to the (1+1) MA where $\mu = \lambda = 1$. Two functions called Race^{con} and Race^{uncon} are defined according to given values for n, δ, and τ. For formal definitions we again refer to Sudholt [858]. The (1+1) MA is efficient on Race^{con}, but inefficient on Race^{uncon}. Now, if the local search frequency is halved, the (1+1) MA suddenly becomes inefficient on Race^{con}, but efficient on Race^{uncon}.

The functions Race^{con} and Race^{uncon}, which we call *race functions*, are constructed in similar ways, so we describe them both at once. First of all, we partition all bit strings into their left and right halves, which form two subspaces $\{0,1\}^{n/2}$ within the original space $\{0,1\}^n$ for even n. Each subspace contains a part of a long path. Except for special cases, the fitness is the (weighted) sum of the positions on the two paths. This way, climbing either path is rewarded and the (1+1) MA is encouraged to climb both paths in parallel.

The difference between the two paths in the left and right halves of the bit string is that they are adapted to the two neighborhoods used by mutation and local search, respectively. In the left half, we have a connected path of predefined length. The right half contains a path where only every third point of the long k-path is present. Instead of a connected path, we have a sequence of isolated peaks where the closest peaks have Hamming distance 3. As the peaks form a path of peaks, we speak of an *unconnected path*. While the unconnected path cannot be climbed by local search, mutation can jump from peak to peak as a mutation of 3 specific bits has probability at least $1/(en^3)$. Concluding, local search is well suited to climb the connected path while mutation is well suited to climb the unconnected path.

Now, the main idea is as follows. Choosing appropriate lengths for the two paths, if the local search frequency is high, we expect the (1+1) MA to optimize the connected path prior to the unconnected path. Contrarily, if the local search frequency is low, the (1+1) MA is likely to optimize the unconnected path prior to the connected one. Which path is optimized first can make a large performance difference. In the special cases where the end of any path is reached, we define separate fitness values for Race^{con} and Race^{uncon}. For Race^{con}, if the connected path is optimized first (i.e., wins the race), a global optimum is found. However, if the unconnected path wins the race, Race^{con} turns into a so-called deceptive function that gives hints to move away from all global optima and to get stuck in a local optimum. In this situation, the expected time to reach a global optimum is exponential, i.e., $2^{\Omega(n^\varepsilon)}$ for some constant $\varepsilon > 0$. For Race^{uncon}, the (1+1) MA gets trapped in the same way if the connected path wins and a global optimum is found in case the unconnected path wins.

The precise result is as follows. The preconditions $\delta \geqslant 36$, $\delta/\tau \geqslant 2/n$, and $\tau = O(n^3)$ require that "enough" iterations of local search are performed during a polynomial number of generations. The reason is that local search must be a visible component in the algorithm for the different local search frequencies to take effect. The condition $\tau = n^{\Omega(1)}$ as well as the choice of the initial search point are required for technical reasons.

Theorem 5.4. *Let $\delta = \text{poly}(n)$, $\delta \geqslant 36$, $\delta/\tau \geqslant 2/n$, $\tau = n^{\Omega(1)}$, and $\tau = O(n^3)$. If the (1+1) MA starts with a search point whose positions on the connected and unconnected paths are 0 and n^5, respectively, then with overwhelming probability*

- *the (1+1) MA with local search frequency $1/\tau$ optimizes Race^{con} in polynomial time while the (1+1) MA with local search frequency $1/(2\tau)$ needs exponential time on Race^{con} and*
- *the (1+1) MA with local search frequency $1/\tau$ needs exponential time on Race^{uncon} while the (1+1) MA with local search frequency $1/(2\tau)$ optimizes Race^{uncon} in polynomial time.*

The proof is quite technical; it requires good estimations for the progress made on the connected and the unconnected path, respectively. This is done separately for generations with and without local search, respectively. Using appropriate values for the lengths of the two paths derived from the analysis, one can show the following with overwhelming probability. With local search frequency $1/\tau$, within n^4

generations on both race functions the end of the connected path is reached first. On Race^{con} the (1+1) MA has then found an optimum, while it has become trapped on Race^{uncon}. With local search frequency $1/(2\tau)$, within $\sqrt{2}n^4$ generations the total progress by local search on the connected path is decreased by a factor of roughly $1/\sqrt{2}$, compared to the previous setting. At the same time, the total progress on the unconnected path by mutation is increased by a factor of roughly $\sqrt{2}$. Summing up the progress values yields that then with overwhelming probability the (1+1) MA has found the end of the unconnected path first and Race^{uncon} is optimized, while the (1+1) MA is trapped on Race^{con}.

An interesting insight gained from the analysis is that just one iteration of local search helps significantly with the location of isolated peaks. Mutation has to flip three specific bits in order to reach the next point on the unconnected path. However, if local search is called after mutation and were it only for one iteration, the next point on the unconnected path is also reached if only two out of the mentioned three bits are flipped. The probability for a successful step is hence roughly a factor of $3n$ larger! So, in contrast to our intuition, local search does indeed help to optimize the unconnected path. Fortunately for our proof, the steps made on the unconnected path in generations with local search are unbiased. Creating the next successor on the unconnected path by mutation and local search has the same probability as creating the closest predecessor on the path. Both operations will be accepted with high probability if $\delta \geqslant 6$ since at least $\delta - 1$ iterations of local search are spent to make progress on the connected path. Also recall that the connected path is weighted with a factor of n in the fitness function. Hence, the progress made on the connected path will dominate the effect of movements on the unconnected path. The search on the unconnected path in generations with local search is hence unbiased and the probability of making large progress due to the random walk behavior can be bounded. However, this only holds under the condition that $\tau = n^{\Omega(1)}$, i.e., if the local search frequency is not too high. Otherwise, the variance of the random walk behavior will indeed have a significant effect on the progress on the unconnected path. The author conjectures that with a very high local search frequency, the effect might even be reversed such that the unconnected path has a larger benefit from local search than the connected path.

5.5 Conclusions

Finding a good balance between global and local search is a crucial step in the design of memetic algorithms. This topic has been addressed explicitly or implicitly in a variety of applications as well as in empirical and theoretical works. An important conclusion is that the optimal balance is determined by many aspects. Finding a good balance involves knowledge on the problem structure as well as a careful consideration of the algorithms' operators, local search neighborhoods, settings of other parameters like the mutation strength, and implementation issues.

The chapter also covered theoretical approaches that are important to memetic algorithm researchers as they show the limits on the efficiency of memetic algorithms.

The time complexity of local search is an interesting and rich topic in its own right and it can help to understand the effect of local search in memetic settings. PLS-completeness results indicate that for many practical problems there probably is no algorithm that can always find a local optimum in polynomial time. These intractability results can help prevent researchers from trying to achieve the impossible. Finally, running time analyses for a particular memetic algorithm have demonstrated superpolynomial or exponential performance gaps even for only slight changes of the parametrization. This rules out a priori design guidelines with polynomially related optimization times, for the considered memetic algorithm. All these findings indicate that finding a good parametrization remains an interesting and challenging topic for the years to come.

Chapter 6
Memetic Algorithms in Discrete Optimization

Jin-Kao Hao

6.1 Introduction

Discrete optimization concerns in essence the search for a "best" configuration (optimal solution) among a set of *finite* candidate configurations according to a particular criterion. There are several ways to describe a discrete optimization problem. In its most general form, it can be defined as a collection of problem instances, each being specified by a pair (S, f) [704], where S is the set of finite candidate configurations, defining the *search space*; f is the *cost* or *objective function*, given by a mapping $f: S \rightarrow R^+$.

Solving the instance (S, f) is to find an $s^* \in S$ such that $f(s^*) \leqslant f(s)$ for all $s \in S$ (this minimization formulation can easily be transformed into a maximization problem). Such a configuration s^* is a globally optimal solution (or simply an optimal solution) to the given instance.

Given its generality, discrete optimization allows many problems of practical and theoretical importance to be conveniently formulated. Examples are the classical problems of general integer programming, permutation problems (e.g., traveling salesman problem, bandwidth minimization, linear arrangement), and constraint satisfaction and optimization problems (satisfiability problems in propositional logic, graph partitioning, k-coloring). Discrete optimization naturally covers practical problems of the environment, renewable energy, distribution, infrastructure design, communications and productivity in the manufacturing and service sectors.

However, discrete optimization problems are known to be difficult to solve in general. Most of them, in particular those of practical interest, belong to the class of NP-hard problems, and thus cannot be efficiently solved to optimality. Over the past decades, important efforts have been made to improve the solution methods and important progresses have been achieved in both exact and heuristic strategies in pursuit of optimal or near optimal solutions.

Jin-Kao Hao
LERIA, Université d'Angers, 2 Boulevard Lavoisier, 49045 Angers Cedex 01, France
e-mail: jin-kao.hao@univ-angers.fr

This chapter concerns the design of Memetic Algorithms (MAs) [615, 617] for finding optimal or high quality near optimal solutions to hard discrete optimization problems.

6.2 Survey of Memetic Algorithms for Discrete Optimization

6.2.1 Rationale

From a fundamental point of view, the task of searching for a best solution in a combinatorial space is all about a suitable balance between "exploitation" and "exploration" for an effective examination of the given search space. The dual concept of exploitation and exploration covers two fundamental and complementary aspects of any effective search procedure. This concept is also known under the term "intensification" and "diversification" introduced within the Tabu Search (TS) methodology [317].

Exploitation emphasizes the ability of a method to examine intensively and in depth specific search areas while exploration is the ability of a method to diversify the search in order to find promising new search areas. Consequently, if the search focuses solely on exploitation, it will confine itself in a limited area, fails to visit other areas of the search space, and may be trapped in poor optima. On the other hand, a method relying heavily on exploration and overlooking exploitation will lack capacity to examine in depth a given area and miss out solutions of good quality. To be effective, a search method thus needs to appropriately conciliate exploitation and exploration. Memetic Algorithms constitute a very interesting framework offering a variety of strategies and mechanisms to achieve this general objective.

MAs are hybrid search methods that are based on the population-based search framework [35, 239] and neighborhood-based local search framework (LS) [393]. Popular examples of population-based methods include Genetic Algorithms and other Evolutionary Algorithms while Tabu Search and Simulated Annealing (SA) are two prominent local search representatives. The basic rationale behind a MA is to combine these two different search methods in order to take advantage of their complementary search strategies. Indeed, it is generally believed that the population-based search framework offers more facilities for exploration while neighborhood search provides more capabilities for exploitation. If they are combined in a suitable way, the resulting hybrid method can then offer a good balance between exploitation and exploration, assuring a high search performance.

Like other metaheuristics, MAs are a general optimization framework that can potentially be applied to various discrete search or optimization problems. Nevertheless, it should be clear that a blind application of MAs (or any other metaheuristics) to a particular problem will not be able to lead to satisfactory solutions. To be effective, the MA framework must be carefully adapted to the given problem and integrate problem-specific knowledge within its search operators and strategies. This is the key point of a successful MA application in practice.

6 Memetic Algorithms in Discrete Optimization

6.2.2 Memetic Algorithms in Overview

Memetic Algorithms [615, 617] are a population-based computational framework and share a number of features with methods like Evolutionary Algorithms [35, 239], and Scatter Search [320]. MAs operate on a set of candidate solutions and use these solutions to create new solutions by applying variation operators such as combinations and local improvements.

From a general perspective, a MA is composed of a number of basic components: a pool of candidate solutions (also called population of individuals) to sample the search space, a combination operator (crossover) to create new candidate solutions (offspring) by blending two or more existing solutions, an improvement operator to ameliorate offspring solutions, and a population management strategy. In addition to these elements, the MA also needs an evaluation or fitness function to assess the quality of each candidate solution as well as a selection mechanism to determine the candidate solutions that will survive and undergo variations.

From an operational perspective, a typical MA starts with an initial population (see §6.3.4) and then repeats cycles of evolution. Each cycle, also called a generation, consists of four sequential steps.

1. *Selection of parents*: Selection aims to determine the candidate solutions that will survive in the following generations and be used to create new solutions. Selection for reproduction often operates in relation with the fitness (quality) of the candidate solutions; high quality solutions have thus more chances to be chosen. Well-known examples of selection strategies include roulette-wheel and tournament. Selection can also be done according to other criteria such as diversity. In such a case, only "distanced" individuals are allowed to survive and reproduce. If the solutions of the population are sufficiently diversified, selection can also be carried out randomly. The selection strategy influences the diversity of the population (see also §6.3.3).

2. *Combination of parents for offspring generation*: Combination aims to create new *promising* candidate solutions by blending (suitably) existing solutions (parents), a solution being promising if it can potentially lead the optimization process to new search areas where better solutions may be found. To achieve this, the combination operator is often designed such that it captures the semantics of the targeted problem to ensure the heritage of good properties from parents to offspring. Additionally, the design of the combination operator should ideally take care of creating *diversified* offspring. From a perspective of exploration and exploitation, such a combination is intended to play a role of strategic diversification with a long term goal of reinforcing the intensification. A carefully designed combination operator constitutes a driving force of a successful MA.

3. *Local improvement of offspring*: The goal of local improvement is to improve the quality of an offspring as far as possible. For this purpose, local improvement takes an offspring as its input (current solution) and then iteratively

Algorithm 10. Memetic Algorithm Template

```
1  Input: |P|; // Size of population P
2  Output: s*; // Best solution found
3  P ← POPGENERATION(|P|); //
4  POPEVALUATION(P); // Fitness evaluation of each individual
5  s* ← best(P); // Record the best solution found so far
6  f* ← f(s*); // Record the fitness of the best solution
7  while Stop Condition is not verified do
8      (p₁...pₖ) ← PARENTSSELECTION(P); // k ⩾ 2 parents are selected
9      s' ← RECOMBINATION(p₁...pₖ); // Offspring generation
10     s ← OFFSPRINGIMPROVEMENT(s'); // Improvement of offspring
           solution by local search
11     P ← POPULATIONUPDATE(s,P); // Population update according
           to a quality-diversity rule
12     (s*,f*) ← BESTSOLUTIONUPDATE(s*,f*,P); // Best solution and its
           fitness are always recorded
13 endw
14 return s*
```

replaces the current solution by another solution taken from a given neighborhood. This process stops and returns the best solution found when a user-defined stop condition is met. Compared with the combination operator, local improvement plays essentially the role of intensifying the search by exploiting search paths delimited by the underlying neighborhood. Like combination, local improvement is another key component and driving force of a MA.

4. *Update of the population*: This step decides whether a new solution should become a member of the population and which existing solution of the population should be replaced. Often, these decisions are made according to criteria related to both quality and diversity. Such a strategy is commonly employed in methods like Scatter Search and many Evolutionary Algorithms. For instance, a basic quality-based updating rule would replace the worst solution of the population while a diversity-based rule would substitute for a similar solution according to a distance metric. Other criteria like recency (age) can also be considered. The policies employed for managing the population are essential to maintain an appropriate diversity of the population, to prevent the search process from premature convergence, and to help the algorithm to continually discover new promising search areas.

The general MA template is described in Algorithm 10 where special attention must be payed to the design of particular components. The stop condition can be a maximum number of cycles (generations), a maximum number of evaluations, a maximum number of cycles without improving the best solution, a solution quality to be reached or a lower-bounded threshold for the population diversity.

We deliberately leave out the mutation operator within this MA template. In some sense, local search can be viewed as a guided macro-mutation operator. However, mutation can also be applied to reinforce population diversity. As a lean design principle, only necessary components are included in a MA, any unjustified and superficial elements must be excluded.

6.2.3 Performance of Memetic Algorithms for Discrete Optimization

The computational performance of a MA depends first on the representation of the solution space (solution encoding) which should preferably be problem dependent and ease the design of efficient search operators.

The performance of a MA depends then on the design of its two key search components: Combination and local improvement operators. Their design should integrate useful problem-specific knowledge of the given problem in order to ensure aggressive exploitation and guided exploration.

The performance of a MA is also conditioned by the way the population is managed to promote and maintain a fertile diversity during the search process. Indeed, much like conventional Evolutionary Algorithms, premature convergence can easily occur if the population loses its diversity. Diversity management is particularly important with MAs because of the specific nature of their aggressive and intensified search strategies. Consequently, it is crucial for a MA to maintain with rigor a "good" population diversity as long as possible.

The interaction between the components of a MA can directly influence the behavior and the performance of the MA. A long or short local search phase after each combination could change the search trajectories. Similarly, a very effective local search procedure may weaken the role of the combination operator while a very strong combination operator may make it less critical to have a highly efficient local improvement procedure.

Finally, the runtime efficiency of a MA depends for a large part on the choice of the data structures employed to implement the different components of the MA. A typical example concerns local improvement procedures that explore the candidate solutions of a neighborhood and represent the most time-consuming part of a MA. In such a situation, it is critical to devise appropriate data structures to enable and streamline a fast neighborhood evaluation (see §6.3.1.3). Otherwise, the computational overheads will jeopardize the search power of the method.

6.3 Special Design Considerations

6.3.1 Design of Dedicated Local Search

Local improvement is one of the most important components of a MA and ensures essentially the role of intensive exploitation of the search space. This is typically achieved either by dedicated local search heuristics (see examples in [460, 523, 524]) or by tailored general neighborhood search methods. In this part, we focus our discussion on adaptation of local search metaheuristics [393], but a large part of the discussion applies to the design of local improvement procedures based on specific heuristics.

6.3.1.1 Local Search Template

Let (S, f) be our search problem where S and f are respectively the search space and optimization objective. A neighborhood N over S is any function that associates to each solution $s \in S$ some other solutions $N(s) \subset S$. Any solution $s' \in N(s)$ is called a neighboring solution or simply a neighbor of s. For a given neighborhood N, a solution s is a *local optimum* with respect to N if s is the best in terms of f among the solutions in $N(s)$.

The notion of neighborhood can be explained in terms of the *move* operator. Typically applying a move mv to a solution s changes s slightly and leads to a neighboring solution s'. This transition from a solution to a neighbor is denoted by $s' = s \oplus mv$. Let $\Gamma(s)$ be the set of all possible moves which can be applied to s, then the neighborhood $N(s)$ of s can be defined by: $N(s) = \{s \oplus mv | mv \in \Gamma(s)\}$.

A typical local search algorithm begins with an initial configuration s in S and proceeds iteratively to visit a series of configurations following the neighborhood. At each iteration, a particular neighbor $s' \in N(s)$ is sought to replace the current configuration and the choice of s' is determined by the underlying metaheuristic and by referring to the quality of the neighboring solution. For instance, a strict Descent algorithm always replaces the current solution s by a *better* neighbor s' while tabu search replaces the current solution by a *best* neighbor s' even if the latter is of inferior quality. Still with simulated annealing, the transition from s to a randomly selected neighbor s' is conditioned by a changing probability.

6.3.1.2 Neighborhood Design

The success of a LS algorithm depends strongly on its neighborhood. The neighborhood defines the subspace of the search problem to be explored by the method. For a given problem, the definition of the neighborhood should structure the search space such that it helps the search process to find its way to good solutions.

The choice of neighborhood is conditioned by the representation (genotype) used to encode the candidate solutions of the search space (phenotype). It may further depend on the structure and constraints of the problem on hand. Here we briefly review some neighborhoods associated to three conventional representations, which have a variety of applications.

- *Binary representation*: With this representation, each solution of the search space is coded by a binary string. Binary representation is very popular in discrete optimization due to the fact that many problems are naturally formulated with binary variables. Typical examples include SAT/Max-SAT, Knapsack, Unconstrained Quadratic Optimization, graph bi-partitioning etc. For these binary problems, two basic neighborhoods are defined by the k-$flip$ and *Swap* move operators. The k-$flip$ move changes the values of k ($k \geqslant 1$) variables. So any neighbor $s' \in N(s)$ has a Hamming distance of k to solution s. A larger k induces a larger (and stronger) neighborhood. Nevertheless, whether a larger neighborhood should be preferred in practice depends on the computational cost to evaluate the neighborhood. *Swap* exchanges the values of two variables that have different values. Note that *Swap* can be simulated by two 1-*flip* moves.

- *Permutation representation*: Here, each solution of the search space corresponds to a permutation $\pi : \{1..n\} \rightarrow \{1..n\}$. Permutation representation has a large range of applications in discrete optimization. Prominent examples include Traveling Salesman Problem, Flow-Shop/Job-Shop scheduling, Linear Arrangement, Bandwidth Minimization etc. Two basic neighborhoods for this representation are available using *Swap* and *Rotation* moves. Given a permutation (solution) π, The *Swap* move exchanges $\pi(i)$ and $\pi(j)$ for some i and j ($i \neq j$). If π' is a neighbor of π by swapping i and j, then $\pi'(k) = \pi(k)$ for $k \neq i, j$, $\pi'(i) = \pi(j)$ and $\pi'(j) = \pi(i)$. The *Rotation* move rotates all the values between $\pi(i)$ and $\pi(j)$ for some $i < j$. Thus, if π' is a rotation neighbor of π obtained with $i < j$, then $\pi'(k) = \pi(k) + 1$ for $i \leqslant k < j$, $\pi'(j) = \pi(i)$, and $\pi'(k) = \pi(k)$ for all other k. Note that $Rotation(i, j)$ can be simulated by $j - i$ successive *Swap* moves starting with $Swap(i, i+1)$.

- *Integer representation*: With this representation, each solution of the search space corresponds to an integer vector whose values are taken from some discrete domains. Integer representation is very useful and convenient for many constraint satisfaction and optimization problems. A common neighborhood is defined by a "one-change" move that consists in replacing the current value of a single variable by a new domain value. The set of candidate variables under consideration for a value change can be identified with a number of rules specific to the problem at hand. For instance, if the search algorithm deals with unfeasible solutions, i.e. some variables are receiving conflicting values relative to some constraints, the set of candidate variables can be constituted of the subset of *conflicting* variables [289, 291, 672]. Such a neighborhood is typically employed in local search algorithms for solving Constraint Satisfaction Problems. More generally, candidate variables for a value change can be identified as those that are critical for improving the objective function or for reaching the feasibility.

These neighborhoods can be applied directly to a given problem if the problem fits well the required representation. A common practice is to adapt a conventional neighborhood with problem-specific knowledge. Moreover, in some situations, it is useful to investigate the possibility of multiple neighborhoods that can be applied at different stages of the search process (see §6.3.1.4 below).

6.3.1.3 Neighborhood Evaluation

Another design issue that arises is the evaluation of a given neighborhood. Indeed, a local search procedure moves iteratively from the current solution to a new solution chosen within the neighborhood. To make this choice, local search needs to know the cost variation (also called the *move value*) between the current solution s and a candidate neighbor $s' \in N(s)$. The move value indicates whether the neighbor s' is of better, worse or equal quality relative to s. Let $\Delta f = f(s') - f(s)$ denote this move value.

- *Incremental evaluation*: Basically, there are two ways to obtain Δf for a neighbor. The trivial way is to calculate $f(s')$ from "scratch" using the objective function[1] f. Doing this way may be expensive if f needs to be evaluated very often or if the evaluation of f itself involves complex calculations. A more efficient alternative aims to derive the value of $f(s')$ from the value $f(s)$ by updating only what is strictly necessary. Indeed, if a neighbor s' is close to its initial solution s, which is true for many neighborhoods, then the evaluation of $f(s')$ can be carried out in this incremental manner. For a number of basic neighborhoods, like those shown previously, such an incremental evaluation is often possible.

- *Full search of neighborhood*: The incremental evaluation can be applied to *all the neighbors* of a given neighborhood relation. In this case, it is generally useful to investigate dedicated data structures (call it Δ-table) to store the move values for all the neighbors of the current solution. Δ-table provides a convenient way to know the quality of each neighbor and enables an efficient search of the full neighborhood. With such a Δ-table, the local search algorithm can decide easily at each iteration which neighbor to take according to its search strategy. For instance, a best-improvement descent algorithm will take the move that is identified by the most negative value in the Δ-table to minimize the objective function. After each move, the Δ-table (often only a portion of it) is updated accordingly using the incremental evaluation technique to propagate the effect of the move. Δ-table is a very useful technique for local search algorithms. This is particularly the case for descent-based methods like Tabu Search where a best neighbor needs to be identified (see examples in [393]).

- *Approximative evaluation*: The practical usefulness of Δ-table depends on both the complexity and the number of updates needed after each move transition. It may happen that, the move value can not be incrementally calculated or the Δ updates need to change a large portion of Δ-table. In this case, it would be useful to replace the initial evaluation function by a (fast) approximative evaluation function [424]. More generally, approximate evaluation is useful if the evaluation function is computationally expensive to calculate or if the function is ill-defined.

- *Order of evaluation*: If the neighborhood is not completely searched, one must decide the order in which the neighborhood is explored. For instance, the first-improvement descent technique moves to any improving neighbor. If there are several improving neighbors, the descent search picks the "first" one encountered

---
[1] For the reason of simplicity, the term "objective function" is used here. A more precise term is "evaluation function", see §6.3.4.

in the order the neighbors are examined. To allow such a method to increase its search diversity, a random order may be preferred [704].

6.3.1.4 Combination of Neighborhoods

Very often, different neighborhoods may be available, enabling alternative ways to explore the search space. In such a situation, it is interesting to consider combined use of multiple neighborhoods. For illustrative purpose, consider two neighborhoods N_1 and N_2. Then one can consider at least three ways to use them in a combined way.

First, *neighborhood union* $N_1 \cup N_2$ includes all the neighbors of the two underlying neighborhoods, so that any member of N_1 and N_2 is a member of $N_1 \cup N_2$. A local search algorithm using this combined neighborhood selects the next neighboring solution among all the solutions in both neighborhoods. This combination has no sense if one neighborhood is fully included in the other one.

With *Probabilistic neighborhood union* $N_1 \oslash N_2$, a neighbor solution in N_1 (or N_2) belongs to $N_1 \oslash N_2$ with probability p (resp. probability $1-p$). A local search algorithm using this combined neighborhood selects at each iteration the next neighbor from N_1 with probability p and from N_2 with probability $1-p$.

Token-ring combination $N_1 \rightarrow N_2$ is time-dependent and defined alternatively either by N_1 or N_2 according to some pre-defined conditions [209]. A local search algorithm using this combined neighborhood cycles through these neighborhoods. It typically starts with one neighborhood until the search stagnates, then changes to the other neighborhood until the search stagnates again to switch back to the first neighborhood and so on.

The advantage of combined neighborhood was already demonstrated a long time ago in [524] for solving the Traveling Salesman Problem. More generally, the issue of transitioning among alternative neighborhoods was discussed with the Tabu Search framework and strategic oscillation design in [312]. More recent examples of local search methods focusing on multiple neighborhoods include Variable Neighborhood Search [363], Neighborhood Portfolio Search [209] and Progressive Neighborhood Search [323]. Examples of studies on neighborhood combinations can be found in [353, 539].

6.3.2 Design of Semantic Combination Operator

6.3.2.1 Solution Combination

Combination is another key component of a MA and constitutes one leading force to explore the search space. The basic idea of combination is very appealing since it provides a very general way of generating new solutions by mixing existing solutions. Contrary to local changes of local improvement, combination can bring into new solutions more useful information, that may be beneficial for a healthy evolution of the search process.

As a first step, it would be tempting to consider the application of a blind (random) crossover operators for solution combinations. Doing this has the advantage of ease of application. However, one question should be asked before this approach is attempted: Is the crossover operator meaningful with respect to the optimization objective? If the answer is negative, the crossover operator is probably not appropriate and the sole role it would play in this case would be to introduce some random diversification in the search process.

In practice, instead of applying blind crossovers, it is often preferable to consider dedicated combination operators that have strong "semantics" with respect to the optimization objective. A semantic combination aims to pass intrinsic good properties from parents to offspring. The design of such a combination operator is far from trivial and in fact represents a challenging issue. Although there are some theoretical guidances, the discovery of such a semantic combination operator in practice relies basically on a deep analysis and understanding of the given problem. Compared with the design of local search procedures, the design of a meaningful combination operator constitutes probably one of the most creative parts of an effective MA.

6.3.2.2 Theoretical Foundations

The schemata theory [389] and the building block hypothesis [325] are often mentioned to explain (partially) the performance of Genetic Algorithms. Intuitively, building blocks are promising patterns of solutions that can be progressively assembled by crossover to get improved solutions. Given that this theory is defined for binary and simple Genetic Algorithm, it is not directly applicable in the context of MAs. Nevertheless, assembling building blocks to generate new solutions remains an appealing idea. In [750, 753], the concept of forma is introduced to generalize the schemata theory. A formal framework is even proposed to try to capture some fundamental aspects of MA in [752]. The forma theory suggests a set of general principles for the design of solution representations and recombination operators. According to this theory, a suitable recombination operator is required to fulfill two conditions called *respect* and *proper assortment*. Intuitively, the *respect* condition advocates the heritage of shared characteristics of parents to offspring, while *proper assortment* ensures the heritage of desirable characteristics of each parent by their offspring. This is in accordance with the general principle of conserving good features through inheritance and discarding bad features developed in Grouping Genetic Algorithms [248].

6.3.2.3 Design of Combination Operator

These abstract considerations only provide us with very general guidances for designing recombination operators. For a particular problem, it is still necessary to find out what are the building blocks (interesting patterns or characteristics) of solutions that can be assembled and inherited through the recombination process. Unfortunately, there is no short-cut to this quest and a fine analysis and deep understanding of the given problem is indispensable to find useful clues.

First, one can analyze the samples of optimal or high quality solutions to possibly identify regular patterns shared by these solutions. Indeed, if such a pattern exists, then the recombination operator can be constrained to conserve the pattern from the parent solutions and to avoid breaking the pattern. Alternatively, the recombination operator can also be encouraged to promote the emergence of favorable building blocks. For instance, such an analysis applied to the Traveling Salesman Problem shows that high quality local optima share sub-tours [523, 524]. This property has been used by several highly successful crossover operators which conserve common edges or sub-tours in offspring solutions [286, 636, 648, 720, 931]. Similarly, for the graph k-coloring problem, an analysis of coloring solutions discloses that some nodes are always grouped to the same color class (i.e. colored with the same color). This characteristic has helped to devise powerful combination operators, as shown in [217, 290] and in [292, 537, 549, 726] with multi-parents.

6.3.2.4 Multi-Parent Combination

Combination may operate with more than two parents. Multiple parent combination is even a general rule for the Scatter Search metaheuristic which uses, in its original form, linear combinations of several solutions to create new solutions [308]. Although there is no theoretical justifications, the practical advantage of multiple parent recombination was demonstrated in several occasions for discrete optimization. For instance, for the graph k-coloring problem, several recent and top-performing algorithms integrate multiple parent combination [292, 537, 549, 726], where color classes from different solutions are assembled to build offspring colorings. More generally, when multiple solutions are used for creating a new solution, one can define special rules to score the solution components of each parent solution and use strategic voting rules to combine components from different parents solutions.

A question that arises for multi-parent combination is how to determine the number of the parents. By using two parents, the offspring is expected to inherit 50% material from each parent. The contribution of each parent to the new solution descreases with an increasing number of parents. If the building blocks from different parents are independent from one another, taking more parents into account would be interesting to build good and diversified offspring. Otherwise, if a building block from a parent is epistatic with respect to the building blocks of other parents, blending more parents means more disruption, and thus should be avoided.

6.3.3 Population Diversity Management

Population diversity is another important issue that should be considered in the design of an effective MA [290, 726, 836]. If the population diversity is not properly managed, the population will converge prematurely and the search process stops with poor local optima. This is particularly true when a small population is used by the MA. In what follows, we first provide some precisions about the nature of diversity and explain how fertile diversity can be promoted and maintained within a population. Note however that diversity is not interesting per se within a MA. The ultimate goal of population diversity is to help the search process not only to avoid premature converge, but also to continually discover interesting new solutions in order to explore non-visited promising search areas. See also Chapter 10.

6.3.3.1 Diversity

Population diversity can be measured by a similarity (or distance) metric applied to the members of the population. The metric can be defined either on the solution representation level (genotype metric) or solution level (phenotype metric) [325]. For instance, pair-wise *Hamming distance* can be used as a genotype metric to measure population diversity. Diversity can also be measured in terms of *entropy* [267] or by the so-called *moment of inertia* [614]. Genotype metric is usually problem independent, and thus may or may not reflect the intrinsic diversity of a population with respect to the given optimization objective.

Population diversity can also be measured at the phenotype level over the solution space. For instance, for partition problems like graph k-coloring, the distance between two partitions can be measured by the so-called *transfer distance* which is the minimum number of elements that need to be moved between classes of one partition so that the resulting partition becomes the other partition [189, 763]. A phenotype metric is defined over the solution space and thus is more likely to measure the real diversity of a population.

In order to observe suitably the population diversity, it is useful to first determine the most appropriate distance or similarity metric with respect to the optimization objective of the given problem. Moreover, if the population diversity needs to be continually monitored, it becomes important to pay attention to the cost of computing the underlaying metric.

6.3.3.2 Promoting and Maintaining Useful Diversity

Population diversity can be promoted and managed at several levels of a MA. One evident possibility is to define specific selection rules to favor the selection of *distanced parents* for mating. Another possibility concerns the variation operators which can be designed in such a way that they favor the generation of diverse and varied offspring. For instance, the "Distance Preserving Crossover" introduced in [286, 588] is constrained to generate an offspring which is at the same distance from both parents. More generally, the path-relinking type of combinations typically construct offspring solutions by considering both the solution quality and its distance to its parent solutions [320] (see also [538] for an example).

Population diversity can also be controlled by the offspring acceptance and replacement strategies. Specifically, this can be done according to both solution diversity and quality. For instance, in [726] a minimum diversity-quality threshold is imposed between the solutions of the population. The acceptance of a new offspring is conditioned not only by its quality, but also by its distance to existing solutions. Similarly, diversity and quality are considered to select the victim solution to be replaced by the offspring.

Other useful ideas for diversity preservation can be found in the areas of Genetic Algorithms. Well-known examples include sharing [327] and crowding [204, 546].

6.3.4 Other Issues

In addition to the components mentioned until now, the design of an effective Memetic Algorithm should take into account a number of other considerations which are briefly discussed in this section.

- *Initial population*: There are basically two ways to obtain an initial population: Random generation and constructive elaboration. While random generation is easy to apply, it can hardly generate initial solutions of good quality. To improve the basic random generation method, a simple sampling technique can be applied. Let P be the population size, then one can generate $K > P$ solutions and then retain only the P "best" ones. Initial generation by construction can be used if some fast greedy heuristics are available for the given problem. Notice that, in this case, the greedy heuristics must be randomized such that each application leads to a different solution. Another issue that can be considered at the initialization stage is to take care of building a diversified population. This can be achieved by controlling the distance between each new solution and the existing solutions of the population. Only distant new solutions are allowed to join the population.

- *Distance*: At several places, MAs may need to measure the distance between two solutions or between a solution and a group of solutions. For instance, parents selection may operate in such a way that the selected parents are sufficiently distant. Similarly, a population management strategy may decide the acceptation or rejection of an offspring by considering its distance to the members of the population. When an operation refers to the notion of distance, it is preferable to employ an appropriate distance metric which is meaningful with respect to the given problem. For instance, for partition problems like graph coloring (see §6.4.1), Hamming distance is not a suitable metric to characterize the difference of two partitions. Instead, transfer distance between partitions should be preferred. Once again, the choice of the distance metric should ideally be correlated with the semantics of the problem on hand.

- *Rich evaluation function*: Evaluation function assesses the quality of a candidate solution with respect to the optimization objective and orients the search method to "navigate" through the search space. A good evaluation function is expected to be able to distinguish each solution from the other solutions and thus to effectively guide the search method to make the most appropriate choice at each iteration. Very often, the initial optimization objective f is directly used as evaluation function. However, such a function may not be sufficiently discriminant to distinguish different solutions. To improve the discriminating power, it is useful to incorporate in the evaluation function additional information, e.g. relative to the structure of the problem instance to be solved. Examples can be found in [248, 431, 772]. Moreover, when constrained optimization problems are considered, some constrains may be hard to satisfy, and thus are relaxed. Among various constraint relaxation techniques, a common practice is to integrate the relaxed constraints into the evaluation function as a (weighted) component or as a part of a multi-component evaluation function (see examples in [316, 902, 903]).

- *Constraints*: The constraints in the considered problem may influence the design of some MA components. For instance, suppose that the MA algorithm is expected to explore only feasible solutions. Then one must decide whether a combination operator is constrained to create only feasible solutions. If infeasible offspring is allowed, it is necessary to consider a dedicated mechanism to repair the broken constraints. Similarly, neighborhood design can take into consideration the constraints to identify eligible moves. For instance, in feasibility search problems, this is often done by identifying problem variables involving violated constraints and restricting the set of authorized moves to those defined on these conflicting variables. Finally, as previously stated, constraints that are difficult to solve can be used in the design of the evaluation function.

- *Connections with Scatter Search and Path Relinking*: As discussed in [311] and [317] (Chapter 9), the MA framework shares ideas with Scatter Search and Path Relinking [313, 320]. These latter methods provide unifying principles for joining solutions based on generalized path constructions (in both Euclidean and neighborhood spaces) and by using strategic design. Solution combination in Scatter Search originated historically from strategies for combining decision rules and combining constraints. In Scatter Search, dispersed new solutions are created from a set of reference solutions by weighted combinations of subsets of the reference solutions that are selected as elite solutions. With Path Relinking, offspring solutions are generated by exploring, within a neighborhood space, trajectories that connect two or more reference solutions. One notices that the reference solutions or subsets of them can be considered as parent solutions for combination while combination resorts to diverse strategies such as attribute voting and weighting.

6.4 Case Studies

In this section, we show two case studies of quite different nature with the purpose of showing how these issues can be effectively implemented in practice. We particularly focus on the design of combination and local search operators.

6.4.1 Graph Coloring Problems

6.4.1.1 Problem Description

Given an integer k and a undirected graph $G = (V, E)$ with a set V of vertices and a set E of edges, a legal k-coloring of G is a partition of V into k distinct color classes such that each color class is composed of pairwise non-adjacent vertices. The graph k-coloring problem (k-COLOR) aims at finding a legal k-coloring for a fixed k while the graph coloring problem (COLOR) determines the smallest k for a given graph G (its *chromatic number* χ_G) such that G has a legal k-coloring. Since COLOR can be handled by solving a series of k-COLOR with decreasing k values, we only consider here k-COLOR.

For a given k-COLOR instance, i.e. an integer k and graph $G = (V, E)$, let $s = \{C_1, C_2...C_k\}$ denote a partition of V into k distinct color classes such that each C_i ($i \in \{1, 2...k\}$) contains all the vertices that are colored with color i. Let S denote all such partitions. For any $s \in S$, define its conflict number $f(s)$ to be the number of pairs of adjacent vertices x and y ($\{x, y\} \in E$) belonging to a same color class of s. Then k-COLOR can be solved by minimizing $f(s)$; $f(s)=0$ implies that s is a legal k-coloring, i.e. all its color classes C_i are conflict-free.

Notice that among the large number of existing heuristic algorithms for k-COLOR, Memetic Algorithms are certainly among the most powerful ones and provide the best results on the well-known DIMACS benchmark instances of this well-known NP-complete problem.

6.4.1.2 Partition Crossovers

In order to design a semantic combination operator, let us try to get an idea about the possible "building blocks" for our problem. The goal of k-COLOR is to determine a set of k *distinct* conflict-free color classes. In this context, color classes can be considered our basic "building blocks". If there are several "good" color classes among some candidate solutions, then these color classes can favorably be recombined to obtain new candidate solutions. This idea was first explored by the Greedy Partition Crossover (GPX) described in [290] and the Union of Independent Sets crossover in [217], which are also related to the design of grouping crossovers described in [248].

Operating with two parent k-colorings s_1 and s_2, GPX builds step by step the k classes $C_1^0, \ldots, C_k^0$ of the offspring s_0. At the first step, GPX creates C_1^0 by choosing a *largest* class from one parent and removes its vertices from both parents s_1 and s_2. GPX repeats then the same operations for the next k-1 steps, but alternates each time the parent considered. If some vertices remain unassigned at the end of these k steps, they are randomly assigned to one of the k color classes. The alternation between the parents aims at a balanced mixture of information from both parents and avoiding the dominance of one parent over the other one during the recombination.

Table 6.1 shows an example with 3 color classes ($k = 3$) and 10 vertices represented by capital letters A,B,$\cdots$,J.

Table 6.1. The Greedy Partition Crossover: An example from [290]

parent $s_1 \rightarrow$	A B C	D E F G	H I J	$C_1^0 := \{D.E.F.G\}$	A B C		H I J
parent s_2	C D E G	A F I	B H J	remove D,E,F and G	C	A I	B H J
offspring s					D E F G		

parent s_1	A B C		H I J	$C_2^0 := \{B.H.J\}$	A C		I
parent $s_2 \rightarrow$	C	A I	B H J	remove B,H and J	C	A I	
offspring s	D E F G				D E F G	B H J	

parent $s_1 \rightarrow$	A C		I	$C_3^0 := \{A.C\}$			I
parent s_2	C	A I		remove A and C		I	
offspring s	D E F G	B H J			D E F G	B H J	A C

The basic idea underlying GPX was also explored with multiple parent combination operators [292, 352, 537, 549, 726]. Using multiple parents for combination is fertile for k-COLOR since this offers more possibilities to obtain good (large) color classes for each step of the recombination operation. By generalizing two parents to multiple parents, refined and additional strategies were also introduced to make the combination process as effective as possible. For instance with the AMaPX operator of [537], in order to favor the creation of *diversified offspring*, each time a color class from a parent is transmitted to the offspring, this parent's k-coloring will not be considered for the next few steps of offspring building. In [726], in order to measure the goodness of the color classes of the parent colorings, the combination operator takes into account the size of each color class, the number of conflicting vertices as well as the degrees of the vertices in the color class.

A question that arises when multiple parents are used is how to determine the number of parents. It is clear that by using more parents, fewer classes will be transmitted from each parent to the offspring and this also implies that the class blending from each parent is also more disrupted. An analysis of the relations between the number of vertices, the number of color classes and the number of parents permits to identify a heuristic rule to fix the right number of parents [726].

In [292], the combination operation is performed within a slightly different context. The algorithm maintains a pool of conflict-free color classes obtained during the search process. From time to time, these color classes are used to generate new k-colorings. Other combination operators using similar ideas are investigated in [217, 352, 549].

6.4.1.3 Local Improvement by Tabu Search

In memetic coloring algorithms, Tabu Search is frequently used for local improvement to ameliorate a new offspring created by the combination operator. For illustration purpose, we use the TS algorithm described in [290] as an example. It uses the constrained "one-change" move described in §6.3.1.2 such that a *neighbor* s' of a given configuration s is obtained by moving a single *conflicting* vertex v from a color class C_i to another color class C_j. When such a move $< v, i >$ is performed, the

6 Memetic Algorithms in Discrete Optimization

couple $<v,i>$ is classified tabu for the next tl iterations. Therefore, v cannot be reassigned to the class i during this period, unless moving v back to the color class i leads to a configuration better than the best configuration found so far (*aspiration criterion*). The tabu tenure tl for a move is variable and depends on the number nb_{CFL} of conflicting vertices in the current configuration: $tl = Random(A) + \alpha * nb_{CFL}$ where A and α are two parameters and the $Random(A)$ function returns a random number from $\{0, \cdots, A-1\}$. To implement the tabu list, it is sufficient to use a $|V| \times k$ table.

The algorithm memorizes and returns the *most recent* configuration s_* among the best configurations found: After each iteration, the current configuration s replaces s_* if $f(s) \leqslant f(s_*)$ (and not only if $f(s) < f(s_*)$). The rational to return the last best configuration is that we want to produce a solution which is as far away as possible from the initial solution in order to better preserve the diversity in the population.

6.4.2 Maximum Parsimony Phylogeny

6.4.2.1 Problem Description

Phylogenetics is the study of evolutionary relationships among various groups of organisms (for example, species or populations). These connections are represented graphically through phylogenetic trees. Computational phylogenetics aims to infer phylogenetic trees from molecular data such as protein or DNA sequences [256]. The main phylogenetic approaches include methods using a distance-matrix, the maximum likelihood or maximum parsimony criterion.

Maximum parsimony phylogeny generally takes as input a multiple sequence alignment which is a matrix M of characters composed of n lines (related to a set S of species, where $|S| = n$) and k columns which represent the characters of the sequences [255]. Each sequence is also called a taxon. Each character of the matrix belongs to an alphabet Σ. A phylogenetic tree T of the given input is a binary tree such that (1) the leaves of T are the set of n species, and (2) each internal node is induced by the sequence of parsimony of its two descendant sequences. Given two sequences $S_1 = <x_1, \cdots, x_k>$ and $S_2 = <y_1, \cdots, y_k>$ with $\forall i \in \{1..k\}, x_i, y_i$ belonging to the power set $\mathscr{P}(\Sigma = \{-, A, C, G, T\})$, the sequence of parsimony $P(S_1, S_2) = <z_1, \cdots, z_k>$ of S_1 and S_2 is given by ([264]) :

$$\forall i, 1 \leqslant i \leqslant k, z_i = \begin{cases} x_i \cup y_i, \text{if } x_i \cap y_i = \emptyset \\ x_i \cap y_i, \text{otherwise} \end{cases} \quad (6.1)$$

The score of the sequence of parsimony defines the "distance" separating its two descent sequences:

$$f_{P(S_1,S_2)} = \sum_{i=1}^{k} c_i \quad \text{where} \quad c_i = \begin{cases} 1, \text{if } x_i \cap y_i = \emptyset \\ 0, \text{otherwise} \end{cases} \quad (6.2)$$

Algorithm 11. The general DiBIP crossover scheme

Input: $T_1, T_2, \delta, \Delta, \oplus, \Lambda$
Output: A child tree T^*

1. Apply the tree-to-distance operator Δ to each parent tree T_i (i=1,2) to obtain the corresponding distance matrix $D_i = \Delta(T_i)$;
2. Apply the matrix operator $\oplus$ to D_1 and D_2 to obtain D^*: $D^* \leftarrow D_1 \oplus D_2$;
3. Apply the distance-to-tree operator Λ to D^* to obtain a child tree: $T^* \leftarrow \Lambda(D^*)$.

Let T be a binary parsimony tree with n leafs or species. T has then $n-1$ sequences of parsimony (internal nodes). Let I denote the set of these internal nodes. The Fitch parsimony score $f(T)$ of T is defined as follows:

$$f(T) = \sum_{i \in I} f_i(T) \qquad (6.3)$$

The aim of the Maximum Parsimony problem (MP) is then to find a most parsimonious phylogenetic tree T^* such that T^* minimizes the parsimony score. Since there are $\prod_{i=3}^{n}(2i-3)$ possible binary trees with n leafs, this problem is a highly combinatorial search problem. The MP problem is computationally difficult since its associated decision problem is equivalent to the NP-complete Steiner problem in a hypercube [277]. MP has been subject of many studies for many years. Among them, neighborhood-based local search and various hybrid algorithms are certainly the most popular solution methods. In what follows, we show a Memetic Algorithm called HYDRA [767], which combines a dedicated tree crossover called DiBIP [322] and a progressive neighborhood local search method [323].

6.4.2.2 Distance-Based Information Preservation Crossover

First, let us notice that conventional tree crossovers known in genetic programming are not suitable here. The Distance-Based Information Preservation crossover (DiBIP) is specifically designed for the MP problem. DiBIP is based on a topological distance between species (leafs) and aims to preserve common properties of parents in terms of this distance between species. For instance, two species that are close (or far) in both parents should stay close (resp. distant) in the offspring. Given two parents trees, the DiBIP crossover is realized in three steps: Calculate a distance matrix for each parent tree, then combine the two resulting matrices to get a third matrix and finally create a child tree from this last matrix.

The general DiBIP crossover scheme is described in Algorithm 11 where T_1 and T_2 denote two parents trees. δ is a distance metric to measure the distance of each pair of species of a tree T, Δ a tree-to-distance operator to obtain a distance matrix of a tree, $\oplus$ a matrix operator to combine 2 distance matrices to produce a new distance matrix, Λ a distance-to-tree operator to construct a tree from a given distance matrix.

A specific DiBIP crossover operator is obtained once δ, Δ, $\oplus$, and Λ are provided. The distance measure δ should be ideally correlated to the evolutionary

6 Memetic Algorithms in Discrete Optimization

changes between species. For instance, 2 species separated in the tree by a small number of evolutionary changes should have a smaller distance than 2 species separated by a large number of changes. The distance measure should additionally be tree-topology dependent. In this sense, the length of the elementary path between 2 species is a possible option while Hamming distance is not suitable here because this metric is totally independent of tree topologies.

Moreover, since we want to preserve representative features of the parents during the crossover operation, a valid matrix operator $\oplus$ should favor such an inheritance from parents to offspring and meet some relation preservation property. For instance, if a pair of species (a,b) is closer than another pair (c,d) in both parents, then this relation should be conserved. Consider the operation $\oplus$ such that for a pair of species (i,j), $(D_1 \oplus D_2)(i,j) = \alpha . \min\{D_1(i,j), D_2(i,j)\} + (1-\alpha) . \max\{D_1(i,j), D_2(i,j)\}$ with $\alpha \in [0,1]$. This indeed defines a valid $\oplus$ operator. Furthermore, this definition offers in fact many possibilities and seems particularly relevant to MP. For instance, the arithmetic average ($\alpha = 0.5$) and the max operator max ($\alpha = 0$) are 2 special cases. At last, let us mention that the arithmetic addition is another simple valid $\oplus$ operator.

We now show a concrete example. Given two species i and j, define their distance δ_{ij} to be the *topological distance*, i.e. the length of the elementary path between the respective ascendants of i and j, (minus 1 if the path contains the root of the tree T). The matrix operator $\oplus$ is the addition $+$ such that $D(i,j) = D_1(i,j) + D_2(i,j)$, which satisfies the relation preservation property previously mentioned. The distance-to-tree operator Λ is a non-deterministic variant of the well-known UPGMA (Unweighted Pair Group Method with Arithmetic Mean) method [833]. Figs. 6.1 and 6.2 show an application of this crossover operator. One observes that the closeness of species in both parents is conserved in the child. This observation applies equally to distant species.

6.4.2.3 Progressive Neighborhood Search

For local improvement, HYDRA uses *Progressive Neighborhood Search* (PNS) which operates with a variable-size neighborhoods [323]. Given a parsimony tree T, a neighboring tree T' is typically obtained by a move that consists in cutting a sub-tree from T and reinserting the sub-tree elsewhere in the initial tree. If a meaningful metric can be defined to measure the distance between the cutting and inserting points, then it would be possible to define neighborhoods of variable sizes. In [323], the topological distance δ shown in Section 6.4.2.2 is used for this purpose. A distance parameter d is introduced to constrain the distance between the pruned edge i and the edge j receiving the insertion such that $\delta_{ij} \leqslant d$.

So, setting $d = \infty$ leads to a large neighborhood where the pruned edge (with its subtree) can be reinserted anywhere in the tree. Consequently, the topological change can be important. This case corresponds in fact to the well-known *Subtree Pruning Regrafting* neighborhood [862] whose size equals $2(n-3)(2n-7)$ [12]. Reversely, setting $d = 1$ gives a small neighborhood where neighboring trees are close to the current tree. This case corresponds to another well-known neighborhood

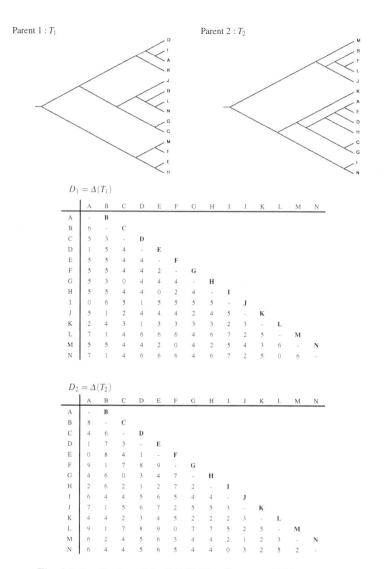

Fig. 6.1. Application of the DiBIP Tree Crossover [322] – The parents

called *Nearest Neighbor Interchange* [922] which swaps two adjacent branches of the tree leading to $(2n-6)$ neighbors [770]. By varying the parameter d, one gets neighborhoods of intermediate sizes.

The *Progressive Neighborhood Search* is based on this parametric neighborhood and its neighborhood changes during the search process by varying the value of d. In the particular MP context, PNS carries out a descent search starting with a large neighborhood (i.e. with large d) and reduces progressively the neighborhood. Indeed, at the beginning of the search, it is possible to obtain strong quality

$D^* = D_1 \oplus D_2$

	A	B	C	D	E	F	G	H	I	J	K	L	M	N
A	-	B												
B	14	-	C											
C	9	9	-	D										
D	2	12	7	-	E									
E	5	13	8	5	-	F								
F	14	6	11	12	11	-	G							
G	9	9	0	7	8	11	-	H						
H	7	11	6	5	2	9	6	-	I					
I	6	10	9	6	11	10	9	9	-	J				
J	12	2	7	10	11	6	7	9	8	-	K			
K	6	8	5	4	7	8	5	5	4	6	-	L		
L	16	2	11	14	15	6	11	13	12	4	10	-	M	
M	11	7	8	9	8	3	8	6	7	5	5	9	-	N
N	13	5	8	11	12	11	8	10	7	5	7	5	8	-

Child : $T^* = \Lambda(D^*)$

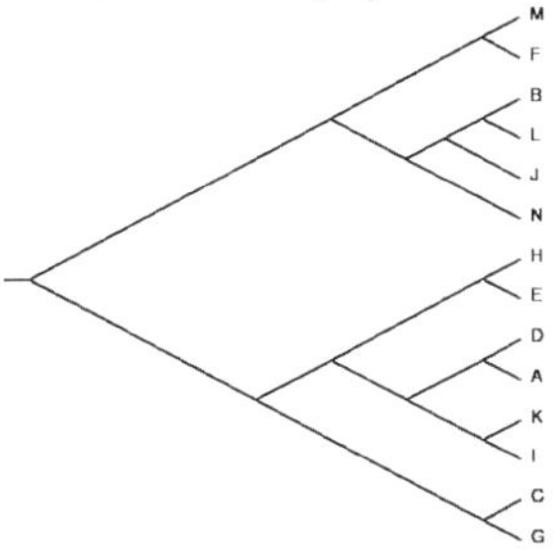

Fig. 6.2. Application of the DiBIP Tree Crossover [322] – The offspring

improvement by important topological modifications of the tree with large d. When the search progresses and the quality of the trees becomes better and better, only small improvements can be expected with small tree modifications. It is thus more judicious to switch to smaller and small neighborhoods to accelerate the search.

One notices that PNS shares some features with Variable Neighborhood Search (VNS) [363]. However, contrary to VNS, the neighborhoods explored by PNS are not systematically of increasing sizes. Within the context of our Maximum Parsimony problem, PNS even progressively reduces its neighborhood.

6.5 Conclusions

In this chapter we have presented the basic concepts of Memetic Algorithms for Discrete Optimization. Focus is given to the key design issues of an effective MA algorithm. We have explained the usefulness of a deep study and understanding of the optimization problem on hand. We have insisted on the importance of a careful adaptation of the general search strategies offered by the MA framework, a suitable incorporation of problem specific knowledge in different components of the MA as well as a logical integration of these components. The pursuit goal is clearly to build an effective MA algorithm that is able to ensure a balanced exploitation and exploration of the search space.

It should be clear that a blind MA application would have little chance to deliver good results for difficult optimization problems. High performance can only be possible by a disciplined and careful specialization of the general MA framework to the targeted problem. It is equally important to apply the "lean design" principle in order to avoid redundant or superficial algorithmic components.

The framework of Memetic Algorithms constitutes an interesting enrichment to the arsenal of existing discrete optimization methods and offers a valuable alternative for tackling hard discrete optimization problems.

Acknowledgements. This work was partially supported by *"Angers Loire Métropole"* and the Region of *"Pays de la Loire"* within the following projects: MILES, BIL, Radapop and LigeRO. I'm grateful to my research collaborators within and outside the Group "Metaheuristics, Optimization and Applications" of the LERIA Laboratory.

Chapter 7
Memetic Algorithms and Fitness Landscapes in Combinatorial Optimization

Peter Merz

7.1 Introduction

Combinatorial optimization problems (COPs) arise in many practical applications in the fields of management science, biology, chemistry, physics, engineering, and computer science. Although the search space is comprised of a finite number of candidate solutions, many of these problems are very complex and thus hard to solve. Often, the search space grows exponentially with the problem size rendering enumeration schemes impractical. Moreover, for many problems it has been shown that they are NP-hard, hence no polynomial time algorithm is known to find optimum solutions. Therefore, effective meta-heuristics are required to find (near) optimum solutions in short time. Memetic algorithms are known to perform well for a wide range of combinatorial optimization problems. Still, an open question is when and why they perform so well. After providing an overview and a common outline of memetic algorithms for combinatorial optimization problems in section 2, we introduce the concept of fitness landscapes in section 3 to address these two questions. In Section 4 and 5 we present case studies of the TSP and the BQP, respectively, in which we show and discuss results from the fitness landscape analysis. Furthermore, we discuss the state-of-the-art meta-heuristics for these problems. Section 6 concludes the chapter.

7.2 MAs in Combinatorial Optimization

The travelling salesman Problem (TSP) is one of the best-known combinatorial optimization problems. Often, new new ideas in meta-heuristics have initially been tested on the TSP and were applied afterwards to other combinatorial problems.

Peter Merz
University of Applied Sciences and Arts Hannover,
Department of Computer Science and Business Administration, 30459 Hannover, Germany
e-mail: `peter.merz@fh-hannover.de`

Hence, it is not surprising that the first memetic algorithms have been developed for the TSP. In the late 80s, several attempts have been made to apply evolutionary algorithms to the travelling salesman problem. Especially, when using recombination, many researchers discovered that it is necessary to use some form of local search within the evolutionary framework [334]. One of the reasons why recombination-based evolutionary algorithms fail to perform well on the TSP is that it is not trivial to recombine some of the edges of two or more TSP tours into a single tour such that all edges are from at least one of the parents and the resulting edge set is a valid TSP tour. Most recombination operators introduce implicit mutations by adding random edges of arbitrary length to ensure feasibility [230, 565, 581, 931]. As a consequence, the offspring tend to be much worse than their parents if the parents have a high fitness. Therefore, researchers considered applying local search to remove these arbitrary long edges from the tours.

7.2.1 Combinatorial Optimization

According to [303], a combinatorial optimization problem P is either a minimization problem or a maximization problem, and it consists of

1. a set D_P of instances,
2. a finite set $S_P(I)$ of candidate solutions for each instance $I \in D_P$, and
3. a function m_P that assigns a positive rational number $m_P(I,x)$ called the solution value for x to each instance $I \in D_P$ and each candidate solution $x \in S_P(I)$.

Thus, an optimal solution for an instance $I \in D_P$ is a candidate solution $x^* \in S_P(I)$ such that, for all $x \in S_P(I)$, $m_P(I,x^*) \leqslant m_P(I,x)$ if P is a minimization problem, and $m_P(I,x^*) \geqslant m_P(I,x)$ if P is a maximization problem.

Due to the fact that the set of candidate solutions is finite, an algorithm for finding an optimum solution always exists. This algorithm, referred to as *exhaustive search*, simply evaluates and compares $m_P(I,x)$ for all $x \in S_P(I)$. Unfortunately, the search space of many combinatorial problems grows exponentially with the problem size, i.e., the number of components in a solution vector x. Thus, this complete enumeration scheme becomes impractical. For a large class of combinatorial optimization problems no alternative algorithms running in polynomial time are known. This phenomenon has led to the development of complexity theory [303], and in particular, to the theory of NP-completeness.

Combinatorial optimization can be considered as a special case of discrete optimization. However, in discrete optimization the search space is not always finite. In contrast to integer programming, combinatorial optimization refers to problems on graphs, matroids and other discrete structures.

7.2.2 MA Outline

Beginning with Brady [81], many researchers have made consequent use of local search in their evolutionary algorithms for the TSP [86, 331, 622, 637, 896]. These

7 Memetic Algorithms and Fitness Landscapes in Combinatorial Optimization

Algorithm 12. The Memetic Algorithm

```
1  begin
2      foreach S in Population do S ← LocalSearch(Init());
3      while not terminated do
4          Offspring ← ;
5          for i ← 0 to crossovers do
6              A ← Select(Population);
7              B ← Select(Population);
8              C ← LocalSearch(Recombine(A, B));
9              Offspring ← OffSpring + C;
10         endfor
11         for i ← 0 to mutations do
12             A ← Select(Population);
13             C ← LocalSearch(Mutate(A));
14             Offspring ← OffSpring + C;
15         endfor
16         Population ← Select(Population, Offspring);
17     endw
18 end
```

approaches can be classified as memetic algorithms although they have not been called so at the time they have been proposed. Some researchers used the term Genetic Local Search [285, 286, 478, 584, 896], others described them as hybrids of evolutionary and local search. Still today, many researchers use the same basic MA framework that is shown in Alg.12. In this framework, local search is consequently applied to all newly created solutions, more precisely to all the members of the initial population created by some initialization operator, and those solutions created by the mutation and recombination operators. In the framework, recombination and mutation are treated independent of each other. In some MAs, mutation is only applied after crossover. However, we concentrate on the framework above since it allows for mutation–only MAs.

When using recombination, selection becomes highly important, since there is a high probability of premature convergence. Due to the fact that local search is expensive, MAs tend to have relatively small population sizes (10–40 individuals). Compared to EAs without local search, the problem of convergence is more severe. When the population only contains very similar solutions, recombination / combined with local search will likely discover the same solutions again and again. It is therefore important to keep diversity in the population. There are several methods to deal with diversity preservation depending on the type of selection: In selection for recombination, one can choose to recombine only those individuals which are sufficiently far away from each other in the search space. Another approach is to consider diversification in the recombination operator itself as has done in HX or DPX [76, 246, 247]. Moreover, in selection for survival duplicates may be removed such that each indivdual is found only once in the population [247] or replacement scheme may be used that replaces similar solutions based on a distance threshold

[285, 286]. Finally, a restart mechanism can be used to diversify the population once convergence has been detected [246, 581].

7.2.3 Related Meta-Heuristics

There are several meta-heuristics which are similar to MAs. Most notably, Scatter Search [313] and Iterated/Stochastic Local Search [533]. While the former incorporates a recombination meachanism as and the MA framework above, the latter can be considered as a special case of the MA above. In that case the population is reduced to 1 and only mutation is used (#recombinations=0,#mutations=1), which simplifies the algorithm significantly. Iterated local search (ILS) was also first proposed for the TSP [427], but has been applied later on to various combinatorial problems [533]. The outline of iterated local search is shown in Alg. 13.

Algorithm 13. Iterated Local Search.
---
1 **begin**
2      $S \leftarrow \text{Init}()$;
3      $S \leftarrow \text{LocalSearch}(S)$;
4   **while** *not terminated* **do**
5          $T \leftarrow \text{Mutate}(S)$;
6          $T \leftarrow \text{LocalSearch}(T)$;
7          $S \leftarrow \text{Select}(S, T)$;
8   **endw**
9 **end**

Iterated local search is also highly similar to some instances of variable neighborhood search (VNS) [604].

7.3 Why and When MAs Work

Although many different meta-heuristics have been proposed for combinatorial optimization problems, only little is known in which cases they are effective. Moreover, every approach comes with a considerable number of parameters and it is often not known which parameter settings are optimum due to the huge parameter space and the computational time required for testing all possible combinations.

In the case of MAs, it would be highly desirable to have guidelines for the development of MAs for new or untackled combinatorial optimization problems. Important questions that arise are: Which local search operator or neighborhood to choose, how many iterations to apply local search, to use recombination or mutation, how to mutate or recombine and so forth. Fitness landscape analysis has been shown to be valuable when trying to find answers to these questions. We therefore provide a short overview of fitness landscapes and statistics methods to analyse them.

7.3.1 The Concept of Fitness Landscapes

The concept of *fitness landscapes* [941], introduced to illustrate the dynamics of biological evolutionary optimization, has been proven to be very powerful in evolutionary theory. As metioned before, the concept has been shown to be useful for understanding the behavior of combinatorial optimization algorithms and can help in predicting their performance. Regarding the search space, i.e. the set of all (candidate) solutions, as a landscape, a heuristic algorithm can be thought of as navigating through it in order to find the highest peak of the landscape; the height of a point in the search space reflects the fitness of the solution associated with that point.

More formally, a fitness landscape (S, f, d) of a problem instance for a given combinatorial optimization problem consists of a set of points (solutions) S, a fitness function $f : S \rightarrow \mathbb{R}$, which assigns a real–valued fitness to each of the points in S, and a distance measure $d : S \times S \rightarrow \mathbb{R}$, which defines the spatial structure of the landscape. A fitness landscape can thus be interpreted as a graph $G_L = (V, E)$ with vertex set $V = S$ and edge set $E = \{(s, s') \in S \times S \mid d(s, s') = d_{min}\}$, with d_{min} denoting the minimum distance between two points in the search space. The diameter $diam\, G_L$ of the landscape is another important property: it is defined as the maximum distance between any two points in the search space.

For binary coded problems ($S = \{0, 1\}^n$), the graph G_L may be a hypercube of dimension n, and the distance measure may be the hamming distance between bit strings. For this landscape, the minimum distance d_{min} is 1 (one bit with a different value), and the maximum distance is $diam\, G_L = n$.

7.3.2 NK-Landscapes

To study rugged fitness landscapes, Kauffman [451, 452] developed a formal model for gene interaction which is called the *NK-model*. In this model, N refers to the number of parts in the system, i.e. genes in a genotype or amino acids in a protein. Each part makes a fitness contribution which depends on the part itself and K other parts. Thus, K reflects how richly cross-coupled the system is; it measures the epistasis, i.e. the richness of interactions among the components of the system.

Each point in the *NK*-fitness landscape is represented by a bit string of length N and can be viewed as a vertex in the N-dimensional hypercube. The fitness f of a point $b = b_1, \ldots, b_N$ is defined as follows:

$$f(b) = \frac{1}{N} \sum_{i=1}^{N} f_i(b_i, b_{i_1}, \ldots, b_{i_K}), \qquad (7.1)$$

where the fitness contribution f_i of the gene at locus i depends on the allele (value of the gene) b_i and K other alleles $b_{i_1}, \ldots, b_{i_K}$. The function $f_i : \{0, 1\}^{K+1} \rightarrow \mathbb{R}$ assigns a uniformly distributed random number between 0 and 1 to each of its 2^{K+1} inputs. The values for $i_1, \ldots, i_K$ are chosen randomly from $\{1, \ldots, N\}$ or from the left and right of locus i. The former is called the random neighbor model while the latter is called the adjacent neighbor model. Since the random neighbor model is NP-hard and the adjacent model is not [929], we focus on the random case. With the *NK* model, the "ruggedness" of a fitness landscape can be tuned by changing the value of *K* and thus the number of interacting genes per locus. Low values of *K* indicate low epistasis and high values of *K* indicate high epistasis.

7.3.3 Analysis of Fitness Landscapes

7.3.3.1 Autocorrelation Analysis

The local properties of the landscape have a strong influence on the effectiveness of a local search, since in a local search the decision which point has to be visited next is based solely on these local properties. The properties can be analyzed with statistical methods such as autocorrelation/random walk correlation analysis. These methods calculate (or estimate) the correlation of neighboring points in the search space with respect to the local search neighborhood. The *random walk correlation function* [845, 846, 928]

$$r(s) = \frac{1}{\sigma^2(f)(m-s)} \sum_{t=1}^{m-s} (f(x_t) - \bar{f})(f(x_{t+s}) - \bar{f}) \tag{7.2}$$

of a time series $\{f(x_t)\}$ defines the correlation of two points s steps away along a random walk of length m through the fitness landscape ($\sigma^2(f)$ denotes the variance of the fitness values). A step denotes here a move from the current solution to a neighboring solution in the fitness landscape. Typical random walk correlation functions for the TSP and *NK*-landscapes are displayed in Fig. 7.1.

Based on this correlation function, the correlation length ℓ [846] of the landscape is defined as

$$\ell = -\frac{1}{\ln(|r(1)|)} \tag{7.3}$$

for $r(1) \neq 0$. The correlation length directly reflects the ruggedness of a landscape. The smaller the value for ℓ, the more rugged the landscape. A landscape is said to be

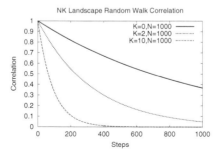

Fig. 7.1. The random walk correlation functions of *NK* landscapes (right) with varying *K*.

smooth if there is high correlation between neighboring points (correlation length is large), and rugged if there is low or no correlation between neighboring solutions (correlation length is small). It can be observed that higher correlation leads to a higher number of iterations in a local search until a local optimum is reached and may also lead to a higher fitness as Kauffman concludes [451]: On the other hand, if the correlation is low, the local search terminates after few iterations in a local optimum with relatively low fitness [451]. The correlation length as a measure of landscape ruggedness can be utilized to compare different neighborhoods for a problem (assumed that the neighborhoods have the same size). The higher the correlation, the more effective the local search.

Alternative landscapes in combinatorial optimization can be found by allowing for infeasible solutions. However, the problem becomes then to find an appropriate penalty function in order to obtain an effective local search. In the graph bipartitioning problem, local search can be performed by exchanging a vertex from one set with a vertex from the other set. An alternative is to move just one vertex to the other set. As a consequence, both sets can have different sizes and hence the solution may be infeasible. Therefore, it is required to introduce a penalty function to penalize infeasible solutions depending on the absolute difference of the sizes of the two sets. The objective becomes

$$f(V_1,V_2) = |e(V_1,V_2)| + \alpha(|V_1| - |V_2|)^2, \qquad (7.4)$$

where V_1, V_2 are the two vertex sets, $e(\cdot,\cdot)$ is the number of edges between the two sets, and α is called the imbalance factor [430]. In [20], the correlation length has been used to determine the perfect imbalance factor α, resulting in the highest random walk correlation. In experiments, it could been verified that the "optimum" imbalance factor leads to the best local search performance. Thus, the correlation length can be used to find the smoothest and hence easiest landscape for a local search for a given problem. In some cases, where the correlation length is problem instance dependent, it may serve as an indicator for the hardness of the instance for a local search. For *NK*-Landscapes as well as for other combinatorial optimization problems such as the quadratic assignment problem, it can be observed that the number of iterations of a local search (the number of moves until a local optimum is reached) decreases for less correlated landscapes and the resulting solution quality becomes worse [451, 581, 587].

7.3.3.2 Fitness Distance Correlation

The effectiveness of the evolutionary meta-search depends on the global properties of the fitness landscape. Since in the MAs discussed in this chapter, the evolutionary variation operators are applied to locally optimum solutions, the distribution of the local optima is the most important global property of a landscape. The distribution can be analyzed with a fitness distance correlation analysis of the local optima – the peaks in the fitness landscape. The fitness distance correlation (FDC) coefficient ρ is defined as

$$\rho(f, d_{opt}) = \frac{\mathrm{Cov}(f, d_{opt})}{\sigma(f)\,\sigma(d_{opt})}, \tag{7.5}$$

where $\mathrm{Cov}(\cdot,\cdot)$ denotes the covariance of two random variables and $\sigma(\cdot)$ denotes the standard deviation. The FDC determines how closely fitness and distance to the nearest optimum in the search space (denoted by d_{opt}) are related. If fitness increases when the distance to the optimum becomes smaller, then search is expected to be easy for selection–based algorithms, since there is a "path" to the optimum via solutions with increasing fitness. A value of $\rho = -1.0$ ($\rho = 1.0$) for a maximization (minimization) problem indicates that fitness and distance to the optimum are perfectly related and that search promises to be easy.

The FDC coefficient has been proposed in [435] as a measure for problem difficulty for genetic algorithms. In [14], a counterexample is presented in which a GA performs well on an uncorrelated landscape. Horjik and Manderick argue why FDC is useful for recombination [394]. [832] propose the NKP model which is a superset of the NK model and show that as K increases the correlation goes down but with no statistically significant effect on the mean fitness of the local optima. A summary of landscape metrics and related issues is provided in [438]. These papers, however, concentrate on evolutionary algorithms without local search.

In respect to MA performance, FDC analysis may reveal a correlation between the fitness and the distance of the local optima to the global optimum. The presence of correlation implies that the fitness increases the closer the local optima are to the global optimum. Therefore, an MA can exploit this feature by 'hopping' from one local optimum to the next local optimum with better fitness until the global optimum is reached. If recombination is used for variation, 'jumping' from one peak to another can be achieved if the recombination is respectful, i.e. if the resulting offspring point lies near both parents and has a distance to its parents that is no greater than the distance between the parents themselves. Compared to other forms of variation, the resulting offspring are closer to other local optima with high fitness. In such cases, it is more likely that the offspring is within a suboptimal 'basin of attraction' (with higher fitness than the parents) rather than jumping into an arbitrary direction.

Besides the FDC, fitness distance scatter plots provide useful information about a fitness landscape. In fact, there are cases in which the FDC can be misinterpreted if the plot is not considered. In Fig. 7.2, typical fitness distance plots are provided. The *NK*-landscape in the left ($N = 1024, K = 2$) is correlated but the landscape ($N = 1024, K = 11$) shown on the right is uncorrelated, the local optima appear to be randomly distributed in the search space. The correlated landscape has a massive central structure, meaning that the optimum is more or less central among other near optimum local optima. This phenomenon can be observed for several other combinatorial optimization problems and is also known as the big valley structure (for minimization problems). The structured landscape is well suited for MA based on recombination while the uncorrelated and structured landscape is not. In fact, for the latter it was shown that variation based on mutation is better than recombination, such as uniform crossover [581]. Thus, in uncorrelated landscapes, jumps in random

7 Memetic Algorithms and Fitness Landscapes in Combinatorial Optimization

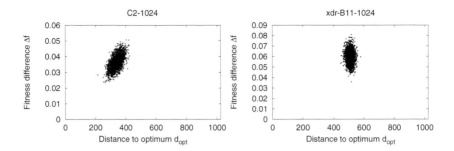

Fig. 7.2. FDC plots for two *NK*-landscapes with $K = 2$ (left) and $K = 11$ (right)

directions with a fixed jump distance are more effective than jumps toward other solutions with high fitness using variable jump distances (respectful recombination).

7.3.3.3 Advanced Fitness Landscape Analysis

Although the analysis techniques described above are valuable, they do not focus on all important aspects of the fitness landscape. Hence, in [583] we proposed some rather simple statistical analysis methods. The first addresses the question how much should be mutated in a mutation-based MA? Intuitively, the idea would be to mutate only the minimum number of components in the solution vector that is sufficient to leave the basin of attraction of the local optimum represented by the current solution. In [583], we computed the average escape rate form local optima depending on the number of mutation steps for various landscapes. Since the higher the number of mutation steps, the higher the chance to escape but also the higher the computational cost in terms of local search iterations to reach a new local optimum, we proposed to calculate the number of local search iterations per escape for various mutation strengths. In Fig.7.3, the escape rates and the number of iterations per escape for *NK*-landscapes with three different values of K are shown on the left, and on the right, respectively. Both escape rates and number of iterations per escape for

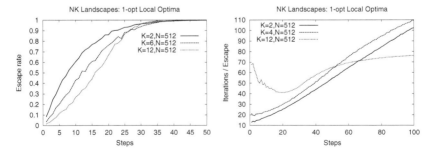

Fig. 7.3. Basins of Attraction of Local Optima in the NK-model

1-opt local optima are shown depending on the number of mutation steps performed to escape. The number of mutation steps is equal to the distance of the mutated solution and the local optimum. The left plot shows that the basins of attraction become larger with increasing K or at least escaping becomes harder: more mutation steps are required to leave the current basin of attraction. This is surprising, since with increasing K the number of local optima increases and we would expect the size of the basins of attraction to decrease. The question arises whether there is an optimum mutation rate in terms of computation costs. At which mutation rate is the number of visited local optima per time unit maximum? The answer is given in the left part of the figure. Clearly, for $K = 12$, there is an optimum at 20 mutation steps. For the landscapes with smaller K, the optimum approaches two mutation steps.

Additionally to this local search escape analysis we proposed random walk analysis starting at local optima. This analysis provides a picture of the search space from a local optimums' perspective. The idea here is that walking away from a local optimum in a random direction may show a different degradation of fitness than walking in the direction of another local optimum. How severe this difference is, depends on fitness distance correlation of the local optima. However, this analysis does not require the knowledge of a global optimum as the FDC analysis does. In Fig.7.4, the results of a random walk analysis for the NK-model is shown for different values of K. The random walks in the direction of another local optimum (simulating recombination) are denoted by 'rec', the random walks in a random direction (simulating mutation) or denoted by 'rw'. In the left, the average fitness values of the solutions on a typical random walk path are displayed over the distance from the starting point of the random walk (parent A). As expected, the fitness decreases when moving away from A. When approaching B the fitness increases as expected. The fitness of the solutions halfway on a random walk between solution A and B tell an interesting story. Let C denote such a point with $d(A,C) = d(A,B)/2$. For $K = \{2,6\}$, this fitness is considerably higher than for the solutions on a random walk in an arbitrary direction indicating that recombination produces much better solutions than mutation. For $K = 12$, this effect can not be observed. Here, the fitness halfway on a random walk between A and B is similar to the fitness on an arbitrary random walk, indicating that recombination is not superior to mutation for this landscape. This is not surprising since the fitness distance correlation analysis reveals that the local optima are randomly distributed in the search space. On the right of the figure, the fitness difference of points on directed and undirected random walks is provided depending on the distance to parent A: The fitness difference $f(C_{rec}) - f(C_{rw})$ for points C_{rec} on a directed random walk and points C_{rw} on an undirected random walk depending on the distance $d(A,C) = d(A,B)/2$ are shown The data is collected over a full run of a memetic algorithm. For $K = 2$, the fitness difference and the distance of the parents are correlated (upper right of the figure). The higher the distance, the better recombination performs compared to mutation (upper left of the figure). Recombination is also always superior to mutation in case of $K = 6$. However, The gain achieved with recombination is highest for distances about 100 and decreases with decreasing distance as well as increasing distance. For $K = 12$, the picture changes. Recombination and mutation perform equally well for

7 Memetic Algorithms and Fitness Landscapes in Combinatorial Optimization   105

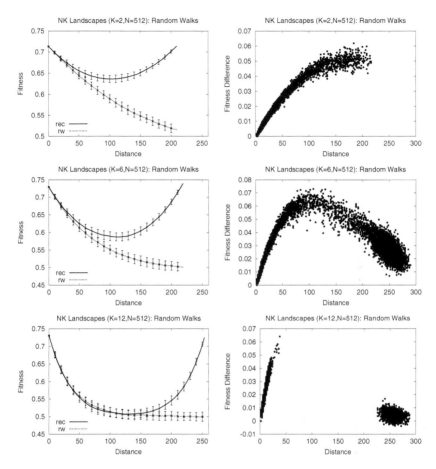

Fig. 7.4. Random Walks Starting from a Local Optimum in the *NK* model.

a distance around 250. This is the case at the beginning of the MA run. Later in the run of the MA, the average distance of the solutions in the population drops due to the effects of mutation with a relatively small mutation rate. For a parents distance smaller than 50, recombination is again superior to mutation.

7.4 Case Study I: The TSP

The travelling salesman problem (TSP) is believed to be the best-known combinatorial optimization problem. Exact methods such as branch & cut as well as many meta-heuristics have been evaluated initially on the TSP. The reason why it has attracted so many researchers is probably that it is very easy to formulate and understand: Given a set of cities and the distances between them, the problem is to find the shortest closed tour that visits each city exactly once. More formally, the tour length

$$l(\pi) = \sum_{i=1}^{n-1} d_{\pi(i),\pi(i+1)} + d_{\pi(n),\pi(1)} \tag{7.6}$$

has to be minimized, where d_{ij} is the distance between city i and city j and π a permutation of $\langle 1,2,\ldots,n \rangle$. Thus, an instance $I = \langle D \rangle$ is defined by a distance matrix $D = (d)_{ij}$, and a solution (TSP tour) is a vector π with $j = \pi(i)$ denoting city j to visit at the i-th step.

A special case of the TSP is the Euclidean TSP. Here, the distance matrix d_{ij} is symmetric, that is $d_{ij} = d_{ji} \quad \forall i,j \in \{1,2,\ldots,n\}$, and the triangle inequality holds: $d_{ij} \leqslant d_{ik} + d_{kj} \quad \forall i,j,k \in \{1,2,\ldots,n\}$. The distance between two cities is defined by the Euclidean distance between two points in the plane. These two assumptions do not lead to a reduction of the complexity; hence the general as well as the euclidian problem is NP-hard. In the following we focus on the Euclidean TSP.

7.4.1 Fitness Landscape

The TSP is also among the first combinatorial optimization problems for which researchers tried to analyze the search space. In order to define the fitness landscape for the TSP, an appropriate distance measure is required.

A suitable distance measure for TSP tours appears to be a function that counts the number of edges different in both tours: Since the fitness of a TSP tour is determined by the sum of the weights of the edges the tour consists of, the distance between two tours t_1 and t_2 can be defined as the number of edges in which one tour differs from the other. Hence

$$d(t_1,t_2) = |\{e \in E \mid e \in t_1 \wedge e \notin t_2\}|. \tag{7.7}$$

This distance measure has been used by several researchers, including [76, 286, 548, 638]. It has been shown that this distance function satisfies all four metric axioms [774].

Alternatively, a distance measure could be defined by counting the number of applications of a neighborhood search move to obtain one solution from the other. In the case of the *2-opt* move, the corresponding distance metric d_{2-opt} is bound by $d \leqslant d_{2-opt} \leqslant 2d$ [548].

7.4.1.1 Autocorrelation Analysis

Stadler and Schnabl [847] performed a landscape analysis of random TSP landscapes considering different neighborhoods: the *2-opt* and the *node exchange* neighborhood. Their results can be summarized as follows.

For the symmetric TSP, both landscapes (based on *2-opt* and *node exchange*) are AR(1) landscapes. The random walk correlation function for random landscapes is of the form

$$r(s) \approx \exp(-s/\ell) = \exp(-b/n \cdot s), \tag{7.8}$$

with n denoting the number of nodes/cities of the problem instance and b denoting the number of edges exchanged between neighboring solutions. Thus, for the *2-opt*

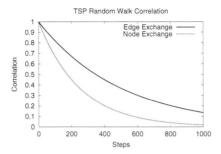

Fig. 7.5. Random walk correlation functions for the TSP based on edge exchange and node exchange.

landscape, the normalized correlation length $\xi = \ell/n$ is $\frac{1}{2}$, for the *node re-insertion* landscape ξ is $\frac{1}{3}$, and for the *node exchange* landscape ξ is $\frac{1}{4}$. This result coincides with experimentally obtained results that *2-opt* local search is much more effective than local search based on *node exchange* or *node re-insertion* [764]. The correlation functions for edge and node exchange are shown in Fig.7.5.

The formula 7.8 implies that a landscape with a strict *3-opt* neighborhood is more rugged than a landscape with a *2-opt* neighborhood. One may conclude that a *2-opt* local search performs better than a *3-opt* local search. However, the opposite is true, since the *3-opt* neighborhood is much greater than the *2-opt* neighborhood and the *3-opt* neighborhood as defined above contains the *2-opt* neighborhood. Therefore, a *3-opt* local search cannot perform worse than a *2-opt* local search in terms of solution quality.

7.4.1.2 Fitness Distance Correlation

The fitness distance correlation of local minima of the TSP has been analyzed in [75, 76] for a single instance and also in [581] for serveral other typical TSP instances. Given the distance measure described above, the results show a strong correlation between tour length and distance to the optimum solution. In fact, the TSP was the first problem to show the deep valley structure, meaning that better local optima tend to be close to the global optimum. Moreover, the global optimum is found in a big valley surrounded by the other local optima and the local optima concentrate in a relatively small part of the search space. A radically different structure would be a random distribution of the local optima in the search space. Hence there would be no correlation between distance to the optimum and tour length. In Fig.7.6, the fitness distance plot of a typical TSP instance is shown together with a fitness distance plot for a completely unstructured landscape of the quadratic assignment problem (QAP). The figure indicates that not all combinatorial problems have a 'nice' landscape as the TSP. The findings also provide an explanation of the success of many of the TSP heuristics. Many meta–heuristics implicitly exploit the fact that local optima are close to each other. An obviously reasonable strategy is to hop from one

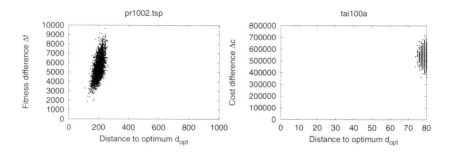

Fig. 7.6. FDC plots for a TSP instance (left) and a QAP instance (right).

local optimum to the next better one until the global optimum is reached. This is, in fact, what MAs and ILS do.

7.4.2 State-of-The-Art Meta-Heuristics for the TSP

The TSP has served as a test-bed for new heuristic approaches including evolutionary algorithms (EA) and memetic algorithms. Consequently, many approaches, both evolutionary and non-evolutionary, have been proposed. Here, we focus on those approaches which are highly effective and scalable. Euclidean TSP instances up to a size of 1,000 can be considered as trivial for most algorithms. In fact, these small problems can usually be solved exactly by Branch & Cut [24] in a few seconds. Therefore, these instances are no longer of interest for heuristics research on the TSP. For instances up to approx. 30,000 cities, very effective heuristics have been proposed most of which are based on the powerful Lin-Kernighan (LK) heuristic [524], a variable k-opt local search. An example is Helsgaun's LK implementation (LKH) [380].

Only few evolutionary algorithms can compete with LKH. One of the best evolutionary approaches is the EA of Nagata using EAX crossover [647] and 2-opt local search. This algorithm finds (near) optimal tours up to a size of 33,000 cities, although with a high runtime. Recently, Nguyen *et al.*[664] have proposed a memetic algorithm which utilizes a variant of the MPX crossover operator [637] and a Lin-Kernighan local search variant with 5-opt moves. Results are reported for instances up to 85,900 cities. The authors claim that their algorithm is more effective than LKH. Moreover, the authors describe an approach for solving the World TSP (approx. 2 million cities) by solving and merging subproblems. But results for other instances in the range from 100,000 to 10 million cities are not reported.

For instances larger than 100,000 cities, only few heuristics have been proposed. For these instances, the DIMACS TSP implementation challenge [428] lists several approaches of which the best are based on the LK heuristic: The multi-level algorithm of Walshaw [912] first reduces the size of a TSP instance stepwise and then applies the (chained) LK heuristic to the smaller problems. The results are inferior to the results obtained by directly applied chained LK or iterated LK heuristics. These

7 Memetic Algorithms and Fitness Landscapes in Combinatorial Optimization 109

heuristics are based on the principle of iterated local search [533], an evolutionary heuristic incorporating local search. The idea is to stepwise improve the current best solution by mutating it and subsequently applying local search. The first iterated local search was the iterated Lin-Kernighan (ILK) heuristic by Johnson [427]. Other variants have been proposed such as the chained Lin-Kernighan heuristic [25, 26]. These ILK heuristics have been applied to instances with up to 10 million cities. The only algorithm within the DIMACS challenge not using LK as a subroutine and still being highly effective for large instances is the dynamic programming approach of Balas and Simonetti [39].

Except for the LKH heuristic, none of the mentioned algorithms provides a lower bound on the optimum solution. To the best of our knowledge, the only evolutionary algorithm computing lower bounds is the one proposed in [556]. However, the approach deals with instances below 2,400 cities only.

7.4.2.1 An ILS Approach for Very Large TSP Instances

In [590] we have presented an iterated local search approach that simultaneously improves lower and upper bounds for a TSP instance to provide a gap for the best solution found. The gap determines the maximum deviation from the optimum solution and therefore provides a quality measure for the obtained TSP tour. The approach differs from exact algorithms like Branch & Cut [179] in that no efficient linear programming (LP) solver is required and it differs from approximation algorithms such as PTAS (polynomial time approximation scheme) [27] in that no general performance guarantee is provided. Instead, the quality is proved for each instance and a particular run: a final gap between lower and upper bound of 1% means that the solution found is at most one percent above the optimum (in practice the real gap is much lower).

Our local search is based on the LK heuristic, hence our iterated local search is called iterated LK. The general outline of our iterated LK is shown in Alg. 14. In contrast to other approaches, our ILK incorporates a lower bound computation.

Algorithm 14. The Advanced Iterated Link-Kernighan Heuristic
---
1 **begin**
2   C ← FindInitialCandidateSet(Instance);
3   Tour ← Init();
4   Tour ← LocalSearch(C, Tour);
5   C ← FindInitialLowerBound(C, TourLength(Tour));
6   **for** *iter* ← *1* **to** *MaxIter* **do**
7   Tbest ← Tour;
8   Tour ← Mutate(Tour);
9   Tour ← LocalSearch(Tour);
10   **if** *TourLength(Tour)* < *TourLength(Tbest)* **then** Tbest ← Tour;
11   **if** *(iter mod 400) = 0* **then** C ← UpdateLowerBound(C, TourLength(Tbest));
12   **endfor**
13 **end**

This computation is interleaved with the optimization algorithm as can be seen in the figure: every 400 iterations of the ILK, the lower bound is improved until there appears to be no more improvement possible (the lower bound computation has converged). The lower bound computation possibly modifies the candidate edge set, which is used by the local search to look for improving moves (edge exchanges).

To find initial solutions (*Init()* in the pseudo code), we use the Quick-Boruvka heuristic [26, 428], and the initial candidate set (*FindInitialCandidateSet(Instance)* in the pseudo code) is based on a subgraph containing the two nearest neighbors for each quadrant of a city [26]. This candidate set has the property of being connected.

The mutation operator used in the algorithm is non-sequential four exchange [524, 588] using a random walk on the candidate set to find edges to be included in the tour. This operator has been proven to be very effective in conjunction with Lin-Kernighan local search [26]. The random walk on the candidate edge set assures that edges with a relatively small length instead of arbitrarily long edges are included.

As mentioned before, we use a variant of the original Lin-Kernighan heuristic for the local search. Compared to the original LK, we use 3-opt moves as submoves instead of 2-opt moves at all levels. We do not use backtracking which simplifies the implementation drastically without affecting the performance. In this aspect our implementation is similar to LKH.

In order to compare with other state-of-the-art approaches, Table 7.1 shows a comparison with the eleven best performing algorithms (out of 90) listed on the DIMACS TSP challenge web page. The summary was produced with the statistics

Table 7.1. Comparison of DIMACS TSP Challenge Results on E1M.0. ILK-PM-.1N denotes our ILK with 1 million iterations and ILK-PM-.1N denotes our ILK with 1,2 million iterations.

% HK	Seconds	Implementation	Reference
0.787	17544.0	ILK-PM-.12N	this paper
0.792	77161.6	ILK-NYYY-N	([663])
0.797	16062.0	ILK-PM-.1N	this paper
0.804	8694.0	ILK-PM-.1N without LB	this paper
0.841	6334.0	ILK-NYYY-Ng	([663])
0.879	42242.5	MLCLK-N	[912]
0.888	3480.2	ILK-NYYY-.5Ng	([663])
0.903	19182.7	BSDP-6	[39]
0.903	19503.1	BSDP-8	[39]
0.903	21358.3	BSDP-10	[39]
0.903	19108.1	CLK-ABCC-N.Sparc	[25]
0.905	19192.3	CLK-ACR-N	[26]
0.910	16008.0	CLK-ABCC-N.MIPS	[25]
0.945	20907.6	MLCLK-.5N	[912]

code from the challenge. Thus the running time reported in the table is normalized to a DEC Alpha processor with 500 MHz in order to allow a comparison of the different approaches. The quality is given as the percentage excess over the Held-Karp (HK) bound. As shown in the table, our algorithm provides a significantly better tour quality than the other approaches. And it does this in a fraction of time of the second best approach which is also an ILK implementation. Note that none of the competitors computes a lower bound. For the 10 million city instance E10M.0, the quality of our approach is 0.75% over the Held-Karp bound compared to the best algorithm of the DIMACS challenge which is 1.63% over the Held-Karp bound! This is due to the fact that the best algorithms for the smaller instances do not scale as well as our approach. While the runtime of our approach without lower bound computation grows linearly with the problem size, the runtime of the others clearly grows faster and and yields in the non-applicability of these algorithms to very large problem instances (>1 million) whereas our approach is still very successful even if the lower bound computation is activated.

More details on the algoritm as well as the results can be found in [590].

7.5 Case Study II: The BQP

In the *unconstrained binary quadratic programming problem* (BQP), a symmetric rational $n \times n$ matrix $Q = (q_{ij})$ is given, and a binary vector of length n is searched for, such that the quantity

$$f(x) = x^t Q x = \sum_{i=1}^{n} \sum_{j=1}^{n} q_{ij} x_i x_j, \quad x_i \in \{0,1\} \, \forall i = 1, \ldots, n \quad (7.9)$$

is maximized. This problem is also known as the *(unconstrained) quadratic bivalent programming problem, (unconstrained) quadratic zero–one programming problem*, or *(unconstrained) quadratic (pseudo-) boolean programming problem* [55]. The general BQP is known to be NP-hard but there are special cases that are solvable in polynomial time [55]. In [926], it has been shown that there are special cases of the BQP, which can be solved efficiently with simple EAs, but there are also cases, for which these EAs have been proven to be ineffective (exponentially growing running time).

The BQP has a large number of applications, for example in capital budgeting and financial analysis problems [506, 571], CAD problems [485], traffic message management problems [300], machine scheduling [10], and molecular conformation [722]. Furthermore, several other combinatorial optimization problems can be formulated as a BQP, such as the maximum cut problem, the maximum clique problem, the maximum vertex packing problem and the maximum independent set problem [414, 707, 708].

There is a close relation between binary quadratic programming and *NK*-landscapes: The objective function of the BQP can be decomposed into n functions.

The fitness of a BQP solution can thus be rewritten as a sum of functions for each site, called the fitness contributions f_i of site i in the genome:

$$f(x) = \sum_{i=1}^{n} f_i(x_i, x_{i_1}, \ldots, x_{i_{k(i)}}), \tag{7.10}$$

$$f_i(x) = \sum_{j=1}^{n} q_{ij} x_i x_j. \tag{7.11}$$

Similar to the *NK*-landscapes defined in [452], the fitness contribution f_i of a site i depends on the gene value x_i and of $k(i)$ other genes $x_{i_1}, \ldots, x_{i_{k(i)}}$. While for *NK*-landscapes $k(i) = K$ is constant for all i, in the BQP $k(i)$ is defined as the number of non-zero entries in the i-th column of matrix Q. Hence, the degree of epistasis in a BQP instance can be easily determined by calculating the density of the matrix Q. It is defined as the number of non-zero entries divided by the number of total entries in the matrix. Thus, the density is between 0 and 1, where 0 means no epistasis and 1 maximum epistasis (every gene depends on the values of all other genes).

7.5.1 Fitness Landscape

Since the BQP is binary-coded and local search for the BQP is based on the *k*-opt neighborhood as defined as

$$\mathcal{N}_{k\text{-}opt}(x) = \{x' \in X \mid d_H(x',x) \leqslant k\} \tag{7.12}$$

where d_H denotes the hamming distance between bit strings and X the set of all bit strings of length n ($X = \{0,1\}^n$), the landscape considered in the search space analysis of the BQP is $\mathscr{L} = (X, f, d_H)$. The graph of this landscape is a hypercube of dimension n in which the nodes represent the (candidate) solutions of the problem. An edge in the graph connects neighboring points in the landscape, i.e. points that have hamming distance 1.

7.5.1.1 Autocorrelation Analysis

Since there are no theoretical results on the autocorrelation function or the random walk correlation function for the BQP, experiments have been conducted in [591] to estimate the correlation length of selected landscapes. The instances were taken from ORLIB [54] and [18, 318]. Here, we summarize the findings: Considering all selected instances, the quotient of n/ℓ varies in tight bounds: the lowest value for n/ℓ is 2.36 and the highest is 2.71. Compared to *NK*-landscapes, this is fairly low since in the *NK*-model $n/\ell \approx K + 1$. For the instances denoted **glov500**, the values are very similar (2.67 ± 0.04) and thus remain constant independent of the density of the problem. For the set **kb-g**, the values for n/ℓ do change with the density of Q, but the values become smaller with increasing density. This is surprising, since in the *NK*-model, the correlation length decreases with increasing epistasis, and the

7 Memetic Algorithms and Fitness Landscapes in Combinatorial Optimization

density can be regarded as a measure of epistasis in the BQP. For the set of instances of size 2500 and a density of 0.1, the values for n/ℓ are constant (about 2.66).

Summarizing, all the instances of the BQP considered here have got a smooth landscape similar to *NK*-landscapes with $K \leqslant 3$.

7.5.1.2 Fitness Distance Correlation Analysis

In a FDC analysis, we studied the correlation of fitness (objective f(x)) and distance to the optimum for local optima with respect to *1-opt* local search. The findings can be summarized as follows. In most cases, the average distance between the local optima and the average distance to the global optimum (best-known solution) are very similar. Moreover, the local optima are located in a small region of the search space: the average distance between the local optima is between a fourth and sixth of the maximum distance (the diameter) between two solutions in the search space. For set **glov500**, the average distance to the optimum is a sixth of the diameter independent of the density of Q. For set **beas2500** the local optima are even closer to the optimum in relation to the maximum distance of two solutions in the landscape: the average distance to other local optima is more than a seventh of the diameter of the landscape. The FDC coefficient varies from -0.56 to -0.81 excluding **glov500-4**. The FDC coefficient for this instance is -0.31.

In Figure 7.7, some scatter plots are provided in which the distance to the global optimum is plotted against the fitness difference $\Delta f = f(x_{opt}) - f(x_{loc})$ for each local optimum found. The figure indicates that the local optima are even closer to each other than for smooth *NK*-landscapes with $K = 2$, revealing the deep valley/massiv central property.

7.5.1.3 Advanced Fitness Landscape Analysis

In order to analyze the structure of the basins of attraction more closely, we conducted several experiments for the BQP [583]. For each problem instance, 1000 local optima were generated and mutated 100 times with a specified mutation rate.

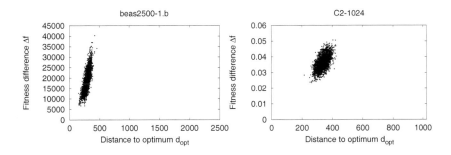

Fig. 7.7. 1-opt Local Optima FDC plots for a BQP instance (left), an *NK*-landscape with $K = 2$ (right)

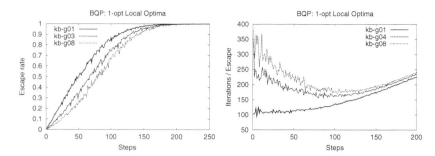

Fig. 7.8. Basins of Attractions of Local Optima: The escape rate (left) and LS iterations/escape

The number of mutation steps was increased from 1 to $n/2$ where n denotes the problem size. Since both problems are binary coded, the number of mutation steps is defined as the number of bit-mutations executed by the mutation operator.

The BQP instances from [319] denoted kb-g01, kb-g02, ... , kb-g10, where the number indicates the density of matrix Q (01 means density 0.1 and 10 denotes density 1.0), were used in the experiments. All instances have a problem size of $n = 1000$. In all cases, 1-opt local search (single bit-flip neighborhood) was used with the best improvement strategy [582, 589]. Selected results are presented in Fig. 7.8 (right). The question arises whether there is an optimum mutation rate in terms of computation costs. At which mutation rate is the number of visited local optima per time unit maximum? The answer is given in Fig. 7.8 (left). In the figure, the number of local search iterations per escape is displayed over the number of mutation steps. For densities greater 0.1, the optimum is around hundred mutation steps and the optimum approaches 2 steps for density 0.1.

To investigate the properties of random walks starting at local optima we conducted several experiments on the problem instances mentioned above. During the run of a memetic algorithm, random walks were performed by selecting two parents A and B and performing a random walk from A to B to simulate recombination as well as a walk with the same length starting at A in an arbitrary direction to simulate mutation. Fig. 7.9 shows the results of the random walk analysis for the BQP on selected instances. Directed random walks have a much higher average fitness than undirected random walks. As the right of the figure indicates, the fitness difference is always positive, independent of the distance between the start and end-points of the random walks. During the whole run of the MA, recombination is clearly superior to mutation since the random walks from one local optimum to the other produce solutions with much higher fitness than random walks starting at the same local optima but in arbitrary direction.

As the FDC analysis reveals, BQP landscapes are structured, even with a high density of matrix Q. Hence, recombination appears to be superior to mutation as the random walk correlation analysis indicates. However, in [591] it has been shown that simple recombination schemes are not very effective in MAs for the BQP. The

7   Memetic Algorithms and Fitness Landscapes in Combinatorial Optimization   115

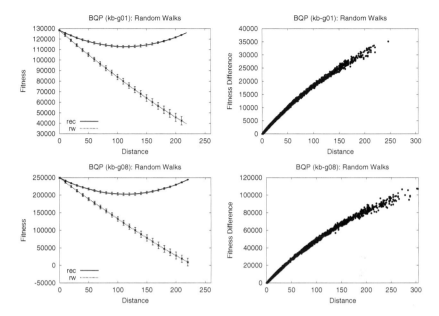

Fig. 7.9. Random Walks Starting from Local Optima of the BQP

reasons become obvious considering the results of the local search escape analysis: the local optima in the BQP have very large basins of attractions leading to the fact that after recombination and local search, one of the parent local optima is rediscovered very often. As a consequence, additional techniques are required to prevent this from happening, as shown in [591].

7.5.2   State-of-the-Art Meta-Heuristics for the BQP

Several (meta–)heuristic approaches have been proposed for the BQP. In the following, we briefly review effective heuristic algorithms capable of finding optimum/best-known or very good near-optimum solutions for the BQP.

Glover, Kochenberger, and Alidaee [319] have proposed a tabu search heuristic for instances of up to 500 variables. Their method consists of a strategic oscillation scheme that alternates between constructive and destructive phases.

Lodi, Allemand, and Liebling [532] proposed a heuristic based on an evolutionary algorithm for the same problem set studied by Glover *et al.* Their heuristic is combined with the local search algorithm that is based on the constructive and destructive phases of the tabu search in [319]. Their crossover operator is similar to uniform crossover [864], utilizing the MinRange algorithm, which is based on the property by Pardalos and Rodgers [706].

Alkhamis, Hasan, and Ahmed proposed a simulated annealing algorithm [11]. Unfortunately, only small problem instances with up to 100 variables were investigated. In [55], Beasley has provided larger BQP test problems with up to 2500 variables as new test problems of the ORLIB [54]. Beasley includes the best-known solutions for each of the provided instances found by tabu search and simulated annealing.

In our genetic local search algorithm [586], a simple local search (called *1-opt*, see below) and a variant of uniform crossover, HUX [246], were employed. For several large instances of [55], they provided new best-known solutions and have shown that their algorithm outperforms the two alternative heuristics reported by Beasley. Furthermore, we developed a greedy heuristic and two local search heuristics called *1-opt* and *k-opt* [589]. We showed that in particular the *k-opt* local search is capable of finding high-quality solutions even for the large-scale problem instances, and they also proposed that these heuristics are well suited as components for metaheuristics, such as MA.

Katayama and Narihisa [446] proposed a new simulated annealing-based heuristic with a reannealing process. The approach was tested on the large instances ranging from 500 to 2500 variables contained in the ORLIB. Although simulated annealing is based on the simple *1-opt* neighborhood structure, better average solution results for the large instances were found as compared to the other heuristics: our genetic local search *et al.* [586] and the heuristics by Beasley [55]. Moreover, the heuristic was considerably faster than the others, and new best-known solutions for several large instances were reported.

7.5.3 A Memetic Algorithm Using Innovative Recombination

In [591], we proposed a memetic algorithm using a new recombination operator that takes the properties of the search space of the BQP into account. Although the landscape is correlated/structured, recombination operators such as HUX or simple uniform crossover are not that effective due to the large basins of attraction of the local optima, as stated above. This is even more true when a powerful variable k-opt local search is used.

The outline of the MA is provided in Alg. 15. The population is initialized (Init()) with the randomized greedy heuristic we proposed in [589]. The local search used is a randomized k-opt local search algorithm proposed in [447, 448], which is based on the k-opt local search proposed in [589]. Similar to the Lin-Kernighan algorithm for the TSP [524], the basic idea of the heuristic is to find a solution by flipping a variable number of k bits in the solution vector per iteration. In each step, a sequence of n solutions is generated by flipping a random bit with positive gain or the bit with the highest associated gain. Furthermore, a candidate set is used to assure that each bit is flipped no more than once. The best solution in the sequence is accepted as the new solution for the next iteration.

To minimize the number of times a local optimum is rediscovered, we have proposed a new variation operator. The basic idea is to utilize a simple local search for introducing new alleles, i.e. alleles not contained in both parents. The crossover can be regarded as innovative, since new alleles are introduced based on the associated gain in fitness. Hence the name *innovative variation*. The operator works as follows:

7 Memetic Algorithms and Fitness Landscapes in Combinatorial Optimization 117

Algorithm 15. BQP-MA

```
1  begin
2      foreach S in Population do S ← LocalSearch(Init());
3      while not terminated do
4          Offspring ← {};
5          for i ← 0 to crossovers do
6              A ← Select(Population);
7              B ← Select(Population);
8              C ← LocalSearch(Recombine(A, B));
9              Offspring ← OffSpring + C;
10         endfor
11         Population ← Select(Population, Offspring);
12         if Converged(Population) then
13             foreach S in Population\Best do S ← LocalSearch(Mutate(S));
14         endif
15     endw
16 end
```

In the first step, the common and the non-common bits of the parents are identified. Then, the contents of parent I_a are copied to the offspring. Variation is now achieved by alternately flipping non-common bits and common-bits: In a loop, a randomly selected non-common bit is flipped with a positive associated gain, if there is at least one such non-common bit. The common bit with the maximum associated gain is flipped afterwards, even if the gain is negative. If a bit has been flipped, it is removed from the set it was contained in (either the common or non-common bit set). The loop is repeated n times where n is the number of non-common bits.

Mutation is only applied when the population is converged. In such a case, we perform a diversification/restart strategy, which is borrowed from [246], in order diversity the population by moving to other points of the search space if no new best individual in the population was found for more than 30 generations. In response to this requirement, the individuals except for the best one in the population are mutated by flipping randomly chosen $n/3$ bits for each individual of length n. After that, each of the mutated individuals is improved by the randomized *k-opt* local search to obtain a renewal set of local optima and the search is started again with the new, diverse population. The performance of our MA is shown in Table 7.2. We have tested our algorithm on several benchmark instances from the literature. The first set kb-g consists of 10 instances of size $n = 1000$ that have been provided by Kochenberger and have been used in the performance evaluation of scatter search [18]. The densities of instances in the problem set are between 0.1 and 1.0. The last two sets beas1000 ($n = 1000$) and beas2500 ($n = 2500$) were first studied by Beasley [55], each of which consists of ten instances with $dens(Q) = 0.1$.

In [591], a detailed comparison with other approaches for the BQP has been made. Summarizing, The MA approach provides a higher average solution quality than other approaches. CPU times have not been reported for all approaches or are not directly comparable to our results. However, our MA is superior or at least

Table 7.2. Computational results of the MA with innovative variation incorporating the randomized *k-opt* local search algorithm for test problem instances from the literature.

		best		MA with Innovative Variation		
Instance	$dens(Q)$	known	best	avg (quality %)	b/30	t1/s (gens)
kb-g01	0.1	131456	131456	131456.0 (0.000000)	30	6.1 (6)
kb-g02	0.2	172788	172788	172788.0 (0.000000)	30	12.8 (9)
kb-g03	0.3	192565	192565	192565.0 (0.000000)	30	11.4 (4)
kb-g04	0.4	215679	215679	215679.0 (0.000000)	30	42.0 (23)
kb-g05	0.5	242367	242367	242367.0 (0.000000)	30	15.6 (5)
kb-g06	0.6	243293	243293	243293.0 (0.000000)	30	69.4 (30)
kb-g07	0.7	253590	253590	253590.0 (0.000000)	30	45.7 (13)
kb-g08	0.8	264268	264268	264268.0 (0.000000)	30	40.2 (12)
kb-g09	0.9	262658	262658	262618.0 (0.015219)	25	140.1 (40)
kb-g10	1.0	274375	274375	274335.4 (0.014423)	15	143.9 (41)
beas1000-1	0.1	371438	371438	371438.0 (0.000000)	30	6.7 (9)
beas1000-2	0.1	354932	354932	354932.0 (0.000000)	30	7.7 (10)
beas1000-3	0.1	371236	371236	371236.0 (0.000000)	30	5.8 (5)
beas1000-4	0.1	370675	370675	370675.0 (0.000000)	30	6.6 (7)
beas1000-5	0.1	352760	352760	352760.0 (0.000000)	30	11.8 (20)
beas1000-6	0.1	359629	359629	359629.0 (0.000000)	30	11.0 (17)
beas1000-7	0.1	371193	371193	371193.0 (0.000000)	30	9.1 (12)
beas1000-8	0.1	351994	351994	351994.0 (0.000000)	30	28.1 (50)
beas1000-9	0.1	349337	349337	349337.0 (0.000000)	30	6.1 (6)
beas1000-10	0.1	351415	351415	351415.0 (0.000000)	30	7.3 (9)
beas2500-1	0.1	1515944	1515944	1515944.0 (0.000000)	30	59.9 (15)
beas2500-2	0.1	1471392	1471392	1471357.8 (0.002322)	26	165.2 (60)
beas2500-3	0.1	1414192	1414192	1414183.1 (0.000629)	29	87.8 (30)
beas2500-4	0.1	1507701	1507701	1507701.0 (0.000000)	30	42.2 (9)
beas2500-5	0.1	1491816	1491816	1491816.0 (0.000000)	30	76.1 (29)
beas2500-6	0.1	1469162	1469162	1469162.0 (0.000000)	30	78.5 (26)
beas2500-7	0.1	1479040	1479040	1479040.0 (0.000000)	30	92.2 (30)
beas2500-8	0.1	1484199	1484199	1484199.0 (0.000000)	30	47.1 (10)
beas2500-9	0.1	1482413	1482413	1482413.0 (0.000000)	30	140.0 (70)
beas2500-10	0.1	1483355	1483355	1483336.9 (0.001218)	28	108.6 (43)

comparable to other state of the art approaches including tabu search and scatter search. In [700], a multi-start tabu search has been proposed that appears to be similarly effective. Again, a direct comparison is not easy. The author, however, fails to compare with the results from [591]. Instead, older results are considered from [586].

7.6 Conclusion

In this chapter, we have discussed fitness landscape analysis as suitable methodology for discovering search space properties relevant for the development of memetic algorithms. We have shown that some combinatorial optimization problems have structured landscapes that have a deep value/massive central structure. This structure is known to be a reason why MAs perform well on certain combinatorial problems. We have argued that the results from the fitness landscape analysis can be used to design local search (autocorrelation analysis) or can help in deciding to use mutation based or recombination variation (fitness distance analysis). In two case studies, we have demonstrated that MAs are highly effective and belong to state-of-the-art meta-heuristics. For the binary quadratic programming problem, we have shown that an advanced fitness analysis can help in designing a recombination operator by assuring that it has a high chance to leave the basin of attraction of the current local optima. This innovative variation operator increases the effectiveness of the MA considerably.

Chapter 8
Memetic Algorithms in Continuous Optimization

Carlos Cotta and Ferrante Neri

8.1 Introduction and Basic Concepts

Intuitively, a set is considered to be discrete if it is composed of isolated elements, whereas it is considered to be continuous if it is composed of infinite and contiguous elements and does not contain "holes".

More formally, to introduce the concept of continuous optimization, some preliminary definitions are required. If we consider sub-sets of real numbers, where the partial order is obviously valid, a set S is said to be *dense* if

$$\forall x_1, x_2 \in S: \exists x_3: x_1 \leqslant x_3 \leqslant x_2. \tag{8.1}$$

If the property above is not satisfied for all the points, the set is said to be *discrete*. When the property is not satisfied for some of the points in D, the set is composed of multiple not interconnected dense sets.

It must be remarked that, since a set of infinite numbers cannot be represented in a machine, in computer science all the sets are in principle discrete. On the other hand, a set where the distance between each pair of consecutive points is not bigger than the machine precision ε can be considered as a dense set. In other words, the definition of a dense set in computer science can be modified in the following way. A set S is said to be *dense in computer science* if

$$\forall x_1, x_2 \in S, x_1 < x_2: \exists x_3 \in S: [(x_3 - x_1) \geqslant \varepsilon] \wedge [(x_2 - x_3) \geqslant \varepsilon]. \tag{8.2}$$

Carlos Cotta
Dept. de Lenguajes y Ciencias de la Computación. Universidad de Málaga,
Campus de Teatinos, 29071 Málaga, Spain
e-mail: ccottap@lcc.uma.es

Ferrante Neri
Department of Mathematical Information Technology, P.O. Box 35 (Agora),
40014 University of Jyväskylä, Finland
e-mail: ferrante.neri@jyu.fi

A multidimensional set composed of the Cartesian product of multiple dense sets S is said to be decision space D. An optimization problem defined on a decision space D is said to be *continuous optimization problem*. More specifically, throughout this chapter we refer to the minimization problem of an objective function $f(x)$, where x is a vector of n design variables in a decision space D. In general, this optimization problem can be subject to a set of constraints. For the sake of simplicity, in this chapter we just consider the minimization within a hyper-rectangular space.

Although the concept of continuous optimization is strictly related to the concept of continuous functions, the two concepts should not be identified. According to the Cauchy definition, a continuous function is characterized by the following property: an infinitesimal change in the independent variable corresponds to an infinitesimal change of the dependent variable. The optimization of a continuous function is always a continuous optimization problem. The reverse statement is not not true. In computer science, even when the function displays discontinuity points still the resulting problem is continuous.

This chapter focuses on continuous optimization problems and on the application of Memetic Algorithms (MAs) in order to solve such problems. Section 8.2 highlights the difference between global and local optimization for continuous problems. Section 8.3 briefly illustrates a set of popular global optimizers which can be used as an evolutionary framework within a MA.

8.2 Global and Local Continuous Optimization

In discrete optimization, solutions are simply characterized by their fitness values. Thus, a solution can either be optimal or suboptimal. In continuous optimization, the situation is different as the position of each candidate solution within the decision space takes a high importance. It can intuitively be seen from the Cauchy definition of continuous function that for a given point its closest points are expected to have a similar performance with respect to the point. In this context, the concept of *neighborhood* is extremely important. For a given point, its neighborhood is that set of points characterized by distance ε from it. This concept is fundamental in continuous optimization because, unlike the discrete optimization case, it make sense to discuss about small movements and search directions. In other words, unlike what happens in the discrete case, in continuous optimization it makes sense to discuss about the *gradient* which can be redefined, along the generic variable x_i, for the "continuous discrete" case of computer science in the following way:

$$\frac{\partial f}{\partial x_i} = \frac{f(x_i + \varepsilon) - f(x_i)}{\varepsilon}. \tag{8.3}$$

If a gradient can be defined, it can be used from a starting point to select the most promising neighbor and thus to identify a promising search direction. The information derived from the knowledge of the gradient values can be obviously exploited within an optimizer. A major difference should be highlighted between the gradient defined above for continuous problems in computer science and the classical gradient in mathematical analysis. While in mathematical analysis a null gradient corresponds to a critical point, i.e., a true local/global optimum, plateau or saddle point, in computer science a null gradient (according to the definition above) in a point means that the entire neighborhood of this point has the same fitness values; thus the point falls within a plateau. From this consideration, it is clear that the null gradient condition cannot be used in computer science to identify the true local/global optima (which is the goal of the optimization) but only plateaus. The detection of local optima should be performed in a different way: a point is a *local minimum*(maximum) if the objective function values of the neighborhood are all higher (lower) than that of the point.

Without a loss of generality, let us consider minimization problems. Usually optimization problems are multimodal, i.e., contain multiple local minima. However, the goal in optimization is to detect the global optimum, that is, in our case, the minimum exhibiting the lowest function value. All the methods that make an explicit or implicit use of gradient information tend to detect the closest local minimum. Thus, an efficient global optimizer should not be based only on gradient information but also on direct fitness comparisons among solutions regardless their position within the decision space. This approach guarantees an extensive search and hopefully allows that algorithms get stuck within local optima. In this context, it is important to define the concept of *basin of attraction*. Two definitions can be given in both a broad and restricted sense. In a broad sense, for a given search strategy, objective function, and starting point(s), a basin of attraction is the set of points which can be reached. However, when in a generic way computer scientists refer to the term basin of attraction without specifying the search strategy, it is meant that the specific search strategy is the classical deterministic hill-descender which perturbs separately each variable. Thus, a decision space can be mapped as a composition of basins of attraction and the goal of global optimization is to detect the globally optimum basin of attractions and avoid the local ones.

MAs in continuous optimization are thus thought as algorithmic structures which require both global and local search components whose coordination make the success of the computational paradigm. These structures are usually composed of an evolutionary framework which has the role of performing the global search and one or more local search algorithms activated within the generation cycle of the external framework.

8.3 Global Optimization Algorithms

While some local search algorithms have been previously illustrated, in this chapter some global search algorithms, which are shown as examples of evolutionary frameworks in MAs, are briefly presented in the following section.

8.3.1 Stochastic Global Search, Brute Force and Random Walk

The simplest (and often not so efficient) way to perform the global optimal search of a black box function is the progressive perturbation of one or more solutions in order to improve upon its performance. The search can be performed by various search rules, for example by generating a new solution within the decision space or by adding a randomized perturbation vector to a trial solution. This class of algorithms is often named Stochastic Global Search or simply Stochastic Search, see [838], and has the crucial importance of being the basic principle behind all the modern computational intelligence optimization algorithms. It must be observed that all the modern algorithms which take their inspiration on the most various natural sources, such as principles of the evolution or collective behavior of animals or even MAs, are at the end stochastic search algorithms which differ one from another on the trial solution generation mechanism or the strategy for retaining the solutions (and selecting the search directions).

In order to clarify this concept let us consider two classical global optimization algorithms which are based on completely opposite search logics. The first, namely brute force, consists of the construction of a regular grid within the decision space and the sample of the points in correspondence to the nodes. This algorithm has been taken into account in this context because it is a global search algorithm based on a fully deterministic generation of solutions. Another famous simple stochastic search is the random walk, see [337]. This algorithm perturbs each coordinate of a candidate solution by means of a Gaussian distribution. It can be immediately observed that the random walk is a highly randomized method as the trial search directions rely only on stochastic perturbations.

As an additional remark, although very different, these two methods are both plagued by the same problem: their performance highly depends on the parameter setting. In the brute force, the selection of step size, and thus amount of points to sample, must be carried out to avoid inefficient search or an unacceptable computational time. Likewise, in the random walk the success of the algorithm heavily depends on the mean value and standard deviation of the perturbation Gaussian. In other words, regardless the degree of randomization in the search logic, when there is no information on the objective function, the parameter setting becomes key point in the algorithmic performance.

8.3.2 Evolution Strategy and Real Coded Evolutionary Algorithms

In 70s, while Genetic Algorithms (GAs) were developed for discrete and combinatorial optimization problems [389], Evolution Strategy (ES) were developed for continuous optimization problems [760, 798]. In ES, each individual is a real-valued vector composed of its candidate solution representation x and a set of self-adaptive parameters σ:

$$(x, \sigma) = (x_1, \ldots, x_n, \sigma_1, \ldots, \sigma_n) \qquad (8.4)$$

8 Memetic Algorithms in Continuous Optimization

In many evolution strategy variants, a set of self-adaptive parameters of a second kind can be added to the solution encoding. At each generation cycle, parent selection relies on pseudo-randomly selecting some solutions to undergo recombination and mutation. In evolution strategies a big emphasis is placed on mutation while recombination sometimes plays a minor role (although it is not simply dismissed as in evolutionary programming) – see [65] for a in-depth treatment of these two operators in ES. The general mutation rule is defined, for the generic i^{th} design variable, by:

$$\sigma_i = \sigma_i e^{N(0,\tau') + N_i(0,\tau)} \tag{8.5}$$

and

$$x_i = N(x_i, \sigma_i) \tag{8.6}$$

where $N(\mu, \sigma)$ is normally a distributed random number with mean μ and standard deviation σ. The update of σ can be performed by means of several rules proposed in literature. The most famous are the 1/5 success rule [760], uncorrelated mutation with one step size, uncorrelated mutation with n step sizes and correlated mutation, for details see [239]. The method shown in Eq. (8.5) corresponds to uncorrelated mutations with n step sizes, and τ, τ' are two parameters (the local and the global learning rate respectively) that can be set as [32]:

$$\tau = 1/\sqrt{2\sqrt{n}} \tag{8.7}$$

$$\tau' = 1/\sqrt{2n} \tag{8.8}$$

The notation $N_i(0, \tau)$ is used to denoted a different random number for each parameter, whereas $N(0, \tau)$ is a common –solution-wise– random number. The general idea is that the solutions are mutated within their neighborhood based on a certain probabilistic criterion with the aim of generating new promising solutions.

The recombination can be discrete or intermediary: discrete recombination generates an offspring solution by pseudo-randomly choosing the genes from two parent solutions, while intermediary recombination generates an offspring whose genes are obtained by calculating a randomly weighted average of the corresponding genes of two parents (other methods are possible though – see Section 8.4).

The parent selection can be performed either in the genetic algorithm fashion by replacing the whole parent population with the best members of the offspring population or by merging parent and offspring populations and selecting the wanted number of individuals on the basis of their fitness values. These strategies are usually known as comma and plus strategy respectively.

In the 90s, a reorganization of the knowledge regarding evolution inspired metaheuristics was performed. This lead to the fact that GAs, ES, Evolutionary Programming and other branches of the field have all been seen as an expression of the same idea and named Evolutionary Algorithms (EAs). These algorithms, characterized by four phases, 1) parent selection, 2) crossover, 3) mutation, 4) survivor selection, can be implemented to both continuous and discrete optimization, by properly

Fig. 8.1. Functioning of the BLX−α recombination operator. Offspring variable z_i is randomly sampled from the interval denoted by a thick line.

representing the solutions and their recombination. The most natural way to represent candidate solutions of a continuous optimization problem is simply to use them "as they are", i.e., have a representation of vectors of real numbers without any conversion (as in classical GAs where all the numbers were converted to binary).

A multitude of recombination strategies among pairs or small groups of solutions have been proposed in literature. The advantages of one strategy with respect to another are, in general, dependent on the problem. A very broadly used recombination strategy is the so called BLX−α crossover, see [246, 382]. For two given parent solutions x and y, their offspring z is generically calculated in the following way:

$$z_i = U[m_i - \alpha I, M_i + \alpha I] \tag{8.9}$$

where α is a parameter, $M_i = \max(x_i, y_i)$, $m_i = \min(x_i, y_i)$, $I = |x_i - y_i|$ and $U[a,b]$ is a uniformly distributed random number in the interval $[a,b]$. Parameter α is thus used to tune the explorative capability of crossover – see Fig. 8.1. A *parent centric* variant of BLX−α is also defined in [536] by sampling each offspring variable from a closed interval of radius $2\alpha I$ centered at any of the corresponding parental variables.

Precisely related to this exploration issue (or more generically to the avoidance of premature convergence), it is worth mentioning another EA variant that is commonly used as the population-based engine of continuous MA, namely the CHC (Cross generational elitist selection, Heterogeneous recombination, and Cataclysmic mutation) algorithm [246]. The main idea of this algorithm is to combine strong selective pressure with incest-prevention strategies and explorative recombination. The incest-prevention strategy amounts to avoiding that two very similar solutions are recombined (since this would likely produce very similar offspring as well, hence leading to diversity loss and potential premature convergence). To do so, a distance parameter δ is maintained, determining the minimal distance that must exist between two solutions if these are to be recombined. This parameter can change dynamically in order to cope with the progressive convergence of the population. As to the selection, it is typically done using the plus strategy of ES. Algorithm 16 shows the pseudocode of the CHC algorithm.

A final evolutionary approach for continuous optimization that deserves being mentioned is the Covariance Matrix Adaptation Evolution Strategy (CMA-ES) [362]. This algorithm falls within the class of estimation of distribution algorithms [534] (EDAs) and has been shown to be extremely efficient when solving continuous optimization benchmarks [28]. CMA-ES is based on generating solutions via a multivariate normal distribution whose mean and covariance matrix is adaptively

8 Memetic Algorithms in Continuous Optimization

Algorithm 16. Pseudo-code of the CHC algorithm

```
1  begin
2      generate initial population P ← {p₁,···,p_μ};
3      initialize distance parameter δ;
4      while ¬ termination-condition do
5          create solutions pairs S ← (pᵢ, pⱼ);
6          P' ← ∅;
7          for (p, p') ∈ S do
8              d ← distance(p, p');
9              if d ⩽ δ then
10                 p'' ← recombine(p, p');
11                 P' ← P' ∪ {p''};
12             endif
13         endfor
14         P ← plus-select(P, P');
15         if P' = ∅ then
16             decrease δ;
17             if δ < 0 then
18                 restart population P;
19                 initialize distance parameter δ;
20             endif
21         endif
22     endw
23 end
```

learnt as in EDAs, i.e., utilizing truncation selection to pick a subset of the best solutions generated in each step, and using these solutions to update the distribution parameters. CMA-ES has a solid theoretical background and several desirable properties such as invariance to several transformations of the objective function and a relatively low number of parameters. Furthermore, it can not only serve as a population-based engine but also as a local searcher if adequately parameterized, e.g., $(1+1)$-CMA-ES [605], We refer to [360] for further details and source code of the CMA-ES algorithm.

8.3.3 Particle Swarm Optimization

Particle Swarm Optimization (PSO) is a population-based optimization metaheuristic introduced in [458], and then developed in various variants for test problems and applications. The main metaphor employed in PSO is that a group of particles makes use of their "personal" and "social" experience in order to explore a decision space and detect solutions with a high performance. More specifically, a population of candidate solutions is randomly sampled within the decision space. Subsequently, the fitness value of each candidate solution is computed and the solutions are ranked on the basis of their performance. The solution associated to the best fitness value detected overall is named global best x^{gb}. At the first generation, each solution x_i

is identified with the corresponding local best solution x_i^{lb}, i.e., the most successful value taken in the history of each solution. At each generation, each solution x_i is perturbed by means of the following rule:

$$x_i = x_i + v_i \tag{8.10}$$

where the velocity vector v_i is a perturbation vector generated in the following way:

$$v_i = \omega v_i + \alpha_1(x_i^{lb} - x_i) + \alpha_2(x^{gb} - x_i) \tag{8.11}$$

where ω is the so-called inertia parameter (the higher this parameter, the longer it takes the particle to change direction), and α_1, α_2 are two parameters that control the attraction the particle feels towards the best-known local/global solutions. These are typically set uniformly at random –within the interval $(0,1)$, i.e., 0 excluded and 1 included– in each step; we denote as $U(0,1)$ as such a uniform distribution. The fitness value of the newly generated x_i is calculated and if it outperforms the previous local best value the value of x_i^{lb} is updated. Similarly, if the newly generated solution outperforms the global best solution, a replacement occurs. At the end of the optimization process, the final global best detected is the estimate of the global optimum returned by the particle swarm algorithm. It is important to remark that in PSO, there is a population of particles which has the role of exploring the decision space and a population of local best solutions (the global best is the local best with the highest performance) to keep track of the successful movements.

In order to better understand the metaphor and thus the algorithmic philosophy behind PSO, the population can be seen as a group of individuals which search for the global optimum by combining the exploration along two components: the former is the memory and thus learning due to successful and unsuccessful moves (personal experience) while the latter is a partial imitation of the successful move of the most promising individual (social experience). In other words, as shown in the formula above, the perturbation is obtained by the vectorial sum of a move in the direction of the best overall solution and a move in the direction of the best success achieved by a single particle. These directions in modern PSO algorithms are weighted by means of random scale factors, since the choice has to turn out to be beneficial in terms of diversity maintenance and prevention of premature convergence.

Many versions and variants of PSO have been proposed in literature in order to attempt to enhance its performance. In order to give a flavor of possible PSO modifications, two examples are here reported. A popular variant is the linearly variable weight factor ω proposed in [809]:

$$\omega = \omega_{max} - (\omega_{max} - \omega_{min})\frac{g}{G} \tag{8.12}$$

where g is the current generation and G is the generation budget. Parameters ω_{max} and ω_{min} are usually set equal to 0.9 and 0.4, respectively.

8 Memetic Algorithms in Continuous Optimization

Algorithm 17. PSO pseudo-code

```
1  begin
2      generate N_p particles and N_p velocities pseudo-randomly;
3      copy the population of particles into the set of local bests: ∀i, x_{i-lb} = x_i ;
4      while budget condition do
5          for i = 1 : N_p do
6              | compute f(x_i);
7          endfor
8          for i = 1 : N_p do
               // ** Velocity Update **
9              generate a vector of random numbers U(0,1);
10             v_i = ωv_i + U(0,1)(x_i^{lb} − x_i) + U(0,1)(x^{gb} − x_i);
               // ** Position Update **
11             x_i = x_i + v_i;
               // ** Survivor Selection **
12             if f(x_i) ≤ f(x_{i-lb}) then
13                 x_i^{lb} = x_i;
14                 if f(x_i) ≤ f(x^{gb}) then
15                     x^{gb} = x_i;
16                 endif
17             endif
18         endfor
19     endw
20 end
```

Another variant is the constriction factor proposed in [129]. Within such a scheme the velocity update is:

$$v_i = \chi v_i + c_1 U(0,1)\left(x_i^{lb} - x_i\right) + c_2 U(0,1)\left(x_i^{nb} - x_i\right) \quad (8.13)$$

where x_i^{nb} is the best within the neighborhood (see for details [129]). The constriction factor χ is defined as:

$$\chi = \frac{2}{\left|2 - \phi - \sqrt{\phi^2 - 4\phi}\right|} \quad (8.14)$$

where $\phi = c_1 + c_2 = 4.1$ and $c_1 = c_2 = 2.05$, see [129]. A pseudo-code showing the main features of the basic PSO is given in Algorithm 17.

8.3.4 Differential Evolution

Differential Evolution (DE) is an interesting optimizer for continuous problems which shares some properties of evolutionary algorithms (e.g., the crossover) and some others of swarm intelligence algorithms (the one-to-one replacement). According to its original definition given in [853], DE consists of the following steps.

An initial sampling of S_{pop} individuals is performed pseudo-randomly with a uniform distribution function within the decision space D. At each generation, for each individual x_i from the S_{pop} in the population, three mutually distinct individuals x_r, x_s and x_t are pseudo-randomly extracted from the population. According to DE logic, a provisional offspring x'_{off} is generated by mutation as:

$$x'_{off} = x_t + F(x_r - x_s) \tag{8.15}$$

where $F \in [0, 1^+[$ is a scale factor which controls the length of the exploration vector $(x_r - x_s)$ and thus determines how far from point x_i the offspring should be generated. With $F \in [0, 1^+[$, it is meant here that the scale factor should be a positive value which cannot be much greater than 1, see [733]. While there is no theoretical upper limit for F, effective values are rarely greater than 1.0. The mutation scheme shown in Eq. (8.15) is also known as DE/rand/1. Other variants of the mutation rule have been subsequently proposed in literature, see [745]:

- DE/best/1: $x'_{off} = x_{best} + F(x_s - x_t)$
- DE/cur-to-best/1: $x'_{off} = x_i + F(x_{best} - x_i) + F(x_s - x_t)$
- DE/best/2: $x'_{off} = x_{best} + F(x_s - x_t) + F(x_u - x_v)$
- DE/rand/2: $x'_{off} = x_r + F(x_s - x_t) + F(x_u - x_v)$
- DE/rand-to-best/2: $x'_{off} = x_r + F(x_{best} - x_i) + F(x_r - x_s) + F(x_u - x_v)$

where x_{best} is the solution with the best performance among individuals of the population, x_u and x_v are two additional pseudo-randomly selected individuals. It is worthwhile to mention the rotation invariant mutation shown in [732]:

- DE/current-to-rand/1 $x_{off} = x_i + K(x_t - x_i) + F'(x_r - x_s)$

where K is is the combination coefficient, which as suggested in [732] should be chosen with a uniform random distribution from $[0, 1]$ and $F' = K \cdot F$. For this special mutation the mutated solution does not undergo the crossover operation (since it already contains the crossover), described below.

Recently, in [733], a new mutation strategy has been defined. This strategy, namely DE/rand/1/either-or, consists of the following:

$$x'_{off} = \begin{cases} x_t + F(x_r - x_s) & \text{if } U(0,1) < p_F \\ x_t + K(x_r + x_s - 2x_t) & \text{otherwise} \end{cases} \tag{8.16}$$

where for a given value of F, the parameter K is set equal to $0.5(F+1)$.

When the provisional offspring has been generated by mutation, each gene of the individual x'_{off} is exchanged with the corresponding gene of x_i with a uniform probability and the final offspring x_{off} is generated:

$$x_{off,j} = \begin{cases} x_{i,j} & \text{if } U(0,1) < CR \\ x'_{off,j} & \text{otherwise} \end{cases} \tag{8.17}$$

Algorithm 18. DE/rand/1/bin pseudo-code

```
1  begin
2      generate N_p individuals of the initial population pseudo-randomly;
3      while budget condition do
4          for k = 1 : N_p do
5              compute f(x_k);
6          endfor
7          for k = 1 : N_p do
               // ** Mutation **
8              select three individuals x_r, x_s, and x_t;
9              compute x'_off = x_t + F(x_r − x_s);
               // ** Crossover **
10             x_off = x'_off;
11             for i = 1 : n do
12                 generate U(0,1);
13                 if U(0,1) > Cr then
14                     x_off[i] = x_k[i];
15                 endif
16             endfor
               // ** Survivor Selection **
17             if f(x_off) ≤ f(x_k) then
18                 save index for replacement x_k = x_off;
19             endif
20         endfor
21         perform replacements;
22     endw
23 end
```

where $U(0,1)$ is a random number between 0 and 1; j is the index of the gene under examination. This crossover strategy is well-known as binary crossover and indicated as "bin". For the sake of completeness, we mention that there exist a few other crossover strategies, for example the exponential strategy see [733]. However in this paper we focus on the bin strategy since it is the most commonly used and often the most promising.

The resulting offspring x_{off} is evaluated and, according to a one-to-one spawning strategy, it replaces x_i if and only if $f(x_{off}) \leq f(x_i)$; otherwise no replacement occurs. For sake of clarity, the pseudo-code highlighting the working principles of DE is shown in Algorithm 18.

8.4 Particularities of Memetic Approaches for Continuous Optimization

In principle the deployment of memetic algorithms on continuous domains can be done using the generic algorithmic template presented in Chapter 4, much like it is done for combinatorial problems – see Chapter 6. This said, continuous optimization problems have several distinctive features that must be considered in order to come up with efficient memetic solvers. Two of the most relevant ones are:

- *The cost of local search:* in many combinatorial domains it is frequently possible to compute the fitness of a perturbed solution incrementally, e.g., let x be a solution and let $x' \in \mathcal{N}(x)$ be a neighboring solution; then the fitness $f(x')$ can be often computed as $f(x') = f(x) + \Delta f(x,x')$, where $\Delta f(x,x')$ is a term that depends on the particular perturbation done on x and is typically efficient to compute (much more efficiently that a full fitness computation). For example, in the context of the traveling salesman problem and the 2-opt neighborhood, the fitness of a perturbed solution can be computed in constant time by calculating the difference between the weights of the two edges added and the two edges removed. This is much more difficult in the context of continuous optimization problems, which are often non-linear and hard to decompose as the sum of linearly-coupled terms. Hence local search usually has to resort to full fitness computations.
- *The underlying search landscape:* the interplay among the different search operators used in memetic algorithms (or even in simple evolutionary algorithms) is a crucial issue for achieving good performance in any optimization domain. When tackling a combinatorial problem, this interplay is a complex topic since each operator may be based on a different search landscape. It is then essential to understand these different landscape structures and how they are navigated – the "one operator, one landscape" view [434]. In the continuous domain the situation is somewhat simpler, in the sense that there exists a natural underlying landscape in D^n (typically $D = \mathbb{R}$), that is induced by distance measures such as Euclidian distance. In other words, neighborhood structures are defined by closed spheres of radius ε in the case of unary operators, and by solid hypercubes in the case of recombination (recall for example the BLX−α operator). The intuitive imagery of local optima and basins of attraction naturally fits here, and allows the designer to exert some control on the search dynamics by carefully adjusting the intensification/diversification properties of the operators used.

These two issues mentioned above have been dealt in the literature on memetic algorithms for continuous optimization in different ways. Starting with the first one (the cost of local search), it emphasizes the need for carefully selecting when and how local search is applied (obviously this is a general issue, also relevant in combinatorial problems, but definitely crucial in continuous ones). Needless to say, this decision-making is very hard in general [494, 857], see also Chapter 5, but some strategies have been put forward in previous works. A rather simple one is to resort to partial Lamarckianism [396] by randomly applying local search with probability $p_{LS} < 1$. Obviously, the application frequency is not the only parameter that can be adjusted to tune the computational cost of local search: the intensity of local

8 Memetic Algorithms in Continuous Optimization

search (i.e., for how long is local improvement attempted on a particular solution) is another parameter to be tweaked. This adjustment can be done blindly (i.e., prefixing a constant value or a variation schedule across the run), or adaptively. For example, Molina et al. [605] define three different solution classes (on the basis of fitness) and associate a different set of local-search parameters for each of them. Related to this, Nguyen *et al.* [665] consider a stratified approach, in which the population is sorted and divided into n levels (n being the number of local search applications), and one individual per level is randomly selected. This is shown to provide better results than random selection. We refer to [40] for an in-depth empirical analysis of the time/quality tradeoffs when applying parameterized local search within memetic algorithms. This adaptive parameterization has been also exploited in so-called *local-search chains* [608], by saving the state of the local-search upon completion of a certain solution for later use if the same solution is selected again for local improvement. Let us finally note with respect to this parameterization issue that adaptive strategies can be taken one step further, entering into the realm of self-adaptation. An overview of the possibilities to this end is provided in Chapter 11.

As to what the exploitation/exploration balance regards, it is typically the case that the population-based component is used to navigate through the search space, providing interesting starting points to intensify the search via the local improvement operator. The diversification aspect of the population-based search can be strengthened in several ways, such as for example using multiple subpopulations [640], or diversity-oriented replacement strategies. The latter are common in scatter search [320] (SS), an optimization paradigm closely related to memetic algorithms in which the population (or reference set in the SS jargon) is divided in tiers: entrance to them is gained by solution on the basis of fitness in one case, or diversity in the other case. Additionally, SS often incorporated restarting mechanisms to introduce fresh information in the population upon convergence of the latter. Diversification can be also introduced via selective mating, as it is done in CHC (see Sect. 8.3.2). A related strategy was proposed by Lozano et al. [536] via the use of negative assortative mating: after picking a solution for recombination, a collection of potential mates is selected and the most diverse one is used. Other strategies range from the use of clustering [806] (to detect solutions likely within the same basin of attraction upon which it may not be fruitful to apply local search), or the use of standard diversity preservation techniques in multimodal contexts such as sharing or crowding. It should also be mentioned that sometimes the intensification component of the memetic algorithm is strongly imbricated in the population-based engine, without resorting to a separate local search component. This is for example the case of the so-called *crossover hill climbing* [432], a procedure which essentially amount to using a hill climbing procedure on states composed of a collection of solutions, using crossover as move operator (i.e., introducing a newly generated solution in the collection –substituting the worst one– if the former is better than the latter). This strategy was used in the context of real-coded memetic algorithms in [536]. A different intensifying strategy was used by [161], by considering an exact procedure for finding the best combination of variable values from the parents (a so-called *optimal discrete recombination*). This obviously requires that the objective function is amenable to the application of an efficient procedure for exploring the dynastic potential (set of possible children) of the solutions being recombined – see also Chapter 12. We refer to [535] for a detailed analysis of diversification/intensification strategies in hybrid metaheuristics (in particular in memetic algorithms).

Acknowledgements. C. Cotta is partially supported by Spanish MICINN under project NEMESIS (TIN2008-05941) and by Junta de Andalucía under project TIC-6083. This research is supported by the Academy of Finland, Akatemiatutkija 130600, Algorithmic Design Issues in Memetic Computing.

Chapter 9
Memetic Algorithms in Constrained Optimization

Tapabrata Ray and Ruhul Sarker

9.1 Introduction

Memetic Algorithms (MAs) are a fairly recent breed of optimization algorithms created through a synergetic coupling of global and local search strategies [615]. While predecessors of MAs, i.e. Genetic Algorithms (GAs) and Evolutionary Algorithms (EAs) have had significant success in solving a number of real life complex optimization problems in the past, their performance can be greatly improved though a hybridization with other techniques [188]. GAs or EAs hybridized with local search strategies are commonly referred as memetic algorithms. These methods are inspired by models of natural systems that combine the evolutionary adaptation of a population with individual learning within the lifetimes of its members. While, the underlying GA/EA provides the ability for exploration, the local search aids in exploitation [492]. The exploitation schemes adopted in MAs include incorporation of heuristics, approximation algorithms, local search algorithms, specialized schemes for recombination etc.

An excellent review of memetic algorithms has been presented by Ong, Lim and Chen [689]. The performance of a MA is largely dependent on the correct choice of the local search strategies (memes), identification of the sub-set undergoing local improvements and the convergence criterion used in local search strategies. In this chapter, first, we discuss constrained optimization and provide a brief review of using memetic algorithms in solving Constrained Optimization Problems (ConOPs). The representations and local search approaches used in memetic algorithms in solving different ConOPs are also described and reviewed. We also present two case studies to demonstrate the use memetic algorithms in solving ConOPs. The first case study is designed to solve constrained numerical optimization problems with traditional representation while the next is designed to solve a combinatorial

Tapabrata Ray · Ruhul Sarker
School of Engineering and Information Technology, University of New South Wales at Australian Defence Force Academy, Canberra ACT 2600, Australia
e-mail: {t.ray,r.sarker}@adfa.edu.au

optimization problem with an alternative representation. In the first case study, a local search is embedded within an evolutionary algorithm to accelerate its rate of convergence. The evolutionary algorithm unlike common EA's preserves a set of marginally infeasible solutions throughout the course of search in an attempt to identify solutions to constraint optimization problems with a higher rate of convergence. The above MA also adopts a conventional representation scheme.

In the true sprit of MAs, the second study of MA is designed to solve job shop scheduling problems through intelligent representation that includes several problem specific recombination schemes to accelerate the rate of convergence. Both case studies show the benefits of using using MAs in solving ConOPs.

9.2 Constrained Optimization

Many real-world design and decision processes require a solution to Constrained Optimization Problems (ConOPs). In general, the ConOPs can be represented mathematically as follows (without loss of generality, minimization is considered here).

$$\begin{aligned}
\text{Minimize} \quad & f(\mathbf{X}) \\
\text{Subject to} \quad & g_i(\mathbf{X}) \geq 0, \quad i = 1, \ldots, m, \\
& h_j(\mathbf{X}) = 0, \quad i = 1, \ldots, p, \\
& L_i \leq x_i \leq U_i, \quad i = 1, 2, \ldots n
\end{aligned} \quad (9.1)$$

where $\mathbf{X} = (x_1, \ldots, x_n)$ is a vector with n decision variables, $f(\mathbf{X})$ is the objective function, $g_i(\mathbf{X})$ is the ith inequality constraint, $h_j(\mathbf{X})$ is the jth equality constraint, each x_i has a lower limit L_i and an upper limit U_i.

Based on the characteristics and mathematical properties, ConOPs can be of many different types. They may contain different types of variables such as real, integer and discrete, and may have equality and/or inequality constraints. The objective and constraint functions could be either linear or nonlinear. The problem may have one or more objectives, and each objective could be either of maximization or minimization. The functions may be either continuous or discontinuous, and either unimodal or multimodal. The feasible space for such problems could be a small fraction of the search space, the entire search space or a collection of multiple disjoint spaces. The optimal solution may or may not lie on constraint boundaries. A classification of optimization problems can be found in [791]. The application of constrained optimization methods is thus wide. A few examples include: planning (resource allocation, logistics, production planning, and scheduling), engineering design (welded beam, pressure vessel, and VLSI chip design), medical science (optimization of beams for radiotherapy, DNA sequencing), and computer science (data base design and data mining).

Researchers and practitioners use both conventional mathematical optimization methods and more recent methods relying on computational intelligence to solve ConOPs. One drawback of conventional optimization methods is the fact that they require specific properties (such as convexity, continuity and differentiability) of the mathematical model and hence require simplifications of the problem via assumptions [792]. In addition, the choice of a method is determined by the problem classification and sub-classification. In contrast, algorithms based on computational intelligence are simple to implement, do not require underlying properties of the model, are amenable to parallelization and can be readily applied to a range of problems.

An EA is one such class of method based on computational intelligence where a population(set) of solutions are iteratively improved in an attempt to identify global optimal solutions. However, they usually require evaluation of numerous solutions prior to convergence resulting in higher computational times and exhibit poor convergence [455]. On the other hand, local search algorithms converge quickly to a local optimum but lack a global perspective. A combination of a population based algorithm and a local search have resulted in a new class of algorithms referred as MAs which capitalizes the benefits of both algorithms simultaneously. For example, a recent study conducted by Hasan *et al.* [378] on a job shop scheduling problem highlighted that better quality of solutions could be obtained using MA with reduced computational effort as compared to genetic algorithms. Boudia and Prins [79] indicated that the solutions produced by memetic algorithms, for an integrated production-distribution problem, made significant savings as compared to others. More recently Singh *et al.* [816] reported the results of their infeasibility empowered memetic algorithm on a set of CEC-2010 constrained optimization benchmarks. It is also important to highlight that MAs are also attractive for dynamic optimization problems where an improved rate of convergence is required along with the ability to search for global optima. Isaacs *et al.* [407] have reported the performance of a memetic algorithm on dynamic bi-objective problems highlighting the benefits over evolutionary algorithms.

While population based methods such as EAs perform well as compared to conventional methods on unconstrained optimization problems, their performance on constrained optimization problems is not exceptionally good. Common search operators of EAs (such as crossover and mutation) are blind to the constraints. As a consequence, the candidate solutions generated by these operators may violate constraints [126]. Hence, mechanisms for constraint handling play an important role on the performance of such algorithms. Over the past decade, various constraint handling techniques have been proposed in the context of evolutionary optimization [126, 133, 195, 597, 908]. These techniques can be grouped as: penalty functions, special representations and operators, repair algorithms, separation of objectives and constraints, and hybrid methods. The purpose of these methods is to find the constraint violations, and use such information to rank and select the individuals for reproduction. Such methods are referred as MAs with conventional representation and are discussed in depth in the following section.

While many MAs adopt conventional representation i.e. the solution represented as a vector of decision variables, there are many which focus on the underlying solution representation scheme and include specialized representation and/or repair methods to deal with constraints efficiently. The details of such methods are discussed under the broad context of MAs with alternative representation. Two case studies are carefully selected to illustrate the behavior of both these classes of MAs.

9.3 Classification of MAs

As observed in the literature, the trend of MAs used for constrained optimization can be represented by the classification shown in Figure 9.1.

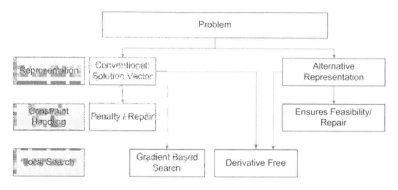

Fig. 9.1. Classification of MAs

Some examples of MAs based on the above classification are given in Table 9.1. It is interesting to observe that MAs, with chromosome representation based on solution vector, use penalty or repair method for dealing with constraint violation. On the other hand, MAs, with alternative chromosome representation, use derivative free local search method, and use either feasible individuals or repair infeasible individuals to deal with constraints. From the review in an earlier section, it is clear that the alternative representation is popular for solving combinatorial optimization problems.

9.4 MAs with Conventional Representation

In this section, we will discuss optimization problem solving, using MAs, where the complete mathematical model is available and a chromosome is represented as a vector of decision variables.

Handoko *et al.* [356] developed a MA where a GA was combined with a gradient based local search to solve nonlinear programming problems. The constraint violation was handled using three simple rules as of Deb [195] : (i) the feasible individual is preferred over the infeasible one; (ii) for two feasible individuals, the individual with better fitness is preferred; and (iii) for two infeasible individuals, the individual with lower constraint violation is preferred. Their experimental results indicated that MAs outperformed conventional algorithms in terms of both quality of solution and the rate of convergence.

9 Memetic Algorithms in Constrained Optimization

Table 9.1. Examples of MAs in literature

	\multicolumn{3}{c}{Representation (as discussed earlier)}	
	Based on solution vector	\multicolumn{2}{c}{Alternative Representation}	
	Constraint handling	\multicolumn{2}{c}{Constraint handling}	
	Penalty/Repair	Penalty/Repair	Ensures feasibility
Gradient based local search	Handoko et al. [356] Singh et al. [816] Kelner et al. [455] Barkat Ullah et al. [44]		
Derivative free local search	Lin and Liang [518] Barkar Ullah et al. [45] Boudia and Prins [79] Park et al. [713]	Hasan et al. [377, 378]	Prins [734, 735] Fallahi et al. [249] Ngueveu et al. [662] Mendoza et al. [579] Marinakis & Marinaki [555]

Singh et al. [816] designed an infeasibility empowered MA for solving constrained optimization problems where an underlying EA was combined with a local search (Sequential Quadratic Programming (SQP)). The constraint violation was tackled using principles of infeasible solution embedding Singh et al. [759] and the results were reported for the series of 18 constrained test problems as introduced in CEC-2010 competition.

Lin and Liang [518] proposed a hybrid algorithm where a GA was combined with an adaptive penalty method and a line search technique (Hooke and Jeeves). The performance of the algorithm on a series of 13 well-known benchmark problems established its robustness.

Kelner et al. [455] proposed a hybrid algorithm as a combination of a GA and a local search strategy based on the interior point method, for solving constrained multi-objective mathematical models. The constraints were handled using the rules proposed by Deb [195]. The efficiency of the algorithm was demonstrated using a number of test problems.

Barkat Ullah et al. [44] proposed an agent based memetic algorithm in which four local search algorithms were used for adaptive learning. The algorithms included random perturbation, neighborhood and gradient search methods. Subsequently, another specialized local search method was designed to deal with equality constraints (Barkat Ullah et al. [45]. The constraints were handled using the rules proposed by Deb [195]. Although the algorithm identified high quality solutions on the set of 13 benchmarks, the computational time was a bit longer than state-of-the-art algorithms (Runarsson and Yao [782]) as the underlying lattice-like environment and orthogonal crossovers consumed a fair amount of time.

Liu et al. [527] developed a memetic co-evolutionary differential evolution algorithm where the population was divided into two sub-populations. The purpose of one sub-population is to minimize the fitness function, and the other is to minimize the constraint violation. The optimization was achieved through interactions between the two sub-populations. No penalty coefficient was used in the method while a Gaussian random number was used to modify the individuals when the best solution remained unchanged over several generations. The results indicate the algorithm being computationally inexpensive in terms of memory requirements and CPU times and efficient when compared with existing state of the art algorithms.

While most of the applications reported above are tested on mathematical benchmarks, several practical applications have also adopted conventional representation. Boudia and Prins [79], Park *et al.* [713], and Berretta and Rodrigues [61] dealt with three different practical problems and in all studies chromosomes were designed using conventional representation. Boudia and Prins [79] considered the problem of cost minimization of a production-distribution system. The moves (local search) used were 2-OPT, relocate a customer, and swap between two customers. A repair mechanism was also applied for constraint satisfaction. The algorithm reported significant savings as compared to two other existing methods. Park *et al.* [713] combined a GA with a tunnel-based dynamic programming scheme (as a local search) to solve highly constrained non-linear discrete dynamic optimization problems arising from long-term planning. The infeasible solutions were repaired by regenerating partial characters. The algorithm successfully solved reasonable sized practical problems which cannot be solved using conventional approaches. A multi-stage capacitated lot-sizing problem was solved by the memetic algorithm proposed by Berretta and Rodrigues [61] using heuristics as local search coupled with usual crossover and mutation operators. The results using the above method were better than those generated using existing heuristics.

9.5 MAs with Alternative Representations

While the above section highlighted a number of successful MAs that have been designed to solve constrained optimization problems using conventional representation schemes, there are also a number of MAs that have been designed to solve problems using alternative representation schemes. Combinatorial problems require many integer (mainly binary) variables and logical constraints to represent them mathematically. Hence, a chromosome design based on the decision variables of the mathematical model as a vector becomes too long. Just to give an idea, let us consider a single variable piecewise linear function or a continuous nonlinear function that can be approximated by a number of piecewise linear functions. To express these functions mathematically for n segments, we need $(n+1)$ real variables, $(n-1)$ binary variables and $(n+1)$ logical constraints. So, $2n$ variables in the chromosome and additional $(n + 1)$ constraints are required to represent the function of a single variable. In alternative chromosome design, one can use just one variable as illustrated in Ray and Sarker [758]. The applications of alternative representations in MAs are briefly reviewed below.

Prins [734] developed a memetic algorithm for solving vehicle routing problems (VRPs) which outperformed most Tabu Search (TS) heuristics (best known algorithms for VRPs at that time) on a number of test instances. The solution was represented using a TSP-like permutation chromosome, without trip delimiters, and local search procedures (like moving or swapping some nodes) were used in lieu of mutation for search. Later, Prins [735] proposed two more memetic algorithms for heterogeneous fleet vehicle routing problems (HFVRPs) that are based on chromosome encoded as giant tours, without trip delimiters. Such chromosomes do not directly represent the decision variables of the corresponding mathematical model of the problem. In both of the above studies, Prins applied an optimal evaluation procedure that splits the tours into feasible trips and assign vehicles to them. As a result, no repair mechanism or penalty method was required. The perturbation was achieved through the relocation of one customer, the exchange of two customers, and 2-OPT moves operated on one or two selected routes. In order to maintain diversity, a distance measure in the solution space was used. The algorithm is one of the most successful algorithms for vehicle fleet mix problem with both fixed and variable costs (VFMP-FV) that has been able to discover six new best solutions to benchmark problems.

El Fallahi *et al.* [249] and Ngueveu *et al.* [662] developed a memetic algorithm for multi-compartment vehicle routing problems (MC-VRPs) and cumulative vehicle routing problems (CCVRPs) respectively. In these algorithms, the chromosome representation and evaluation procedure are similar to Prins [734]. However, the moves (local search) in the first algorithm are based on 2-OPT, relocate and I-interchange and the second include 2-OPT, relocation of one customer and exchange of two customers. Mendoza *et al.* [579] proposed a memetic algorithm for a variant of MC-VRPs with a different representation known as the genetic vehicle representation (GVR). In GVR, each permutation contains an ordered set of customers representing a route. This representation allows the straightforward application of the selected crossover, mutation and local search operators designed to work on independent routes. The authors used relocate and 2-OPT as the local search schemes.

Marinakis and Marinaki [555] proposed a memetic algorithm for the solution of VRPs. The MA makes use of a GA framework with an expanding neighborhood search. Although, significantly better solutions were reported on two sets of benchmark instances, there is no comparison on computational time.

Hasan *et al.* [377, 378] developed a memetic algorithm for solving job-shop scheduling problems. They used job pair-relation based genotype representation, priority rules as local search, and a repair mechanism for changing the infeasible individuals into feasible. It is generally accepted that the time taken per generation of MA would be higher than that of GA. However Hasan *et al.* [378] proved that MA, as compared to GA, not only improves the quality of solutions but also reduces the overall computational time. The proposed MA improved the average of the best solutions over GA by 2.623%, while reducing the computational time by 40.57% on average per problem. It is also important to take note that these are based on 40 well-known series of benchmark problems.

9.6 Numerical Case Studies

Two case studies are discussed in depth in the following sub-sections.

9.6.1 Case Study 1: Infeasibility Empowered Memetic Algorithm for Constrained Optimization Problems: MA with Conventional Representation

In this section we present an Infeasibility Empowered Memetic Algorithm (IEMA) which is a combination of Infeasibility Driven Evolutionary Algorithm(IDEA) and a local search based on Sequential Quadratic Program (SQP). IDEA is a derived variant of EAs in which a small proportion of marginally infeasible solutions are preserved to accelerate the rate of convergence. While most EAs rank feasible solutions above infeasible solutions, IDEA ranks solutions based on the original objectives along with additional objective representing constraint violation measure. In addition, "good" infeasible solutions are ranked higher than the feasible solutions, and thereby the search proceeds through both feasible and infeasible regions, resulting in greater rate of convergence to optimal solutions. The studies reported in [759, 817] indicate that IDEA has better rate of convergence over conventional EAs for a number of constrained single and multi-objective optimization problems. The following subsections provide the background of IDEA and necessary details of IEMA.

9.6.1.1 Infeasibility Driven Evolutionary Algorithm (IDEA)

A generalized single-objective optimization problem can be formulated as shown in (9.1). It is a usual practice to convert the equality constraints to inequality constraints using a small tolerance (*i.e.* $h(\mathbf{x}) = 0$ is converted to $|h(\mathbf{x})| \leq \varepsilon$). Hence, the discussion presented here is with regards to presence of inequality constraints only.

To effectively search the design space (including the feasible and the infeasible regions near constraint boundaries), the original single objective constrained optimization problem is reformulated as bi-objective unconstrained optimization problem as shown in (9.2).

$$\begin{aligned} \text{Minimize} \quad & f_1'(\mathbf{x}) = f_1(\mathbf{x}) \\ & f_2'(\mathbf{x}) = \text{violation measure} \end{aligned} \quad (9.2)$$

The additional objective represents a measure of constraint violation, which is referred to as "violation measure". It is based on the amount of relative constraint violations among the population members. Each solution in the population is assigned m ranks, corresponding to each m constraints. The ranks are calculated as follows. To get the ranks corresponding to i^{th} constraint, all the solutions are sorted based on the constraint violation value of ith constraint. Solutions that do not violate the constraint are assigned rank 0. The solution with the least constraint violation value gets rank 1, and the rest of the solutions are assigned increasing ranks in the

Algorithm 19. Infeasibility Driven Evolutionary Algorithm (IDEA)

1 **begin**
    // Given population size N number of generations $N_G > 1$
    and Proportion of infeasible solutions $0 < \alpha < 1$
2     $N_{inf} \leftarrow \alpha * N$;
3     $N_f \leftarrow N - N_{inf}$;
4   set $pop_1 \leftarrow$ Initialize();
5   Evaluate(pop_1);
6   **for** $i = 2$ *to* N_G **do**
7         $childpop_{i-1} \leftarrow$ Evolve(pop_{i-1});
8   Evaluate($childpop_{i-1}$);
9         $(S_f, S_{inf}) \leftarrow$ Split($pop_{i-1} + childpop_{i-1}$);
10   Rank(S_f);
11   Rank(S_{inf});
12         $pop_i \leftarrow S_{inf}(1:N_{inf}) + S_f(1:N_f)$
13   **endfor**
14 **end**

ascending order of their constraint violation values. The process is repeated for all the constraints and as a result each solution in the population gets assigned m ranks. The violation measure is the sum of these m ranks corresponding to m constraints.

The main steps of IDEA are outlined in Algorithm 19. IDEA uses simulated binary crossover (SBX) and polynomial mutation operators to generate offspring from a pair of parents selected using binary tournament as in NSGA-II [200]. Individual solutions in the population are evaluated using the original problem definition (9.1) and the infeasible solutions are identified. The solutions in the parent and offspring population are divided into a feasible set (S_f) and an infeasible set (S_{inf}). The solutions in the feasible set and the infeasible set are ranked separately using the non-dominated sorting and crowding distance sorting [200] based on 2 objectives as per the modified problem definition (9.2). The solutions for the next generation are selected from both the sets to maintain infeasible solutions in the population. In addition, the infeasible solutions are ranked higher than the feasible solutions to provide a selection pressure to create *better* infeasible solutions resulting in an active search through the infeasible search space.

A user-defined parameter α is used to maintain a set of infeasible solutions as a fraction of the size of the population. The numbers N_f and N_{inf} denote the number of feasible and infeasible solutions as determined by parameter α. If the infeasible set S_{inf} has more than N_{inf} solutions, then first N_{inf} solutions are selected based on their rank, else all the solutions from S_{inf} are selected. The rest of the solutions are selected from the feasible set S_f, provided there are at least N_f number of feasible solutions. If S_f has fewer solutions, all the feasible solutions are selected and the rest are filled with infeasible solutions from S_{inf}. The solutions are ranked from 1 to N in the order they are selected. Hence, the infeasible solutions selected first will be ranked higher than the feasible solutions selected later.

Algorithm 20. Infeasibility Empowered Memetic Algorithm (IEMA)

1 **begin**
   // Given population size N number of generations $N_G > 1$
   and Proportion of infeasible solutions $0 < \alpha < 1$
2   $N_{inf} \leftarrow \alpha * N$;
3   $N_f \leftarrow N - N_{inf}$;
4   pop_1 = Initialize();
5   Evaluate(pop_1);
6   **for** $i = 2$ *to* N_G **do**
7     $childpop_{i-1} \leftarrow$ Evolve(pop_{i-1});
8   Evaluate($childpop_{i-1}$);
9     $(S_f, S_{inf}) \leftarrow$ Split($pop_{i-1} + childpop_{i-1}$);
10   Rank(S_f);
11   Rank(S_{inf});
12    $pop_i \leftarrow S_{inf}(1:N_{inf}) + S_f(1:N_f)$;
13   **x** $\leftarrow$ Random solution in pop_i;
14   **x**$_{best}$ $\leftarrow$ Local_ search (**x**);
      // **x**$_{best}$ is the best solution found using local search
      from **x**
15   Replace worst solution in pop_i with **x**$_{best}$;
16   Rank(pop_i);
17   *Rank the solutions again in pop_i*
18 **endfor**
19 **end**

9.6.1.2 Infeasibility Empowered Memetic Algorithm (IEMA)

The proposed algorithm IEMA is constructed using IDEA as the baseline algorithm. For single objective problems, a local search can be a very efficient tool for optimization. However, its performance is largely dependent on the starting solution. The proposed algorithm tries to exploit the advantages of both these approaches, i.e. 1) searching near the constraint boundaries by preserving marginally infeasible solutions during the search, and 2) the effectiveness of local search to expedite the convergence in potentially optimal regions of the search space. Hence, we refer to the proposed algorithm as Infeasibility Empowered Memetic Algorithm (IEMA).

The proposed IEMA is outlined in algorithm 20. In IEMA, during each generation, apart from the evolution of the solutions in IDEA, a local search is done from a random solution in the population, for a prescribed number of function evaluations (set to 2000 here). Sequential Quadratic Programming (SQP) [729] has been used in the presented studies for the local search. Thereafter, the worst solution in the population is replaced by the best solution found from the local search. The ranking of solutions is done in the same way as done in IDEA. The injection of good quality solutions found using the local search guides the population towards potentially optimal regions of the search space. The evolved solutions in turn act as good starting solutions for the local search in subsequent generations.

9   Memetic Algorithms in Constrained Optimization   145

9.6.1.3   Results on CEC-2010 Benchmark Problems

- **Experimental setup:** The performance of IEMA is presented for one of the most recent difficult set of constrained optimization benchmarks, i.e. that of IEEE CEC-2010, constrained optimization competition. Twenty five runs of the proposed algorithm IEMA are done on each of the test problems C01 - C18 [550]. The parameters used for IEMA are same for each problem, i.e. no tuning of parameters is done across the problems. The parameters are listed in Table 9.2. A maximum of 2000 function evaluations are allotted to the local search within each generation.

Table 9.2. Parameters used for IEMA

Parameter	Value
Population Size	200
Max. FES	for 10D problems: 200000
	for 30D problems: 600000
Crossover Probability	0.9
Crossover index	15
Mutation Probability	0.1
Mutation index	20
Infeasibility Ratio (α)	0.9

- **PC configuration:** All the runs are made on a cluster with the compute nodes DL140G3 5110 NHP Sata, with following configuration:
  1. Processor - Dual-core Intel Xeon 5110
  2. RAM - 4GB
  3. Operating system - Redhat Linux

  IEMA algorithm is implemented in Matlab 2008a.

- **Summary of results:** The results for 10D problems are shown in Table 9.3, whereas the results for 30D problems are listed in Table 9.4. To determine the median, following procedure is adopted. All the runs in which a feasible solution was found are sorted based on the best function value obtained. Thereafter, all the runs in which no feasible solutions are found are sorted based on the mean constraint violation of the best (infeasible) solution found. Feasible runs are ranked above infeasible runs. In the sorted list, the 13th solution is reported as the median solution (only if the median is feasible). The best, mean and worst runs reported in the tables are based only on the runs in which at least one feasible solution was found. The number of such feasible runs are also reported in the tables for each problem. The median value, if infeasible is also not reported.

From Table 9.3, it is observed that for 10D problems, IEMA is able to achieve all (25) feasible runs for 12 problems out of 18. The best value obtained for many problems are much better than the median and worst values, indicating a possibility of highly multimodal objective functions. This also results in a correspondingly high value of standard deviation (std), as seen from the table.

For 30D problems (Table 9.4), the results are worse as compared to the 10D problems. For 4 out of 18 functions, no feasible solution was identified. Among the remaining 14 functions, all 25 runs were feasible for 11 problems. Once again, the results are seen to have a high standard deviation value as in 10D case, and the best values found are much better than the median/worse values for some of the problems.

Table 9.3. Performance of IEMA on 10D problems

	C01	C02	C03	C04	C05	C06
Best	-0.74731	-2.27771	1.46667e-16	-9.98606e-06	-483.611	-578.662
Median	-0.74615	-2.27771	3.2005e-15	-9.95109e-06	-483.611	-578.662
Mean	-0.743189	-2.27771	6.23456e-07	-9.91135e-06	-379.156	-551.47
std	0.00433099	1.82278e-07	1.40239e-06	8.99217e-08	179.424	73.5817
Feasible	25	25	25	25	24	24
	C07	C08	C09	C10	C11	C12
Best	1.74726e-10	1.00753e-10	1.20218e-09	5.4012e-09	-0.00152271	-10.9735
Median	1.9587e-09	3.94831e-09	333.32	42130.4	-0.00152271	-0.199246
Mean	3.25685e-09	4.0702	1.95109e+12	2.5613e+12	-0.00152271	-0.648172
std	3.38717e-09	6.38287	5.40139e+12	3.96979e+12	2.73127e-08	2.19928
Feasible	25	25	23	19	24	24
	C13	C14	C15	C16	C17	C18
Best	-68.4294	8.03508e-10	9.35405e-10	4.44089e-16	9.47971e-15	2.23664e-15
Median	-68.4294	1.29625e-08	26.1715	0.0320248	2.59284e-12	6.78077e-15
Mean	-68.0182	56.3081	1.57531e+08	0.0330299	0.00315093	1.61789e-14
std	1.40069	182.866	6.04477e+08	0.0226013	0.0157547	3.82034e-14
Feasible	25	25	25	25	25	25

- **Convergence plots:** The convergence plots for C09, C10, C14, C15, C17 and C18 are shown in Figure 9.2. The plots show the feasible solutions only, for the best runs corresponding to these problems. The objective values have been plotted in log scale in order to aid visualization.
- **Time complexity:** The time complexity of the algorithm is shown in Table 9.5. $T1$ and $T2$ are as defined in [550]. $T1$ represents the average (across C01-C18) time taken for evaluating the problem 10000 times, whereas $T2$ represents the average time taken across C01-C18 by the algorithm IEMA to run through 10000 FES.

9 Memetic Algorithms in Constrained Optimization   147

Table 9.4. Performance of IEMA on 30D problems

	C01	C02	C03	C04	C05	C06
Best	-0.821883	-2.28091	-	-	-286.678	-529.593
Median	-0.819145	-2.27767	-	-	-	-
Mean	-0.817769	-1.50449	-	-	-270.93	-132.876
std	0.00478853	2.14056	-	-	14.1169	561.042
Feasible	25	25	0	0	4	2
	C07	C08	C09	C10	C11	C12
Best	4.81578e-10	1.12009e-09	7314.23	27682	-	-
Median	6.32192e-10	0.101033	7.91089e+06	1.1134e+07	-	-
Mean	8.48609e-10	17.7033	2.98793e+07	1.58342e+07	-	-
std	4.84296e-10	40.8025	4.50013e+07	1.68363e+07	-	-
Feasible	25	25	25	25	0	0
	C13	C14	C15	C16	C17	C18
Best	-68.4294	3.28834e-09	31187.6	6.15674e-12	9.27664e-10	1.37537e-14
Median	-67.6537	7.38087e-09	7.28118e+07	1.26779e-10	5.67557e-06	2.12239e-14
Mean	-67.4872	0.0615242	2.29491e+08	0.00163294	0.0883974	4.73841e-14
std	0.983662	0.307356	4.64046e+08	0.0081647	0.15109	6.5735e-14
Feasible	25	25	25	25	25	25

Table 9.5. Time complexity of IEMA (in seconds)

	$T1$	$T2$	$(T2-T1)/T1$
10D problems	2.57636	9.05104	2.51312
30D problem	2.57854	13.2825	4.1512

9.6.2 Case Study 2: MA with Alternative Representation

The job-shop scheduling problem (JSSP) is a well-known practical planning problem in the manufacturing sector. A classical JSSP is a combination of N jobs and M machines. Each job consists of a set of operations that has to be processed, on a set of known machines, and where each operation has a known processing time. A schedule is a complete set of operations, required by a job, to be performed on different machines, in a given order. In addition, the process may need to satisfy other constraints such as (i) no more than one operation of any job can be executed simultaneously and (ii) no machine can process more than one operation at the same time. The objectives usually considered in JSSPs are the minimization of makespan. The total time between the starting of the first operation and the ending of the last operation, is termed as the "makespan". We first develop a traditional GA for solving JSSPs. We then proposed three versions of memetic algorithms using three new

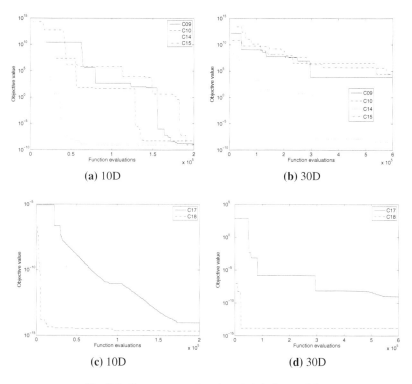

Fig. 9.2. Convergence plots (*y*-axis is in log scale)

priority rules for improving the performance of traditional GA, namely: partial reordering (PR), gap reduction (GR) and restricted swapping (RS). The performances of our proposed algorithms are analyzed by solving 40 well-known benchmark problems. The chromosome representation, priority rules and the performance analysis are briefly discussed below.

9.6.2.1 Chromosome Representation

In this study, we do not solve the mathematical model of the job shop problem. Instead we develop GA and MA for solving the problem directly. We select the job pair-relationship based representation for the genotype, as in [649, 946], due to the flexibility of applying genetic operators to it. In this representation, a chromosome is symbolized by a binary string, where each bit stands for the order of a job pair (u,v) for a particular machine m. This binary string acts as the genotype of individuals. The corresponding phenotype represents the job sequence for each machine. Further details on the chromosome design can be found in Hasan *et al.* [377].

9.6.2.2 Priority Rules

The priority rules developed for this study are as follows.

- **Partial Reordering (PR):** In this rule, we identify the machine which is the deciding factor for the makespan and the last job (say J^*) that is to be processed by that machine. That machine can be termed as the bottleneck machine in the chromosome under consideration. Then we find the machine (say M^*) that is required by the first operation of the identified job J^*. The re-ordering rule then suggests that the first operation of the identified job (J^*) must be the first task on machine M^* if it is not already scheduled. If we move the job J^* from its current position to the 1st position, we may need to push some other jobs currently scheduled on machine M^* to the right. In addition, it may provide an opportunity to shift some jobs to the left on other machines. The overall process helps to reduce the makespan for some chromosomes.
- **Gap Reduction (GR):** After each generation, the generated phenotype usually leaves some gaps between the jobs. Sometimes, these gaps are necessary to satisfy the precedence constraints. However, in some cases, a gap could be removed or reduced by placing a job from the right side of the gap. For a given machine, this is like swapping between a gap from the left and a job from the right of a schedule. In addition, a gap may be removed or reduced by simply moving a job to its adjacent gap at the left. This process would help to develop a compact schedule from the left and continuing up to the last job for each machine. Of course, it must ensure no conflict or infeasibility before accepting the move.
- **Restricted swapping (RS):** For a given machine, the restricted swapping rule allows swapping between the adjacent jobs if and only if the resulting schedule is feasible. This process is carried out only for the job which takes the longest time for completion.

9.6.2.3 Implementation

First, we implement a simple GA for solving JSSPs. We use simple two point crossover and bit flip mutation as reproduction operators. We then implemented three versions of MAs by introducing the priority rules as local search techniques as follows:

- MA(PR): Partial re-ordering rule with GA,
- MA(GR): Gap reduction rule with GA, and
- MA(GR-RS): Gap reduction and restricted swapping rule with GA

In both GA and MA, we apply elitism in each generation to preserve the best solution found so far, and also to inherit the elite individuals more than the rest. In performing the crossover operation, we use the tournament selection that chooses one individual from the elite class of the individuals (*i.e.* the top 15%) and two individuals from the rest. This selection then plays a tournament between the last two and performs crossover between the winner and the elite individual. We rank the individuals on the basis of the fitness value. From our extensive parametric analysis, we have chosen the crossover and mutation rate as 0.45 and 0.35 respectively. We set the population size to 2500 and the number of generations to 1000. Note that JSSPs usually require a higher population size. For example, Pezzella *et al.* [721] used a population size of 5000 even for 10×10 problems. In our approach, GR is applied to every individual. On the other hand, we apply PR and RS to only 5% of randomly selected individuals in every generation. To test the performance of our proposed algorithms, we have solved the 40 benchmark problems designed by Lawrence [509] and have compared the results.

9.6.2.4 Result and Analysis

Each problem was run 30 times and Table 9.6 compares the performance of four algorithms we implement [GA, MA(PR), MA(GR), and MA(GR-RS)] in terms of the % average relative deviation (ARD) from the best result published in the literature, the standard deviation of % relative deviation (SDRD), and the average number of fitness evaluations required. From Table 9.6, it is clear that the performance of the MAs are better than the GA, and MA(GR) is better than both MA(PR) and GA. The addition of RS to MA(GR), which is known as MA(GR-RS), has clearly enhanced the performance of the algorithm. Out of the 40 test problems, both MA(GR) and MA(GR-RS) obtained exact optimal solutions for 23 problems. In addition, MA(GR-RS) obtained optimal solutions for another 4 problems and substantially improved solutions for 10 other problems. In general, these two algorithms converged quickly, which can be seen from the average number of fitness evaluations.

Table 9.6. Comparing our four algorithms for 40 test problems

Algorithm	Optimal Found	ARD (%)	SDRD	Average # of generations	Average # of Fitness eval.(10^3)	Average Computational time (s)
GA	15	3.591	4.165	270.93	664.90	201.60
MA(PR)	16	3.503	4.192	272.79	660.86	213.42
MA(GR)	23	1.360	2.250	136.54	356.41	105.87
MA(GR-RS)	27	0.968	1.656	146.63	388.58	119.81

As shown in Table 9.6, the addition of the local search techniques to GA (for the last two MAs) not only improves the quality of solutions significantly but also helps in converging to the solutions with a lower number of generations and a lower total number of fitness evaluations. However, as the local search techniques require additional computation, the computational time per generation for all three MAs is higher than GA. For example, the average computational time taken per generation by the algorithms GA, MA(PR), MA(GR) and MA(GR-RS) are 0.744, 0.782, 0.775 and 0.817 seconds respectively. Interestingly, the overall average computational time per test problem solved, for the algorithm MA(GR-RS), is the lowest among the four algorithms implemented. As of Table 9.6, for all 40 test problems, the algorithm MA(GR-RS) improved the average of the best solutions over GA by 2.623%, while reducing the computational time by 40.57% on average per problem. This clearly demonstrates the strength of MAs.

9.7 Summary and Conclusions

This chapter provides a review of various memetic algorithms that have been proposed over the years to deal with constrained optimization problems. Details of two distinct and widely different classes of MAs are presented in the chapter. The first MA adopts a conventional representation scheme and employs a population based global search and a SQP for local search. The population based global search component of MA explicitly maintains a fraction of marginally infeasible solutions in a quest to accelerate its rate of convergence. The second MA and its variants on the other hand is designed to efficiently solve job shop scheduling problems. The algorithm employs specialized representation, recombination and local search strategies/heuristics in an attempt to improve the rate of convergence. The examples clearly highlight the potential benefits that can be realized through the use of MAs and the range of local learning schemes that can be used to further enhance its performance.

Acknowledgements. The authors would like to acknowledge the help received from Dr. Kamrul Hasan for providing results for the Job Shop Scheduling problem, and Mr. Hemant Kumar Singh for benchmarking IEMA and formatting of this manuscript. The efforts of Dr. Amitay Isaacs to generate several parts of the code is also acknowledged.

Chapter 10
Diversity Management in Memetic Algorithms

Ferrante Neri

10.1 Introduction

In Evolutionary Computing, Swarm Intelligence, and more generally, population-based algorithms diversity plays a crucial role in the success of the optimization. Diversity is a property of a group of individuals which indicates how much these individuals are alike. Clearly, a group composed of individuals similar to each other is said to have a low diversity whilst a group of individuals dissimilar to each other is said to have a high diversity. In computer science, in the context of population-based algorithms the concept of diversity is more specific: the diversity of a population is a measure of the number of different solutions present, see [239].

Ideally, a population-based algorithm is supposed to work in high diversity conditions during the early stages of the optimization, then progressively lose diversity in proximity to the global basin of attraction, and eventually lose all diversity when the global optimum is detected. The latter condition means that the entire population is characterized by a unique genotype, i.e. the global optimum. The described functioning happens very rarely in practice since the algorithm tends either to prematurely converge to a suboptimal solution or to stagnate. Premature convergence is an undesired condition, which very often jeopardizes the functioning of Evolutionary Algorithms (EAs), consisting of a diversity loss in the presence of a sub-optimal (and unsatisfactory) candidate solution, see [246]. Stagnation is typical of Swarm Intelligence Algorithms (SIAs) but is present also in some EA structures. An algorithm stagnates when it does not succeed at enhancing upon its individual with the best performance while the diversity is still high. In other words, the algorithm repeatedly explores less promising areas of the decision space and thus does not manage to register improvements.

Due to their different structures, EAs and SIAs require different and complementary techniques for handling diversity. More specifically, in EAs a mechanism which

---

Ferrante Neri
Department of Mathematical Information Technology, P.O. Box 35 (Agora),
40014 University of Jyväskylä, Finland
e-mail: ferrante.neri@jyu.fi

preserves diversity and thus inhibits premature convergence is beneficial, while such an approach in SIAs can be detrimental and turn into stagnation behavior.

In Memetic Algorithms (MAs), since their earliest definition in [621] and early original works in Memetic Computing (MC), see [615] and [622], the problem of diversity is taken into account and implicitly analyzed. Since MAs perform the search by employing multiple search logics, diversity is preserved by studying the decision space under complementary perspectives, see [489]. This means that if the search logic within the evolutionary framework fails at detecting new promising solutions, the local search components give an extra chance to the algorithm to detect fresh and promising genotypes. This is probably one of the main reasons contributing to the success of MAs.

However, as remarked in [239], MAs by themselves are not a "magic solution" to optimization problems, and the employment of multiple search logics does not guarantee the prevention of premature convergence or stagnation. For example, a MA based on an evolutionary framework and employing local search components can naturally lose diversity since the application of the local search to a set of points belonging to the same (sub-optimal) basin attraction would produce the convergence of a part of the population to the corresponding local optimum.

In order to prevent MAs from premature convergence and stagnation, several approaches attempting to handle population diversity in MAs have been proposed during recent years. This chapter deals with diversity in MAs and presents a survey of techniques recently proposed in literature for handling diversity and coordinating the various algorithmic components contained within MAs. Section 10.2 gives a short survey on the topic. Section 10.3 focuses on Fitness Diversity adaptation and, presents various diversity metrics and the related adaptation techniques.

10.2 Handling the Diversity of Memetic Algorithms: A Short Survey

Most of the MAs proposed in the literature employ an evolutionary framework (and not a swarm intelligence framework). Thus, most of the work on diversity attempts to preserve diversity and prevent premature convergence.

A classical and straightforward approach has been proposed in [246] where a generational Genetic Algorithm (GA) employing truncation selection is proposed. The algorithm randomly pairs parents; but only those string pairs which differ from each other by some number of bits (i.e., a mating threshold) are allowed to reproduce. In this way, diversity is preserved by inhibiting the presence of duplicates. A similar approach has been proposed with reference to an engineering problem in [863] and [205].

In [640] the problem of diversity is handled by employing a structured population. A distributed GA and a local search algorithm process the entire population. The sub-population evolves independently and thus preserves the diversity of the entire population.

10 Diversity Management in Memetic Algorithms

In [648] a local search crossover is integrated within the evolutionary framework. The basic idea of this local search crossover is to remove and replace genes in a selected parent solution on the basis of its common and different edges with the other parent solution. As a result, the offspring is genotypically different from the parents and diversity is preserved.

In [581] a specific crossover for preserving the diversity is proposed. This crossover keeps constant the Hamming distance (i.e. the number of genes in a candidate solution at which the corresponding symbols are different) between parents and offspring. Moreover, in [581] a restarting mechanism is proposed. This simple (and sometimes efficient) mechanism consists of resampling the individuals of the population in the presence of diversity loss and possible premature convergence.

In [491] a MA composed by a GA and an adaptive local search algorithm is proposed. This adaptive local search is inspired by Simulated Annealing, see [468] and [122], and is supposed to improve upon the available genotypes when the population is diverse and to increase the diversity when the population is approaching the convergence condition. The diversity preservation logic proposed in [491] can be summarized in the following way: a solution which is slightly worse than the starting one can be accepted under the condition that it increases the diversity in the population. More formally, for a given minimization problem and for a given candidate solution x, a newly generated solution x' replaces x according to the following probability:

$$P = \begin{cases} 1 & \text{if } f(x') \leqslant f(x) \\ e^{\frac{k|f(x')-f(x)|}{|f_{min}-f_{avg}|}} & \text{otherwise} \end{cases} \quad (10.1)$$

where f_{min} and f_{avg} are, respectively, minimum and average fitness values among the population individuals and k is a normalization constant. This technique measures the diversity by means of the fitness value and is strongly related to the fitness diversity adaptation which will be extensively discussed in Section 10.3.

In [492] the encoding of memetic information (in the mentioned paper mutations for some problems and local search algorithms for another problem) is performed within the solutions. A probabilistic criterion manages the transmission of the memes and thus search strategies from parents to offspring. In [492], multiple local search algorithms are employed, de facto composing a multimeme algorithm, see [496] and [489]. The resulting algorithmic structure is supposed to prevent premature convergence by offering multiple search perspectives of the decision space. The main algorithmic philosophy is that the combination and coordination of a set of various search logics enhances the chance of obtaining a high performance or, more modestly, at least overcome the bottlenecks resulting from the specific limitations of a certain search structure. For example the employment of a local search algorithm employing a steepest descent pivot rule can be efficient in the proximity of the global optimum when it is important to finalize the search by exploiting the neighborhood while a random walk algorithm can support the evolutionary framework to detect new promising directions when the search still has not detected a promising direction. If a MA employs both these local searches, it might be able to handle both the situations. In addition, the adaptation is supposed to allow the algorithm to decide itself the most proper local search on the basis of the situation. The employment and thus coordination of multiple local search algorithms within a MA is a crucially important topic in Memetic Computing and is somehow the "hearth" and the reason for success/unsuccess of a MA. Some examples of studies on this specific topic are reported in [411], [493], [683], [830] and references therein.

In [806] a MA for clustering is proposed. Two modified selection schemes based on fitness assignment concur at global and local levels to preserve diversity and to prevent premature convergence. In [715], a MA for solving multimodal problems is presented. The concept of fitness sharing is extended to the local search algorithms, thus defining Baldwinian sharing. In practice, the algorithm employs a sharing technique (i.e. a normalization of the fitness values based on the Euclidean distances to affect the sorting/selection and thus prefer a population composed by spread out points) in order to guarantee that diversity is preserved.

In [536] a real-coded MA is proposed. Within this MA two mechanisms for preserving the diversity are employed. The first mechanism, namely negative assortative mating, consists of selecting genotypically distant parents in order to obtain an offspring which does not look similar to either generating parent. The second mechanism, namely Breeder Genetic Algorithm (BGA) mutation [639], is a mutation operator which promotes the generation of distant genes within the solutions by employing an ad-hoc probability distribution function.

In [873] the problem of diversity is handled by using multiple search logics and a structured population. Two adaptive systems for preserving diversity are also presented. Both mechanisms rely on the fact that the frequency of the local search helps to preserve diversity. According to the first adaptive system, at the beginning of the optimization process the sub-populations already contain enough diversity and therefore do not need additional search moves coming from the local search; hence the local search algorithms are activated with a low frequency. Subsequently, since the population naturally tends to progressively lose diversity, the local search is activated with a higher frequency. More specifically, the frequency γ of local search activation is given by the following heuristic rule:

$$\gamma = \frac{1}{\sqrt{2\pi}\sigma} \exp\left(-\frac{1}{2}\left(\frac{gen - \mu}{\sigma}\right)^2\right) \eta \qquad (10.2)$$

where μ and σ are mean value and standard deviation of a Gaussian distribution, gen is the generation number, and η is a scaling factor.

The second adaptation system is more complex and less intuitive compared to the first one. In order to explain this mechanism, let us consider a (sub-)population S of individuals. The population can be partitioned into Q groups $S_1, S_2, \ldots, S_Q$ where each group contains individuals characterized by the same fitness value. With reference to the generic $j-th$ group, we can define the ratio p_j as:

$$p_j = \frac{|S_j|}{\sum_{i=1}^{Q} |S_i|} \qquad (10.3)$$

where with $|*|$ is indicated the cardinality of the set, i.e. how many individuals belong to a given group. On the basis of the described partitioning, Shannon's information entropy, see [775], is defined as:

$$E = -\sum_{j=1}^{Q} p_j \log(p_j). \quad (10.4)$$

For a given population the entropy can be considered as a fitness-based diversity measure. In [873] the entropy variation is used to determine the amount of local search to be employed. More specifically the diversity frequency at the generation *gen* is given by:

$$\beta(gen) = 1 + \frac{E(gen) - E(gen-k)}{E(gen-k)} \quad (10.5)$$

where $E(gen)$ and $E(gen-k)$ (where $gen \geq k$) are the population entropy measure at the $gen-th$ and $(gen-k)-th$ generation, respectively.

10.3 Fitness Diversity Adaptation

Fitness Diversity Adaptive MAs are a class of algorithms which, like other works e.g. [491] and [873], measure fitness diversity in order to estimate the population diversity. This choice is done considering that for multi-variate problems the measure of genotypical distance can be excessively time and memory consuming and thus the adaptation might require an unacceptable computational overhead. Obviously, fitness diversity could not give an efficient estimation of population diversity, since it can happen that very different points take the same fitness values, e.g. if the points lay in a plateau. However, this fact does not effect the decision mechanism of the adaptive system for the following reasons.

The MAs employing Fitness Diversity Adaptation (FDA) are composed of an evolutionary framework and a list of local searchers. The coordination of the local search is carried out by the fitness diversity. More specifically, when the diversity is low one or more explorative local searchers, e.g. Nelder-Mead Simplex [653], are activated in order to offer an alternative search logic, and possibly to detect new promising search directions and increase the diversity. If this mechanism fails and the algorithm keeps losing diversity and converging to some areas of the decision space an exploitative local search algorithm, e.g. Rosenbrock Algorithm [776], attempts to quickly perform the exploitation of the most promising basin of attraction and thus quickly complete the search. If the fitness diversity is low, the candidate solutions in the population have a similar performance. This fact can mean either that the solutions are concentrated within a small region of the decision space, or that the solutions are distributed over one or more plateaus or over two or more basins of attraction having a similar performance. It can easily be visualized that all the listed situations are undesirable and that the activation of an alternative search move can increase the chances to detect "fresh" genotypes. In other words, although the FDA does not guarantee a proper estimation of the population diversity, it is an efficient index to estimate the correct moment of the evolution which would benefit from a local search application.

Although the fitness diversity mechanism sounds reliable at first, it hides two practical issues when the algorithmic design is performed. The first issue is how to measure the diversity while the second is how to use the diversity information for coordinating the local and global search. The following subsections address these two problems.

10.3.1 Fitness Diversity Metrics

Before analyzing the various metrics presented in the literature for measuring diversity a comment on the approach is necessary. As highlighted in [657], there is no "best" metric in general but there is a "most suitable" metric dependent not only on the problem (i.e. the fitness landscape) but also on the nature of the evolutionary framework. For example, an efficient diversity metric for Evolution Strategy (ES) would likely be inadequate to measure the diversity of Differential Evolution (DE). This consideration can be seen as a consequence of the No Free Lunch Theorem [940].

The first fitness diversity metric has been introduced in [104] and then used in [659]. This metric is given by:

$$\xi = \min\left\{\left|\frac{f_{best} - f_{avg}}{f_{best}}\right|, 1\right\}, \qquad (10.6)$$

where f_{best} and f_{avg} are respectively best and average fitness values over the individuals of the population. Measurement ξ can be seen as the answer to the question "How close is the average fitness to the best one?". Thus, if the average fitness value is as good as the best, the diversity is low and $\xi \approx 0$. On the contrary, if the fitness values are very distant the diversity metric is saturated to 1 and the diversity can be considered to be high. In this way, the metric ξ can say whether the local search activation is suitable ($\xi \approx 0$) or unnecessary ($\xi = 1$). This metric proved to lead to a high algorithmic performance in some cases but suffers from robustness, as shown in [657]. The main drawback of this metric is that it is dependent on the codomain width: adding a constant value to the fitness function would lead to an important variation of the diversity values. However, this diversity metric is very efficient in the specific cases it has been used: for multivariate and complex fitness landscapes having a limited range of variability in the fitness values (e.g. $[0, 10]$) and the minimum around zero (e.g. for error minimization in engineering problems).

The second fitness diversity metric has been introduced in [888] and used also in [889]. The metric is:

$$\nu = \min\left\{1, \frac{\sigma_f}{|f_{avg}|}\right\}, \qquad (10.7)$$

where $|f_{avg}|$ and σ_f are respectively the average value and standard deviation over the fitness values of individuals of the population. Also the parameter ν can vary between 0 and 1 and can be seen as a measurement of the fitness diversity and distribution of the fitness values within the population. In other words, v is the answer to the question "How sparse are the fitness values within the population?". As well as ξ, v is codomain dependent and works with a limited range of variability. Unlike ξ, v depends on the standard deviation and thus on the fitness distribution over all individuals of the population. In addition, v is less sensitive than ξ to fitness diversity variations and would not consider high diversity a situation where one individual has a performance significantly better than the others. For this feature if ξ is efficient on an ES framework employing the plus strategy, v can be employed for SIAs and DE i.e. for those algorithms which normally work in high diversity conditions, see [889].

The third fitness diversity metric has been introduced in [658] for a specific medical application. This metric consists of the following:

$$\psi = 1 - \left| \frac{f_{avg} - f_{best}}{f_{worst} - f_{best}} \right| \qquad (10.8)$$

where f_{best}, f_{avg} and f_{worst} are respectively best, average and worst fitness over the individuals of the population. The parameter ψ can be seen as the answer to the question "If we sort all fitness values over a line, which position is occupied by the average fitness?". The metric ψ also varies between 0 and 1. It can be noticed that, unlike the two metrics previously presented, ψ is not codomain dependant, i.e. its value does not depend on the range of variability of the fitness values. Due to its structure, this metric is very sensitive to small variations and thus is especially suitable for fitness landscapes containing plateaus and low gradient areas. Parameter ψ has been successfully employed within memetic frameworks which employ plus strategy in the spirit of the ES.

In [106] the following parameter is used:

$$\chi = \frac{|f_{best} - f_{avg}|}{max|f_{best} - f_{avg}|_k} \qquad (10.9)$$

where f_{best} and f_{avg} are the fitness values of, respectively, the best and average individuals of the population. $max|f_{best} - f_{avg}|_k$ is the maximum difference observed (e.g. at the kth generation), beginning from the start of the optimization process. It is clear that χ varies between 0 and 1; it scores 1 when the difference between the best and average fitness is the largest observed, and scores 0 when $f_{best} = f_{avg}$ i.e. the entire population is characterized by a unique fitness value. Thus, χ is the answer to the question "How much better is the best individual than the average fitness of the population with respect to the history of the optimization process?".

Besides considering it as a measurement of the fitness diversity, χ is an estimation of the best individual performance with respect to the other individuals. In other words, χ measures how much the super-fit outperforms the remaining part of the population. More specifically, the condition $\chi \approx 1$ means that one individual has a performance far above the average, thus one super-fit individual is leading the search. Conversely, the condition $\chi \approx 0$ means that performance of the individuals is comparable and there is not a super-fit. Due to its nature, χ is suitable for guessing the state of convergence in a population of a SIA or a DE. In [106], χ has been defined for coordinating the search components of a MA based on a DE framework. This choice was carried out by taking into account the fact that a DE structure works well when one individual is better than the others since it has the role of guiding the search. However, its performance should not be excessively good with respect to the others; otherwise, it would be unlikely for another individual to succeed at outperforming the leading individual. As a general guideline, a DE population containing a super-fit individual needs to exploit the direction offered by the super-fit in order to eventually generate a new individual that outperforms the super-fit. Conversely, a DE population made up of individuals with comparable fitness values requires that one individual that clearly outperforms the others be generated in order to have a good search lead. A similar analysis can be carried out for Particle Swarm Optimization (PSO) and other SIAs.

In [887] another fitness diversity metric has been introduced. This metric is given by:

$$\phi = \frac{\sigma_f}{|f_{worst} - f_{best}|} \quad (10.10)$$

where σ_f is the standard deviation of fitness values over individuals of the populations, and f_{worst} and f_{best} are the worst and best fitness values, respectively, of the population individuals.

Analogous to the other fitness diversity indexes listed above, ϕ varies between 0 and 1. When the fitness diversity is high, $\phi \approx 1$; on the contrary when the fitness diversity is low, $\phi \approx 0$. The index ϕ can be seen as a combination of ν in formula (10.7) and ψ in formula (10.8) because it represents the distribution of fitness values over individuals of the population with respect to its range of variability. In other words, ϕ is the answer to the question "How sparse are the fitness values with respect to the range of fitness variability at the current generation?". The index ψ was also designed for DE based algorithms. Employment of the standard deviation in the numerator in formula (10.10) is due to the fact that a DE framework tends to generate an individual with performance significantly above the average (as mentioned for the metric χ) and efficiently continues optimization for several generations. In this sense, an estimation of the fitness diversity of a DE population by means of the difference between best and average fitness values can return a misleading result and each value must be taken into account. Regarding the denominator in formula (10.10), a normalization to the range of variability of the current population makes the index co-domain invariant (unlike ν in formula (10.7)) and its estimation is not affected, for example by adding an offset to the fitness function. Thus, the index ϕ can be successfully employed, within a DE framework, on problems of various kinds.

Finally, another fitness diversity index inspired also by the entropy study in [873] has been proposed in [481]. The population is sorted according to the fitness values. Thus an interval $[f_{min}, f_{max}]]$ having width d can be detected. Let us indicate with n_1 the number of individuals falling within $[f_{min}, f_{min} + \frac{d}{3}]$ and with n_3 the number of individuals falling within $[f_{max} - \frac{d}{3}, f_{max}]$. Indicating with N_p the number of individuals of the population and assuming that we want to solve a minimization problem, the diversity is then estimated as:

$$\tau_3 = 0.5 + \frac{n_1 - n_3}{2N_p}. \tag{10.11}$$

In other words, this metric subdivides the population into three quality classes and measures the diversity as a difference of the cardinality of the classes. Metric τ_3 has been used for an ES framework but it might be suitable also for different frameworks. It has successfully been applied to a chemical engineering problem characterized by a highly multi-variate function but likely not a very multi-modal fitness landscape. It must be remarked that although τ_3 also varies between 0 and 1, the interpretation of the parameter is different from the other diversity metrics. The maximum diversity condition occurs when $\tau_3 = 0.5$, which corresponds to maximum distribution of the performance over the individuals of the population. The conditions $\tau_3 \approx 0$ and $\tau_3 \approx 1$ mean that a few individuals have a very high performance with respect to the others and that a few individuals have a very low performance with respect to the others, respectively. In order to visualize this approach, it may be useful to imagine a ring where value 0 and 1 are contiguous. In this sense, this metric measures the balance among the three performance regions. This sophisticated way to measure diversity has the drawback that the metric can suffer from abrupt changes in proximity to 0 and 1 and very slow changes in proximity to 0.5, in correspondence of the same variations within the population. This can make the adaptation rather complicated to handle.

In order to summarize the features of the diversity metrics listed in this section, a synoptical scheme is shown in Table 10.1.

Table 10.1. Diversity Metrics: Synoptical Scheme

Diversity Metric	Framework	Landscape Features	Drawbacks
ξ	EAs	Highly Multi-modal Lanscape	Non scalable
ν	SIAs, DE	Flexible	Non scalable
ψ	EAs	Plateaus, Flat Landscapes	Very sensitive
χ	SIAs, DE	Flexible	Very DE and PSO tailored
ϕ	SIAs, DE	Flexible	Very sensitive
τ_3	EAs	Large Scale not too Multi-modal	Abrupt and Slow Variations

10.3.2 Coordination of the Search: The "Natura non Facit Saltus" Principle

At each generation, when a diversity metric is calculated the problem that follows is how to use such information in order to perform the coordination of global and local search. As mentioned before, let us consider that the MA employs an evolutionary framework and two local search algorithms, the first having explorative features, the second having exploitative features. The goal is to activate the explorative local search algorithm when the population has lost part of its diversity and to activate the exploitative local search algorithm when the population has lost most of its diversity and is approaching a convergence condition. In order to obtain this effect three adaptive schemes have been proposed in the literature.

The first scheme, used in [104], [659], and [658], employs a threshold mechanism for the application of local search. More specifically, when the control parameter surpasses a given threshold, the corresponding local search algorithm is activated. This mechanism can be seen as a probabilistic scheme where the probability of the local search activation, dependent upon the control parameter, is a step function which takes the value 1 within the threshold limits and 0 elsewhere. Although this kind of scheme has proven to be efficient for various applications (see e.g. [657]), the continuous variation of the fitness diversity in an evolutionary algorithm is not in accordance with this step function. In other words, if the fitness diversity metrics measure the necessity of the algorithm increasing/decreasing the local search within the memetic framework, the intensity of the local search is supposed to be related to the variation of the diversity metrics. On the contrary, a step function suggests that the local search is abruptly introduced within the search at its maximum intensity and can thus be too crude an approximation of the exploration/exploitation necessity of the MA.

In order to introduce smooth variation in the intensity of the local search application, two more schemes have been proposed in [106] and [889], respectively. More specifically, the step function has been replaced with a continuous function within the memetic frameworks under examination. Thus, the probability of local search activation is given by a function of the fitness diversity.

Indicating with λ the fitness diversity metric, the first function is the beta distribution function, see [106]:

$$p(\lambda) = \frac{1}{B(s,t)} \cdot \frac{(\lambda-a)^{(s-1)}(b-\lambda)^{(t-1)}}{(b-a)^{(s+t-1)}} \tag{10.12}$$

where a and b are, respectively, the inferior and superior limits of the distribution; $B(s,t)$ is the beta function; $s = 2$ and $t = 2$ are the shape parameters. Parameters a and b must be set on the basis of the algorithm under consideration. The latter parameters play the same role as the thresholds in the previous scheme.

The second, used in [889], is the exponential distribution:

$$p(\lambda) = e^{\frac{-(\lambda-\mu_p)}{2\sigma_p^2}} \qquad (10.13)$$

where μ_p and σ_p are the parameters characterizing the intensity application range of the local search.

In order to better explain the three coordination scheme, let us consider the Fast Adaptive Memetic Algorithm (FAMA) proposed in [104]. This algorithm is based on an ES framework and two local search algorithms. The first local search, playing an explorative role, is the Nelder-Mead Algorithm (NMA) [653] and the second playing an exploitative role, is the Hooke-Jeeves Algorithm (HJA) [391]. For a proper functioning of FAMA, we desire that the NMA be activated when the diversity becomes low in order to give an alternative search logic, and that the HJA be activated in very low diversity condition. Since FAMA employs the ξ metric, this statement can be rephrased as: the NMA is activated when $0.05 < \xi < 0.5$ and the HJA when $\xi < 0.2$. By keeping the same amount of local search, if the beta distribution function is employed then $a = 0$ and $b = 0.68$ for the NMA and, $a = 0$ and $b = 0.3$ for the HJA. Fig. 10.1 gives a graphical representation of the local search coordination, dependent on the diversity metrics, for the FAMA. The diagram shows the step functions (as in the original implementations) in the upper part, the related beta distribution functions in the central part, and the related exponential distributions in the lowest part.

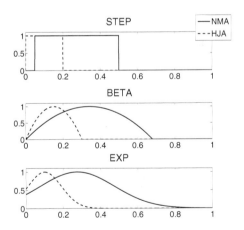

Fig. 10.1. Coordination of the local search for the FAMA

It should be remarked that the scaling of beta and exponential functions is done taking into account the fact that the areas below each trend are the same i.e. the overall balance between global and local search is the same for the original and proposed versions of each algorithm. For the sake of clarity, activation of a local searcher is performed by sampling (by means of a uniform distribution) a pseudo-random number ε in $[0,1]$ and then comparing it with $p(\lambda)$; if $\varepsilon < p(\lambda)$ the corresponding local search is performed.

Numerical results reported in [890] show that the employment of continuous functions is beneficial and succeeds at improving upon the step scheme for a constant amount of local and global search. This fact has been expressed as the "natura non facit saltus" principle. The Latin expression "natura non facit saltus", i.e. nature does not make (sudden) jumps, is a principle of classical physics, claimed since Aristoteles' time until the formulation of the quantum mechanic theory, which states that variation of physical phenomena is continuous, thus not containing "jumps". This concept has been extended to Memetic Computing and more specifically to the local search coordination, dependent on a fitness diversity index. The local search activation should not be abruptly started on the basis of some conditions but should slowly be increased and decreased around a suitable diversity condition.

10.4 Conclusion

This chapter analyzes the problem of diversity in Memetic Computing. The problem of diversity loss is very relevant in Evolutionary Computation since a premature diversity loss can lead to a premature algorithmic convergence into undesired areas of the decision space. Dually, some algorithms could fail at generating new genotypes despite a high diversity and thus stagnate. In Memetic Computing this problem is even more important because the local search application might cause the convergence to the same (or a very similar) point starting from a set of solutions belonging to the same basin of attraction. However, since Memetic Algorithms employ different search logics, if a proper coordination of the algorithmic components is carried out, a successful optimizer can be designed. Modern Memetic Algorithms use different local search algorithms for preserving a proper diversity which promotes the enhancements in the search, and they propose adaptive techniques for coordinating the various algorithmic components.

Several schemes for handling diversity have been illustrated. The employment of structured population has been widely used since it implicitly allows a preservation of diversity. However, distributed algorithms by themselves are not enough to prevent stagnation and premature convergence. Therefore, an adaptive system can support the memetic framework. A control mechanism based on Shannon's entropy can be an efficient countermeasure. Fitness diversity adaptation also provides an efficient diversity control system since a diversity metric is used to coordinate the local search. Although this approach is promising it hides two problems: how to measure the diversity and how to use this information within a memetic framework. In accordance with the No Free Lunch Theorem, there is no optimal diversity metric, but rather its design should take into account the problem and the evolutionary/swarm intelligence structure under consideration. A synoptical table compares the metrics and gives some hints on how to use some diversity metrics proposed in the literature. Regarding the coordination of the algorithmic components, it has been observed that an efficient Memetic Algorithm should contain both explorative and exploitative local search algorithms. The explorative local search algorithm(s) assist the framework to detect novel promising search directions when the diversity is decreasing, while the exploitative one(s) perform an extensive search within already detected basins of attraction when the population has lost most of its diversity. To pursue this aim three control functions are illustrated in this chapter. The first function is a step function, i.e. local search is activated simply by means of threshold comparison. Although this approach is efficient, it has a wide margin of improvement if instead of a step function a continuous function is preferred. Two probability distribution functions have been considered. Previous studies observed that, while keeping constant the amount of local and global search, a Memetic Algorithm employing continuous functions outperforms on a regular basis the corresponding algorithm employing the step function. This fact was previously named the "natura non facit saltus" principle for Memetic Algorithms.

Acknowledgements. This work is supported by Academy of Finland, Akatemiatutkija 130600, Algorithmic Design Issues in Memetic Computing.

Chapter 11
Self-adaptative and Coevolving Memetic Algorithms

James E. Smith

11.1 Introduction

Results from applications of meta-heuristics, and Evolutionary Computation in particular, have led to the widespread acknowledgement of two facts. The first is that evolutionary optimisation can be improved by the use of local search methods, creating so-called Memetic Algorithms. The second is that there is no single "best" choice of memetic operators and parameters- rather the situation changes according to both the problem and the particular stage of search. This has created a growing interest in "Adaptive" Memetic Algorithms which combine a portfolio of local search operators with some method to choose between them. Here we describe techniques which extend these ideas to allow the behaviours of the local search operators to adapt during the search process. In the first case these maybe thought of as Self-Adaptive, so that each member of the evolving population encodes for both an initial solution to a problem, and a learning mechanism which acts on that solution to improve it. More generally, we show that these can be treated as separate co-evolving populations of "genes" and "memes". Following a review of related work, we next describe a framework for meme-gene self-adaptation and co-evolution. This is followed by a summary of the "proof-of-concept" and of findings concerning representation and scalability with self-adaptive memes. Next the paper considers in more depth issues relevant to co-evolution such as credit assignment, and the ratio of population sizes - which can be thought of as the memetic "load" that an evolving population can support.

James E. Smith
Department of Computer Science, University of the West of England,
Bristol, BS16 1QY, UK Name
e-mail: james.smith@uwe.ac.uk

11.2 Background

The performance benefits which can be achieved by hybridising Evolutionary Algorithms (EAs) with Local Search (LS) operators, so-called *Memetic Algorithms* (MAs), have now been well documented across a wide range of problem domains such as optimisation of combinatorial, non-stationary and multi-objective problems (see [493] for a review, and [376] for a collection of recent algorithmic and theoretical work). Typically in these algorithms, a LS improvement step is performed on each of the products of the generating (recombination and mutation) operators, prior to selection for the next population. There are of course many variants on this theme, but these can easily be fitted within a general syntactic framework [493].

In recent years it has been increasingly recognised that the choice of LS operator will have a major impact on the efficacy of the hybridisation. Of particular importance is the choice of move operator, which defines the neighbourhood function, and so governs the way in which new solutions are generated and tested. For example Krasnogor and Smith used Polynomial Local search (PLS) theory to show that the worst-case runtime of an MA is not improved over the underlying EA if the LS neighbourhood function does not differ from those of the EAs variation operators [494]. However, points which are locally optimal with respect to one neighbourhood structure will not in general be so with respect to another, unless of course they are globally optimal. It therefore follows that even if a population only contains local optima, then changing the LS move operator (neighbourhood) may provide a means of progression in addition to recombination and mutation. This observation has led a number of authors to investigate mechanisms for choosing between a set of predefined LS operators which may be used during a particular run of a meta-heuristic such as an EA.

11.2.1 MAs with Multiple LS Operators

There are several recent examples of the use of multiple LS operators within evolutionary systems. Ong *et al.* [683] present an excellent recent review of work in the field of what they term "Adaptive Memetic Algorithms". This encompasses Krasnogor's "Multi-Memetic Algorithms" [486, 487, 491, 492, 496], Smith's COMA framework [821, 824, 825, 830], Ong and Keane's "Meta-Lamarkian MAs [680], and Hyper-Heuristics [96, 97, 169, 456]. In another interesting related algorithm, Krasnogor and Gustafson's "Self-Generating MAs" use a grammar to specify for instance when local search takes place [488, 490]. Essentially all of these approaches maintain a pool of LS operators available to be used by the algorithm, and at each decision point make a choice of which to apply. There is a clear analogy between these algorithms and Variable Neighbourhood Search [363], which uses a heuristic to control the order of application of a set of predefined LS operators to a single improving solution. The difference here lies in the population based nature of MAs, so that not only do we have multiple LS operators but also multiple candidate solutions, which makes the task of deciding which LS operator to apply to any given one more complex.

Ong's classification uses terminology developed elsewhere to describe adaptation of operators and parameters in Evolutionary Algorithms [237, 238, 386, 829]. This categorises algorithms according to the way that these decisions are made. One way (termed "static") is to use a fixed strategy . Another is to use feedback of which operators have provided the best improvement recently. This is termed "Adaptive", and is further subdivided into "external", "local" (to a deme or region of search space), and "global" (to the population) according to the nature of the knowledge considered. Finally they note that LS operators may be linked to candidate solutions (Self-Adaptive). We will adopt this terminology, and also make use of the general term "meme" to denote an object specifying a particular local search strategy.

11.2.2 Self-adaptation in EAs

We are concerned with meta-heuristics which maintain two sets of objects - one of genes and one of memes. If these sets are adaptive, and use evolutionary processes to manage what may now be termed populations, then we can draw some immediate parallels to other work. If the populations are of the same size and selection of the two is tightly coupled (to use the notation of [22]) then this can be considered as a form of Self Adaptation. The use of the intrinsic evolutionary processes to adapt mutation step sizes has long been used in Evolution Strategies [799], and Evolutionary Programming [268]. Similar approaches have been used to self-adapt mutation probabilities [31, 828] and recombination operators[793, 827] in genetic algorithms (GAs) as well as more complex generating operators combining both mutation and recombination [826]. More recently Smith and Serpell have showed that self-adaptation can very effectively govern both the choice and parameterisation of different mutation operators for GAs with permutation representations [807].

11.2.3 Co-evolutionary Systems

If selection is performed separately for the two populations, with memes' fitness assigned as some function of the relative improvement they cause in the "solution" population, then we have a co-operative co-evolutionary system. Following initial work by Husbands and Mill [399] this metaphor has gained increasing interest. Paredis has examined the co-evolution of solutions and their representations [709]. Potter and DeJong have also used co-operative co-evolution of partial solutions in situations where an obvious problem decomposition was available [727]. Both reported good results. Bull [90] conducted a series of more general studies on co-operative co-evolution using Kauffman's static NKC model. In [92] he examined the evolution of linkage flags in co-evolving "symbiotic" systems and showed that the strategies which emerge depend heavily on the extent to which the two populations affect each others fitness landscape. In highly interdependent situations linkage of the two species' chromosomes was preferred –which in our context is equivalent to memes self-adapting as part of the solutions' genotypes. Bull also examined the effect of various strategies for pairing members of different populations for evaluation [91]. This showed mixed results, although the NKC systems he investigated used fixed interaction patterns. This work has recently been revisited and extended by Wiegand *et al.* with very similar findings [933]. Wiegand's work also focused attention on the number of partners with which a member of either population should be evaluated, which draws attention to the trade-off between accurately estimating the value of an object (solution or meme), and using up evaluations doing so. Parker and Blumenthal's "Punctuated Anytime Learning with samples" [714] is another recent approach to the pairing problem by using periodic sampling to estimate fitness, but this is more suited to approaches where the two populations evolve at different rates. Closely related to this, Bull, Holland and Blackmore have examined the effect of changing the relative speed of evolution of populations which they termed "genes" and "memes" [93]. Their results showed that as the relative speed of meme evolution increased a point was reached beyond which gene evolution effectively ceases. However, the NKC systems they use severely limit the types of interaction permitted to an abstraction rather different from most MA applications.

There has also been a large body of research into competitive co-evolution (see [710] for an overview). Here the fitnesses assigned to the two populations are directly related to how well individuals perform against the other population - what has been termed "predator-prey" interactions. Luke and Spector [541] have proposed a general framework within which populations can be co-evolved under different pressures of competition and co-operation. This uses speciation both to aid the preservation of diversity and as a way of tackling the credit assignment problem.

In all the co-evolutionary work cited above, the different populations only affect each other's perceived fitness, unlike the COMA framework where the meme population can directly affect the genotypes within the solution population. This raises the question of whether the modifications arising from Local Search should be written back into the genotype (Lamarckian Learning) or not (Baldwinian Learning). Although the pseudo-code and the discussion below, assumes Lamarckian learning, this is not a prerequisite of the COMA framework. However, even if a Baldwinian approach was used, COMA differs from the co-evolutionary systems above because there is a selection phase within the local search, so that if all of the neighbours of a point defined by the meme's rule are of inferior fitness, then the point is retained unchanged within the population. If one was to discard this criterion and simply apply the rule (possibly iteratively), the system could be viewed as a type of "developmental learning" akin to the studies in Genetic Code e.g. [443] and the "Developmental Genetic Programming" of Keller and Banzhaf [453, 454].

11.3 A Framework for Self-adaption and Co-evolution of Memes and Genes

In this section we describe a conceptual framework designed to support a wide range of algorithms for meme adaptation.

11.3.1 Specifying Local Search

The primary factor that affects the behaviour of the LS is the choice of neighbourhood generating function. This can be thought of as defining a set of points $n(i)$ that can be reached by the application of some move operator to the point i. One representation is as a graph $G = (v, e)$ where the set of vertices v are the points in the search space, and the edges relate to applications of the move operator i.e $e_{ij} \in G \iff j \in n(i)$. The provision of a scalar fitness value, f, defined over the search space means that we can consider the graphs defined by different move operators as "fitness landscapes" [433]. Merz and Freisleben [585] present a number of statistical measures which can be used to characterise fitness landscapes, and have been proposed as potential measures of problem difficulty. They show that the choice of move operator can have a dramatic effect on the efficiency and effectiveness of the Local Search, and hence of the resultant MA.

The second component of Local Search is the choice of pivot rule, which can be *Steepest Ascent* or *Greedy Ascent*. In the former the "termination condition" is that the entire neighbourhood $n(i)$ has been searched, whereas the latter stops as soon as an improvement is found. Note that one can consider only a randomly drawn sample of size $N <<| n(i) |$ if the neighbourhood is too large to search.

The final component is the "depth" of the Local Search. This lies in the continuum between only one improving step being applied to the search continuing to local optimality. Studies with MAs e.g. [366] have shown it affects the performance both in terms of time taken and of quality of solution found.

11.3.2 Adapting the Specification of Local Search

The aim of this work is to provide a means whereby the definition of the LS operator used within a MA can be varied over time, and then to examine whether evolutionary processes can control that variation so that beneficial adaptation takes place. Accomplishing this aim requires the provision of four major components, namely:

- A means of representing a LS operator in an evolvable form i.e. as a meme.
- A means of assigning fitness to memes.
- A choice of population structures and sizes, selection and replacement methods for managing the meme population.
- A set of experiments to permit evaluation and analysis of the system.

The pseudo-code in Algorithm 21 illustrates the algorithmic framework of a COevolutionary Memetic Algorithm (COMA) developed to support this research. Note that although this pseudo-code assumes synchronous evolution, this need not in general be the case. The representation of the memes is a tuple <*Pivot, Depth, Pairing,Move*>, which can readily encompass all of the other requirements identified above. The representation of the tuple elements leads naturally to the choice of evolutionary variation operators. The *Pivot,Depth* and *Pairing* elements can be easily mapped onto integer or cardinal representations. The latter element, is one

Algorithm 21. Pseudo-Code Definition of COMA algorithm

```
   // Given populations P of μ_s solutions and M of μ_m memes
1  initialise P and M randomly ;
2  set generations ← 0;
3  set evaluations ← 0;
4  while run_termination condition is satisfied do
      // Create μ_s solution offspring and store parent ids
5     for i ← 1 to μ_s do
6        set FirstParent[i] ← Select_One_Parent(P);
7        set SecondParent[i] ← Select_One_Parent(P);
8        set Offspring[i] ← Recombine(FirstParent[i],SecondParent[i]);
9        Mutate(Offspring[i]);
10       set i ← i +1;
11    endfor
      // Create mu_m meme offspring according to pairing
12    for i ← 1 to μ_m do
13       set Pairing ← Get_Pairing(M,i);
14       if Pairing = SelfAdaptive then
15          set MemeParent1[i] ← FirstParent[i];
16          set MemeParent2[i] ← SecondParent[i];
             // note this requires μ_m = μ_s.
17       endif
18       else if Pairing = Fitness_Based then
19          set MemeParent1[i] ← Select_One_Parent(M);
20          set MemeParent2[i] ← Select_One_Parent(M);
21       endif
22       else
23          set MemeParent1[i] ← RandInt(1,μ_m);
24          set MemeParent2[i] ← RandInt(1,μ_m);
25       endif
26       set NewMemes[i] ← Recombine(MemeParent1[i],MemeParent2[i]);
27       Mutate(NewMemes[i]);
28       set i ← i+1;
29    endfor
      // Apply local search to Offspring Using Memes
30    for i ← 1 to μ_s do
31       set original_fitness ← Get_Fitness(Offspring[i]);
32       if Pairing = SelfAdaptive then
33          set meme ← i;
34       endif
35       else
36          set meme ← Select_Random(NewMemes);
37       endif
38       set Neighbours ← Apply_Rule_To_Offspring(Offspring[i],NewMemes[meme]);
39       Evaluate_Fitness(Neighbours);
40       set Offspring[i] ← Apply_Pivot_Rule(Neighbours);
41       set Δfitness ← Get_Fitness(Offspring[i]) - original_fitness;
42       Update_Meme_Fitness(NewMemes[meme], Δfitness);
43       set evaluations ← evaluations +1 + |Neighbours|;
44       set i ← i +1;
45    endfor
46    set P ← Offspring;
47    set M ← NewMemes;
48 endw
```

of {*Self-Adaptive, Random, Fitness_Based*} and determines how memes are created and applied to solutions. As is illustrated in the *If..Else* section of the pseudo-code, a range of behaviours from self-adaptive, through collaborative co-evolution to random meme drift can be obtained.

This framework is designed it be generic in the way that move operators are described - for example they could be GP-like expressions as per [288]. However while such richness tends to lead to complexity of expression suitable for practical applications, it can make analysis of evolved behaviour more difficult. Therefore for the initial development work a simpler format was used together with well-understood test problems. In what follows, move operators are encoded as *condition:action* pairs, which specify one pattern to be looked for in the problem representation, and another to replace it. The neighbourhood of a point i then consists of i itself, plus all those points where the substring denoted by *condition* appears in the representation of i and is replaced by the *action*. To give an example, a rule 1#0 $\rightarrow$ 111 matches the binary string 1100111000 in the first, second, sixth and seventh positions, and the neighbourhood is the set {1100111000, 1110111000, 1111111000, 1100111100, 1100111110}.

Note that the string is not treated as toroidal, and the neighbours are evaluated in a random order so as not to introduce positional bias into the local search when greedy ascent is used. Although this representation at first appears very simple, it has the potential to represent highly complex moves via the use of symbols to denote not only single/multiple wild-card characters (in a manner similar to that used for regular expressions in Unix) but also the specifications of repetitions and iterations. Further, permitting the use of different length patterns in the *condition* and *action* parts of the rule gives scope for *cut* and *splice* operators working on variable length solutions.

11.4 Test Suit and Methodology

A range of well understood test problems were used to examine the performance of various self-adaptive and coevolutionary MAs. Some of these are "standard" testbed functions for EAs, others were specifically designed to probe and evaluate certain behaviours. The initial systems only used rules where the *condition* and *action* patterns were of equal length and were composed of values taken from the set of permissible allele values of the problem representation, augmented by a "don't care" symbol (#) which is allowed to appear in the *condition* part of the rule and optionally in the *action* where it is treated as *invert*. In practise, each rule was augmented by a value *rule_length* specifying the number of positions in the pattern string to consider. This permitted not only the examination of the effects of different fixed rule sizes, but also the ability to adapt its value via mutation.

11.4.1 The Test Suite

The first set of problems used are composed of 16 subproblems of Deb's 4-bit fully deceptive function [35]. The fitness of each subproblem i is given by its unitation $u(i)$, that is the number of bits set to "one":

$$f(i) = \begin{cases} 0.6 - 0.2 \cdot u(i) & : \quad u(i) < 4 \\ 1 & : \quad u(i) = 4 \end{cases} \tag{11.1}$$

In addition to a "concatenated" version (4-Trap), a second "distributed" version (Dist-Trap) was used in which the subproblems were interleaved i.e. sub-problem i was composed of the genes $i, i+16, i+32, i+48$. This separation ensured that in a single application even the longest rules allowed in these experiments would be unable to alter more than one element in any of the sub-functions. A third variant of this problem (Shifted-Trap) was designed to be more "difficult" than the first for the COMA algorithm, by making patterns which were optimal in one sub-problem, sub-optimal in all others. Since unitation is simply the Hamming distance from the all-zeroes string, each sub-problem can be translated by replacing $u(i)$ with the Hamming distance from an arbitrary 4 bit string. There were 16 sub-problems so the binary coding of each ones' index was used as basis for its fitness calculation.

The second test function was Watson's Hierarchical-if-and-only-if (H-IFF) function, a highly epistatic problem designed to examine the virtues of recombination. At the bottom level, fitness is awarded to matching pairs of adjacent bits in a solution s, i.e.

$$f_1 s = \sum_{i=0}^{l/2-1} 1 - XOR(s_{2i}, s_{2i+1}) \tag{11.2}$$

and this process is applied recursively, so that a problem of size $l = 2^k$ has k levels. In each ascending level the number of blocks is reduced by a factor of two, and the fitness awarded for each matching pair is increased by a constant factor, in our case 2. This problem has a number of Hamming sub-optima, and two global optima corresponding to the $u(i) \in \{0,1\}$. Problem sizes $l \in \{16, \ldots, 512, 1024\}$ were used, corresponding to 3 to 10 levels. Note that for $l > 16$ the length of the blocks to be identified at the highest levels far exceeded the maximum rule length.

The Maximum satisfiability (Max-SAT) problem is a classical combinatorial optimisation problem, consisting of a number of Boolean variables and a set of clauses built from those variables. A full description and many examples can be found in [392]. For each length $\{50, 100, 250\}$ the first 25 were taken from the sets of uniformly randomly created satisfiable instances around the phase transition (in terms of hardness) where there are approximately 4.3 clauses per variable.

11.4.2 Experimental Set-Up and Terminology

A generational genetic algorithm, with deterministic binary tournament selection for parents and no elitism was used. Population size μ_s was 250 unless otherwise stated. One Point Crossover (with probability 0.7) and bit-flipping mutation (with a bitwise probability of 0.01) were used on the problem representation. These choices were taken as "standard", and no attempt was made to tune them to the particular problems at hand. Mutation was applied to the rules with a allele-wise probability of 0.0625 - the inverse of the maximum rule length allowed to the adaptive version. If the rule length was adaptive, they were randomly initialised in the range [1,16], and during mutation, a value of +/- 1 is randomly added with probability 0.0625, subject to staying in range.

For each problem, 20 runs were made, each continuing until the global optimum was reached, subject to a maximum of 500,000 evaluations. Two performance metrics were considered; the Success Rate (SR) which is the number of runs finding the global optimum, and the Average Evaluations to Success (AES) which is the mean time taken to locate the global optimum on successful runs. The reason for the large cut-off value was to try and avoid skewing results as can happen with an arbitrarily chosen lower cut-off, rather than to be indicative of the amount of time available for a "real world" problem. Note that since one iteration of a local search may involve several evaluations, this allows more generations to the GA, i.e. algorithms are compared strictly on the basis of the number of calls to the evaluation function. Any observed differences in performance were tested for statistical significance using ANOVA and pairwise post-hoc testing using the Tukey's Least Significant Difference (LSD) and Tamhane's T2 tests at the 95% confidence level.

The variants of self- and co-adaptive algorithms that can be instantiated within this framework are denoted as *CXY* where *X* denotes the pairing and is one of *L* (Linked, or self-adaptive), *R* (Random drift) or *T* (Tournament - variants of fitness based coevolution). *Y* denotes the pivot function and is one of *G*reedy, *S*teepest or *A*daptive. Rule lengths are adaptive unless denoted by a numeric prefix. Depth of search is one unless indicated by a suffix -*L* (to local optima) or -*A*daptive.

11.5 Self-adaptation of Fixed and Varying Sized Rules

11.5.1 Self-adapting the Choice from a Fixed Set of Memes

The first experiments in this line of research explored the ability of evolutionary mechanisms to correctly select between a number of fixed memes. This can be achieved trivially within the COMA framework by the use of appropriate initialisation for the meme population, setting the meme recombination probabilities to zero and defining the mutation operator so that it chose between the set of fixed memes rather than operating "within" each meme. In [492] experiments were run on a range of TSP problems using MAs with one of set of ten memes which varied in both their move operators and depth. When the search progress was plotted together, it could clearly be seen that the optimal choice of meme was dependant on the state of the search as well as on the individual TSP instance. Next the population members of "multimeme" algorithm were allowed to self-adapt the choice of which meme to use. The results showed that the progress tracked that of the currently best-performing meme from the "static" MAs, ultimately outperforming each of them. The evolved patterns of meme usage closely matched what might have been "designed" with hindsight, with periods of one meme dominating alternating with periods of broader usage as local optima were reached, then escaped from.

The concept of self-adapting the choice from a fixed set of memes was also successfully demonstrated by Krasnogor *et al.* for protein structure alignment [496].

11.5.2 Self-adaptation of Meme Definitions

Initial experiments were restricted to considering a simple system, and examining first whether the system was able to evolve useful rules for the "trap" problems - first when the rule length naturally matched the structure of the problem, and then whether the system was able to adapt to an appropriate rule length for different problems. For this reason it was decided to avoid the various issues concerning population management, pairing strategies and credit assignment, and instead work with a single improvement step, a fully linked self-adaptive system and a greedy pivot rule. These choices were coded into the chromosomes at initialisation, and variation operators were not used on them. The algorithms used (and the abbreviations which will be used to refer to them hereafter) are as follows:

- A "vanilla" GA with no Local Search (GA).
- A simple bit-flipping MA (SMA-G).
- COMA using a random rules, i.e. with the learning disabled (CRG).
- COMA with self-adaptive memes, greedy pivot and adaptive rule lengths (CLG).
- COMA using fixed length memes (1-CLG,...,10-CLG),

Experiments were run with population sizes (μ_s, μ_m) of 100, 250 and 500.

11.5.3 Results on Trap Functions

The results on 4-Trap showed that the GA, SMA, and 1-CLG algorithms frequently failed to find the optimum but the other COMA variants, always did. On these problems there was a clear benefit to using adaptive neighbourhood local search, although since the CRG algorithm also found the optimum on every run, it cannot be concluded from the Success Rates that learning was taking place. Considering the AES, the GA, SMA and 1-CLG algorithms took significantly longer to locate the optimum. For a population of 500 2-CLG joined the significantly slower group.

In short, it could be observed that for fixed rule lengths of between 3 and 9, and for the adaptive version, the COMA system derived performance benefits from evolving LS rules according to both metrics on this function.

For the Shifted-Trap function, the performances of the GA and SMA were not significantly different from those on 4-Trap because these algorithms solved the subproblems independently and so were "blind" to whether the optimal string for each was different. The COMA results exhibited the same pattern of behaviour noted above; fast, reliable problem solving for all but 1-CLG and 2-CLG, and even for these two the AES results were statistically significantly better than GA or SMA.

On Dist-Trap, GA, SMA and CRG never located the global optimum, regardless of population size. While the Success Rate for COMA was less than for the other problems (typically 10-15/20 for $\mu = 100$ and 15-20/20 for $\mu = 250$), the same pattern was observed of better performance (SR and AES) for the adaptive version and fixed rule lengths in the range 3-5, tailing off at the extremes of the length range.

11.5.4 Analysis of Results and Evolution of Rule Base

The deceptive functions used were specifically chosen because GA theory suggests they are best solved by finding and mixing optimal solutions to sub-problems. Thus the GA failed to solve the function when the crossover operator was not suited to the representation (Dist-Trap). Considering the action of a single bit-flipping LS operator on these "trap" subproblems, a search of the Hamming neighbourhood of a solution will always lead towards the sub-optimal solution when the unitation is 0,1 or 2, regardless of pivot rule. Additionally, the greedy search of the neighbourhood will lead towards the deceptive optimum 75% of the time when the unitation is 3. This explains the poor results of the SMA, and 1-CLG algorithms.

The behaviour of the CLG algorithm was examined by plotting the population mean against time of the rule length, the specificity of the condition (the proportion of values set to #), and the unitation of the action. These results are shown in Figure 11.1.

For the 4-Trap function, the system rapidly evolved medium length (3 – 4), general (specificity < 50%) rules whose action was to set all the bits to 1 (mean unitation 100%). Closer inspection of the evolving rule-base confirmed that the optimal subproblem string was being learnt and applied.

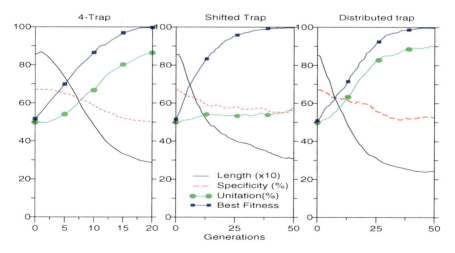

Fig. 11.1. Analysis of Evolved Rules on three problems with different properties

For the Shifted-Trap function, where the optimal sub-blocks are all different, the rule length decreased more slowly from its initial mean value of 8. The specificity also remained higher, and the mean unitation remained at 50%, indicating that different rules were being maintained. This was borne out by closer examination of the evolved rule sets.

The behaviour on Dist-Trap was similar to that on 4Trap, albeit over a longer time-scale. The algorithm could not possibly be learning specific rules about sub-problems, since no rule was able to affect more than one locus of any subproblem. Rather, the system learnt the general rule of setting all bits to 1. The rules were generally shorter than for 4Trap, which means that the number of potential neighbours was higher for any given rule. The high incidence of #s meant that the rule length defined a maximum radius in Hamming space for the neighbourhood, rather than a fixed distance from the original solution. These two observations, together with the longer times to solution, suggest that when the system was unable to find a single rule that matched the problems' structure, a more diverse search took place using a more complex neighbourhood which slowly adapted itself to the current solution population. Full details of these experiments and analysis may be found in [821].

11.5.5 Benchmarking the Self-adaptive Systems

In order to test these hypotheses about how the memes self-adapt in different ways a further set of experiments was run using a wider range of problems, with 50 runs per problem-length to tease out statistically significant differences. For the first two sets of results, both steepest and greedy ascent pivot rules were tried, for the final, MAX-SAT problem, the pivot rule was also allowed to adapt under mutation.

11.5.5.1 Exploiting Search Space Regularities Gives Scalability

The hypothesis memes adapt to identify and exploit any regularities in the problem space was tested by varying the lengths of two problems. The first of these comprised multiple concatenated copies of (11.1) with lengths in the range {40, 60, 80,..., 200}. As expected from above, the results for SMA-G were extremely poor. The next worse algorithm was CRS. The SR steadily decreased 50 (100%) at length 40 to 5 at length 100 and zero above that. All the other algorithms showed SR of 49 or 50 up to length 160, but only the CLS (39) and CLG (50) solved the 200-bit problem. This provides evidence that learning is taking place in the meme populations. The AES results were revealing. The GA was faster than CLG and CLS but the increase in AES with length was worse than linear. The AES results of the successful COMA variants, and analysis of the evolving rule bases, supported the hypothesis of discovering and exploiting regularities. In this case it meant identifying a rule giving the optimal solution to the sub-problems, and then applying it to each sub-problem in the string in successive generations. as shown in Figure 11.2 CLG was the fastest algorithm, followed by CLS, and all three were near-linear. For example, a linear regression of AES to length for CLG fitted the data with a correlation co-efficient of 0.97.

11 Self-adaptative and Coevolving Memetic Algorithms

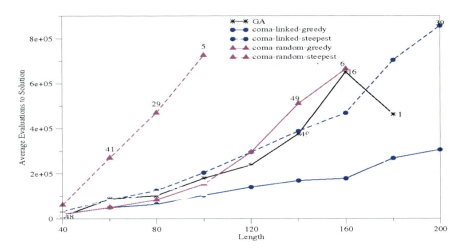

Fig. 11.2. Efficiency of different algorithms on 4 Trap functions with varying length. Annotations beside points show where Success Rates were less than 50/50.

On the H-IFF problems all of the MAs had higher Success Rates than the GA, and again the CLG and CLS were significantly the best. For example, out of 50 runs with $l = 128$ the SR values were 0 (GA, CRG, CRS), 4 (MA), 38 (CLG), 43 (CLS). Only the CLG (10) variant solved the 256 bit problem. As on other problems: the greedy ascent versions found the optimum faster (lower AES) than the equivalent steepest ascent versions but not as reliably (lower SR). ANOVA on the MBF results confirmed that the performance was statistically significantly different with 95% confidence. Post-hoc analysis showed that the CLG and CLS variants had a higher mean best fitness than the others but did not significantly differ.

11.5.5.2 Escaping Local Optima by Changing Neighbourhoods

Shifted-Trap, Dist-Trap and MAX-3SAT were used to examine the behaviour when there were no regularities that could be exploited. On the Dist-Trap function, only the CLS and CLG algorithms ever located the global optimum, and both always did, CLG significantly faster than CLS. On the Shifted-Trap function, the success rates were 39/50 (CRS) 45/50 (SMA-G) and 50/50 (all others). There was no significant difference in the mean times to solution.

On MAX-SAT the GA, steepest/greedy simple MAs (SMA-S, SMA-G), and self-adaptive COMA algorithms with greedy, steepest and adaptive pivot strategies (CLG, CLS, CLA) were run ten times on each instance. Table 11.1 shows the number of success from 250 runs. Full experimental details, and some results omitted here for brevity, may be found in [830].

As can be seen, for the 50 variable instances the simple MAs have the highest success rates, and the GA the worst. For the longer instances all methods are much less successful, and many instances are not solved by any algorithms. SMA-G and

Table 11.1. Success Rates (out of 250) for different length MAX-3SAT problems.

Algorithm	Length 50	Length 100
GA	125	21
SMA-S	154	0
SMA-G	153	25
CLS	141	0
CLG	135	25
CLA	144	8

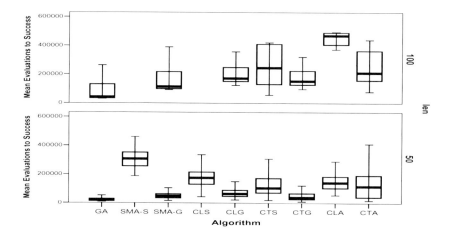

Fig. 11.3. Box plots of AES for 100 (top) and 50 (bottom) variable MAX-3SAT instances.

CLG show the same performance For the shorter instances the steepest ascent strategy is on average better, but there are differences between individual instances. For the longer instances the cost of searching the entire neighbourhood every iteration becomes prohibitive, so that SMA-S and CLS solve no instances. Analysis shows that the adaptive variant CLA performs on a par with whichever of the S or G variants is better for each instance, suggesting successful adaptation.

The AES results show that the GA is the fastest algorithm followed by a close grouping of SMA-G, CLG then CLA, with the CLS algorithm taking more time and having a higher variance. The adaptive pivot variants both fall between their respective greedy and steepest counterparts, both in terms of mean and variance. GA was significantly faster, and the SMA-S significantly slower than the other algorithms, CLS/CLA/CLG did not significantly differ. Analysis of the mean best fitness showed that the CLA algorithm came between the two fixed strategies, but again the ordering of CLG/CLS, and the magnitude of the difference between them, was instance dependant.

11.5.6 Summary of Self-adaptive Results

The results above highlight the problem of choosing the appropriate local search operator which provided the original rationale for the development of COMA. For example, although the Memetic Algorithm with a simple bit-flipping hill climber had the highest Success Rates and Mean Best Fitness on the Max-3SAT problems, it's performance on the other problems was derisory, and frequently worse than the simple GA. In contrast the self-adaptive MAs exhibited better performance than the GA or SMA over a wide range of problems, according to different metrics. Fuller details of the experiments with binary-coded problems may be found in [821, 822, 830], and details of successful application to a protein structure prediction problem may be found in [822, 825]. Overall adapting the pivot rule (CLA) is outperformed by whichever is better of steepest or greedy ascent, but the difference is often marginal, and more importantly the choice (CLS or CLG) is problem dependant.

11.6 Extension to True Co-evolution: the Credit Assignment Problem

Having established the basic principle of evolving memes which coded for LS rules as a means of enhancing optimisation performance in MAs, the next series of experiments used a full co-evolutionary model. Experiments reported in [824, 825, 830] showed that a major factor determining successful adaption was the credit assignment mechanism used to award fitness to a solution. The results also showed that with meme fitness dependant simply on the improvement caused, the choice of pivot and pairing strategies are intertwined.

Unsurprisingly, the greedy variants almost always used less evaluations than the steepest ascent equivalents on successful runs. However, for some problems (but not all) the extra noise introduced by using a greedy ascent was sufficient to "fool" the simple credit assignment mechanism. Thus a good rule will only get a low fitness if the first match only leads to a small improvement, whereas larger improvement (and hence fitness) might be seen if it was applied elsewhere in the solution. Another related source of noise is the choice of partner.

In [831] a number of methods were examined to try and overcome the difficulties of the greedy strategy by reducing the amount of noise present. Meme parent selection used binary tournaments based on fitness defined in the following ways.

- Simple "global-adaptive" scheme where the fitness of a meme was the improvement it caused when applied to a solution (CT). Note that even if a meme perfectly encapsulates the problem structure it can achieve zero fitness if it happens not to match or change the solution it is paired with.
- COMA with a "memory" (CTD). Inspired by Paredis' "Life Time Fitness Evaluation" (LTFE) [710] this uses a time-decaying fitness function of the form:

$$meme\_fitness' = meme\_fitness \cdot \alpha + improvement\_caused \qquad (11.3)$$

A newly created meme takes the average of its parent's fitnesses. After initial experiments a decay factor of $\alpha = 0.5$ was used.
- A modification to the COMA algorithm so that two solution parents contribute to create two offspring solutions via recombination, and similarly for the memes. Each meme is then tested against both of the solutions and the fitness assigned is either the mean (**CT2M**) or better (**CT2B**) of the two improvements noted. In Wiegand's terminology this is a *collaboration poolsize* of two. Each solution takes the better of the two neighbours found for it.

11.6.1 Results: Reliability

Table 11.2 shows the Success Rates achieved with the different algorithms on each function and problem length. The results not just for the 3 Trap variants but also the H-IFF show the clear advantage of Adaptive Memetic Algorithms over both the static counterpart (SMA-G) and a simple Genetic Algorithm (GA). The global-random scheme (CRG) shows lower Success Rates than the other COMA algorithms on most problems. The self-adaptive scheme (CLG) also has lower success rates on the longer H-IFF problems and the SAT problems that the co-evolutionary

Table 11.2. Success Rates of Algorithms on Different Functions

Function	Len	CRG	CLG	CT2BG	CT2MG	CTDG	CTG	GA	SMAG
4Trap	20	10	10	10	10	10	10	10	10
	40	10	10	10	10	10	10	10	6
	60	10	10	10	10	10	10	10	3
	80	10	10	10	10	10	10	10	0
	100	10	10	10	10	10	10	10	0
	120	10	10	10	10	10	10	8	0
	140	9	10	10	10	10	10	3	0
	160	10	10	10	10	10	10	1	0
	180	2	10	7	9	10	10	0	0
	200	0	10	2	4	10	10	0	0
Shifted Trap	64	10	10	10	10	10	10	10	3
Dist-Trap	64	0	10	10	10	10	9	0	0
H-IFF	16	10	10	10	10	10	10	10	10
	32	10	10	10	10	10	10	5	10
	64	2	9	9	10	10	8	4	8
	128	0	3	5	8	6	7	0	0
	256	0	0	6	4	4	3	0	0
SAT	50	131	134	146	152	136	145	114	153
	100	28	21	24	26	16	25	38	27
Total		272	307	319	333	312	327	243	230

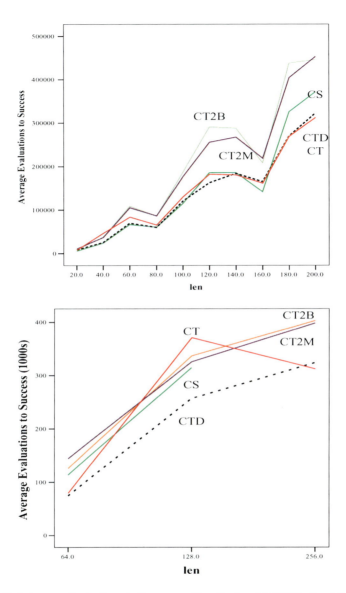

Fig. 11.4. Average Evaluations to Success on Trap (top) and H-IFF (bottom) functions.

variants. Comparing the four different fitness schemes for coevolution, no clear pattern emerges:

- On the 4-trap functions, the meme evolution task is to identify rules of the form #### → 1111, and then maintain and exploit them through repeated application. Here the simpler schemes based on a single pairing (CTG, CTDG) find the optimum slightly more often for the longer instances in the time allowed.

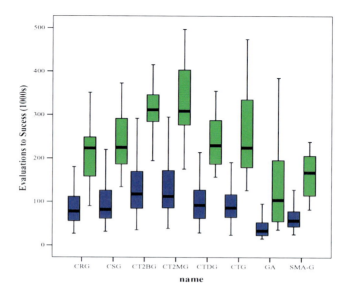

Fig. 11.5. Box-plots of Evaluations to Success on SAT functions. Lighter boxes are for 50-variable instances, darker ones for 100 variables.

- On the Shifted-Trap and Dist-Trap functions, where it is necessary to maintain a *diverse* rule-set in the meme population, algorithms perform the same (100% Success), except the simple CTG (9/10 on Dist-Trap).
- On the H-IFF and SAT problems the fitness schemes based on a collaboration poolsize of 2 are more successful, the averaging version (CT2MG) especially so. Notably the CTDG scheme with memory and a collaboration poolsize of 1 is markedly less successful than the others on the SAT functions.
- Overall the CT2MG algorithm has the highest success rate.

11.6.2 Results: Efficiency

Figures 11.4 and 11.5 illustrate the change in the mean time to locate the optimum for the Trap, H-IFF and SAT functions used with different length instances. The results for the GA, SMAG and CRG are omitted from the first two for the sake of clarity as they are so poor. On the Trap functions the results with collaboration poolsize 1 (CTG, CTDG, CLG) are obtained faster than with the poolsize of 2 (CT2MG, CT2BG), the difference being increasingly statistically significant for the longer instances. This is a natural result of the overhead of testing each meme against two solutions - since the solution just takes the better of the two improvements to be the result of its Lamarkian learning, the other evaluations are "wasted" from that point of view. This explains the lower SR for CT2MG/CT2BG on longer 4-Trap problems. However on the H-IFF function the CTG approach is not only less successful than the CTDG approach, but takes more evaluations when it does find the optimum. This can be explained by the fact that the algorithm needs to make a decision between the all '1's solution and the all '0's solution, and the use of a memory helps make this decision consistent between generations. On the SAT problems, where there is no regular problem structure to be learnt and exploited, the CT2M/B G schemes again significantly take longer.

11.7 Varying the Population Sizes

The results in the previous section clearly demonstrate the advantages of a credit assignment mechanism that does not rely solely on the improvement caused when a meme is applied to a single solution. In general those schemes that make use of multiple collaborations (to use Wiegend's terminology) - either explicitly within the same generation, or via a memory - have higher success rates, but this is sometimes at the expense of significantly increased run-times. The memory-based approach (CTDG) is faster, but can be mislead as shown by the lower SR results for the SAT functions. We hypothesise that this is because the meme population is not converging in these runs, so the use of fitness inherited from both parents is more "noisy".

One obvious way to assess memes in the context of multiple solutions (points in the search space) without "wasting" evaluations is to reduce the size of the meme population μ_m relative to μ_s. To investigate this, a series of experiments were run using different size meme populations. After some brief initial experimentation, the following changes were made to the parameter settings, with the results shown in Table 11.3:

- The "#" character in a "action" string is taken to mean "invert the current value".
- The solution population size was increased to 500 and self-adaptive mutation was applied using the scheme outlined in [820, 823, 850].
- The tournament size in the meme population was increased from 2 to 5. This effectively reduces the size of the meme population since less fit memes have a smaller probability of being selected as parents.
- The fitness of each meme is assigned by summing improvement that meme caused in different solutions divided by the number of calls to the evaluation function used. However multiple copies of memes were allowed so this potentially provides a mix of what Schoenauer *et al.* have termed "extreme" and "average" value rewards in the context of adaptive operator selection in EAs [260].

The results of these experiments are presented in Table 11.3 and Figure 11.6, and can be summarised as follows:

- Overall the COMA algorithms are clearly more effective (higher SR) than the GA and SMA.
- Although not shown for reasons of clarity, the coevolutionary memetic algorithms are also overall more efficient (lower AES) than the GA or SMA.

Table 11.3. Success Rates of different functions as number of memes is varied

Algorithm	H-IFF 16	32	64	128	256	512	1024	Total	Trap 40	80	120	160	200	Total	Max-Sat 50	100	Total
CTA-pop-10	50	48	45	24	23	15	5	210	48	46	34	34	32	194	196	40	236
CTA-pop-50	50	50	49	46	39	31	302	50	47	49	48	44	238	229	49	278
CTA-pop-100	50	50	50	44	37	29	302	50	50	50	49	49	248	232	54	286
CTA-pop-200	50	50	50	46	42	37	29	309	50	50	50	50	50	250	257	62	319
CTA-pop-400	50	50	50	48	41	36	29	304	50	50	50	50	50	250	260	56	316
GA	50	33	2	0	0	0	0	85	30	3	0	0	0	33	100	15	115
CTG-pop-10	50	49	42	34	30	17	9	231	50	39	39	36	32	196	212	45	267
CTG-pop-50	50	50	47	47	41	34	33	302	50	50	47	45	45	237	224	49	273
CTG-pop-100	50	50	49	49	41	37	31	307	50	50	50	50	48	248	247	49	296
CTG-pop-200	50	50	50	46	44	38	33	311	50	50	50	50	50	250	247	61	308
CTG-pop-400	50	50	50	49	41	36	38	314	50	50	50	50	50	250	257	57	314
SMA-G	50	49	24	1	0	0	0	124	38	6	0	0	0	44	246	48	294

11 Self-adaptative and Coevolving Memetic Algorithms

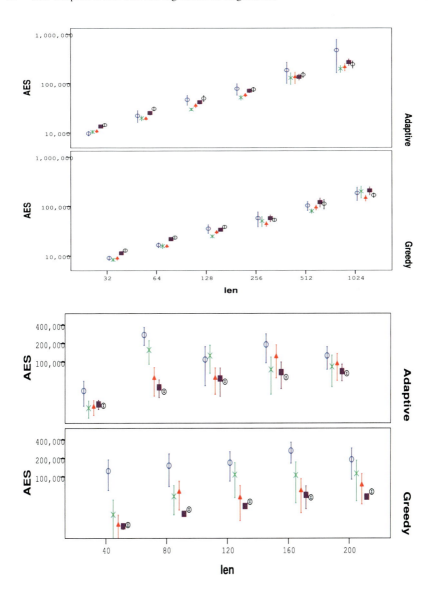

Fig. 11.6. Average Evaluations to Success on Trap (bottom) and H-IFF (top) functions as a function of length and number of memes. Error bars represent 95% conficence intervals for mean, grouping within each length is (l to r) 10,50,100,200,400 memes.

- On average there is little difference in effectiveness or efficiency between the fixed (CTG) and adaptive (CTA) pivot rules.
- Adapting the pivot rules creates more reliably efficient methods - the 95% confidence intervals for the AES are smaller for the CTA than for the corresponding CTG algorithms.

- The algorithms with low numbers of memes ($\mu_m \in \{10, 50\}$) are less effective. This may well arise from premature convergence or loss of diversity in the meme population, which could be ameliorated by reducing the selection pressure or increasing the mutation rate.
- The variation in efficiency reduces as the number of memes is increased - for similar reasons to the previous observation.
- The algorithms with 200 memes are the most effective (highest overall SR, especially on H-IFF and MAX-SAT) whilst not being significantly less effective than the algorithms with 400 memes (AES values not significantly different with 95& confidence).

Of particular interest is the relationship between the time taken to solve problems, and their length. As can be seen there appears to be a linear trend in Figure 11.6 - although the logarithmic scale should be noted. This is most evident for the H-IFF functions where a wider range of lengths is used. Using the SPSS tool to fit a curve to results for CTA, pooling the results for 200 and 400 memes reveals that a relationship of the form $AES = 233.3 \cdot len^1.018$ accounts for over 80% of the variation in solution times.

11.8 Conclusions

This chapter describes a conceptual framework within which self-adaptive and co-evolutionary memetic systems can be examined. Starting with systems which self-adapt the choice of which meme to use from a fixed set, and then moving through self-adaptation of the meme behaviours to a full co-evolutionary system, experimental results show progressively enhanced problem-solving behaviour using a variety of mechanisms.

The extension to co-evolution showed that the credit assignment mechanisms is critical, and selection within the meme population can be affected by noise arising from a number of sources. Mechanisms such as the use of multiple partners, or memory have been examined. The most promising appears to be a decoupling of the two populations with fewer memes than solutions.

Along the way the meme definitions have become progressively richer - permitting wildcards, inversion, and length adaptation in the pattern matching, and adapting the choice of pivot function. The stage is now well prepared for the use of richer definitions such as GP-like functions, which may be application specific as used elsewhere e.g. evolving MAX-SAT solvers using primitive elements derived from other heuristics [288, 461].

Chapter 12
Memetic Algorithms and Complete Techniques

Carlos Cotta, Antonio J. Fernández Leiva, and José E. Gallardo

12.1 Introduction

As mentioned in previous chapters in this volume, metaheuristics (and specifically MAs) have a part of their raison d'etre in practically solving problems whose resolution would be otherwise infeasible by means of other non-heuristic approaches. Such alternative non-heuristic approaches are complete methods that –unlike heuristics– do guarantee that the deviation from optimality of the solution they will provide is somehow bounded (and as a particular case, that the optimal solution will be found). These methods are eventually limited by the curse of dimensionality, yet they may still constitute a very interesting resource either from the application point of view, or from the lessons that can be learnt from them. Indeed, in some sense these approaches could be considered complementary to metaheuristics rather that mere "rivals". Even more so in the case of MAs, whose philosophy has been since its inception much more flexible and integrative rather than dogmatic or exclusive.

This said, despite the eclosion of metaheuristics as powerful optimization techniques during the 80s and 90s, inter-breeding between the fields of provably problem-solving and heuristic problem-solving was relatively limited until the last decade (some seminal works dating back from the mid 90s – e.g., [165]). The last years however have witnessed a remarkable increase in the number of works trying to combine ideas from these two areas. Certainly, MAs have also played an important role in this cross-fertilization of search paradigms. Along this chapter we will review some of the lines of research that have emerged in this regard. To this end, we will begin by briefly revisiting complete techniques to highlight their strengths and weaknesses, and what they have to offer to metaheuristics. Subsequently, we will outline some of the efforts that have been made in the literature to classify

---

Carlos Cotta · Antonio J. Fernández Leiva · José E. Gallardo
Dept. Lenguajes y Ciencias de la Computación, Universidad de Málaga,
ETSI Informática, Campus de Teatinos, 29071 Málaga, Spain
e-mail: {ccottap,afdez,pepeg}@lcc.uma.es

hybrid approaches. Although these classifications are usually general and intended to cover more than just complete-heuristic combinations, they will provide a framework within which actual combinations of MAs with exact techniques (or from the broad interpretation of memetic algorithms, MAs incorporating exact techniques) can be studied. This will be done in Sections 12.4 and 12.5.

12.2 Background

Complete techniques are those whose results can be proved to be at bounded distance from the optimum. From a very general point of view, these techniques can be further subdivided into techniques that guarantee finding the optimal solutions, i.e., exact techniques, and techniques that only provide a fixed or adjustable bound (that is, a bound that can be reduced by spending more computational effort), i.e., approximation techniques. Curiously, and this is something that may be worth some further analysis from a sociological and/or philosophical point of view, the community of researchers working on approximation theory has been traditionally more skeptical with respect to the value of metaheuristic optimization. Conversely, it is also true that the usefulness of approximation algorithms has not been always appreciated by the metaheuristic community, in part due to the inherent limitations of the former in many practical contexts – see for example [224] for a glimpse of the computational complexity of PTAS (a polynomial time approximation scheme, probably one of the jewels of the crown in approximation theory) for several common problems.

Focusing thus on exact techniques, such as for example branch and bound [507], dynamic programming [57], branch and cut [654], etc. these are characterized by the fact that they guarantee finding optimal solutions at the cost of a non-polynomial growth of computation time (and often memory consumption too). Their limitations are those emanating from the theory of computational complexity, such as the conspicuous P vs NP question. It must be noted however that such classical (unidimensional) hardness characterizations are not necessarily correlated with practical performance. A much more interesting characterization can be obtained from the field of parameterized complexity [221], in which hardness is approached from a multidimensional perspective, factoring out some *parameter(s)* from the input and trying to isolate the problem's difficulty in them. If this can be done –formally, if the complexity of the problem can be shown to be polynomially related to the input size (once the parameter is factored out), and the degree of the polynomial is unrelated to the value of the parameter– the problem is said to be fixed-parameter tractable (FPT). FPT problems can be solved for small values of the parameter using the arsenal developed by the parameterized-complexity community – e.g., [667]. Hard problems can be nevertheless detected from a parameterized perspective, and for such problems metaheuristics are fully in order.

There are many ways in which the hybridization of metaheuristics in general (and MAs in particular) with exact techniques can be fruitful: exact techniques can, for example, reduce their resource consumption if they obtain valuable input from metaheuristics (e.g., improved bounds); on the other hand, metaheuristics routinely

12 Memetic Algorithms and Complete Techniques

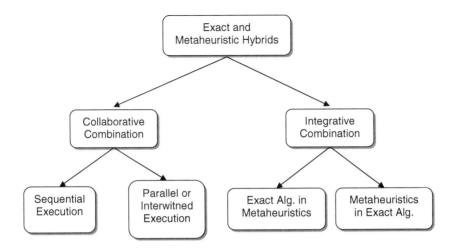

Fig. 12.1. Puchinger and Raidl's classification of exact-heuristic hybrid algorithms.

use search mechanisms —recombination, mutation, etc.— in which exact techniques can play an important role to intensify the search. Furthermore, hybridization of MAs and exact techniques can be defined at several nested levels thus providing multiple ways of boosting each other's performance. In the following we will survey some successful hybridization models reported in the literature along these lines just depicted. Subsequently, we will overview several approaches to classify these hybrid models.

12.3 Classification of Hybridization Approaches

Several taxonomical attempts have been proposed to classify hybrid optimization algorithms. For example, Talbi [872] proposed a mixed hierarchical-flat classification scheme. The hierarchical component captured the structure of the hybrid, whereas the flat component specified the features of the algorithms involved in the hybrid. More precisely, the hierarchical portion of the taxonomy firstly distinguished between low-level (a given function of a metaheuristic is replaced by another metaheuristic) and high-level (combined algorithms are self-contained) hybridization. Secondly, it was distinguished between relay hybridization (a set of metaheuristics is applied in a pipeline fashion) and teamwork hybridization (cooperative optimization models). Cotta [147] proposed another related taxonomy with the dichotomy *strong* vs. *weak* as its root. This distinction referred to whether problem-knowledge was placed in the core of the algorithm, affecting its internal components (e.g., representation and/or genotype-phenotype mapping, operators, etc.), or in the combination of different search algorithms that retained their identity. This terminology is consistent with the classification of problem-solving strategies in artificial intelligence as *strong* and *weak* methods [595].

A much more interesting classification for the purposes of this chapter is that proposed by Puchinger and Raidl [741]. This classification is specifically intended for exact-metaheuristic combinations, and establishes two main categories for such hybrid algorithms:

- *Collaborative combinations*, where an exact algorithm and a metaheuristic method exchange some information, but none of them are part of the other, and
- *Integrative combinations*, where one technique is a subordinate of the other, i.e., there is a *master* algorithm that uses the other one.

These two categories can be further refined depending on the particular of the combination as shown inf Figure 12.1. Thus, a collaborative combination can be sequential or parallel/intertwined, depending on how the control flow passes from one algorithm to the other. Similarly, an integrative combination can be subdivided in models in which an exact technique plays the role of master (i.e., the metaheuristic is embedded in an exact technique), and models in which the opposite is true.

As mentioned before, this latter classification fits nicely into context of this chapter, so we will consider it in order to survey existing hybrid approaches combining MAs and exact techniques.

12.4 Integrative Combinations

One basic form of integrative collaboration consists of endowing a memetic algorithm with an exact technique (ET) so that this ET is a subordinate of the MA. The most common implementation consists of combining an EA with a procedure to perform a complete local search (which can consider the whole neighborhood and in this sense can be viewed as an exact technique). This is usually done after evaluation, although it must be noted however that the integration does not simply reduce itself to this particular scheme. In fact, the purpose of using an ET inside a MA is to provide specific knowledge that can help to a better optimization process. For instance, Algorithm 22 shows a general picture of where an ET can be incorporated inside an MA.

As it can be seen, during the initialization of the population some complete method may be used to generate high quality initial solutions. Of course, this complete method may only consider a subset of the search space, a relaxed version of the problem, or may perform just a truncated search, since otherwise the problem would just be solved at that stage (not to mention the computational cost). An example of relaxed initialization using complete techniques can be found in [148], where a backtracking algorithm is used to create feasible initial solutions for a protein structure prediction problem (thus relaxing optimality to mere feasibility). Another related approach will be discussed in next subsection in the context of collaborative models, and considers a variant of a B&B algorithm –namely beam search– to initialize the population of a MA with the aim of improving its performance. In addition to this an exhaustive LS could be applied to improve the individuals generated initially with the aim of providing a first population of better quality. This procedure

Algorithm 22. Pseudocode of a basic MA based on a integrative collaboration with an exact technique ET

```
1  for i ∈ {1,..., POPULATION SIZE} do
2      pop[i] ← RANDOM-SOLUTION();
3      if Rand[0,1] < p_ET then // ET is applied with probability p_ET
4          EXACT-TECHNIQUE (pop[i]); // Usually ET = Local
                Improvement
5      endif
6  endfor
7  i ← 0;
8  while i < MaxEvals do
9      RANK-POPULATION (pop); // sort population according to
                fitness
10     parent_1 ← SELECT (pop);
11     if Rand[0,1] < p_X then // recombination is done
12         parent_2 ← SELECT (pop);
13         child ← RECOMBINE (parent_1, parent_2); // RECOMBINE might be an
                Exact Technique
14     else
15         child ← parent_1;
16     endif
17     child ← MUTATE (child, p_M); // p_M is the mutation probability
                per gene
18     if Rand[0,1] < p'_ET then // ET is applied
19         EXACT-TECHNIQUE (child); // Usually Local Improvement
                applied here
20     endif
21     pop[μ] ← child; // replace worst
22 endw
23 return best solution in pop;
```

is intimately related to the idea of local branching by Fischetti and Lodi [263], and to Congram's Dynasearch [138, 139].

Another proposal that can be devised from the general schema shown above is the use of an ET as a recombination operator. Recombination or mutation operators can be intelligently designed so that specific problem knowledge is used in order to improve the offspring. For instance, Cotta *et al.* [165] used a problem-specific B&B approach for the Travelling Salesman Problem based on 1-trees and the Lagrangean relaxation [910], to build a hybrid recombination operator. More precisely, the B&B was used in order to build the best possible tour within the (Hamiltonian) subgraph defined by the union of edges in the parents. This recombination procedure was costly, but provided better results than blind edge recombination. This model was later extended to a more general operator termed dynastically optimal recombination (DOR) [164]. The term refers to the *dynastic potential*, which in the framework of Forma Analysis [750] denotes the set of children attainable from a certain set of parents. DOR thus consists of finding the best children in this dynastic potential, i.e., that with the best combination of parental features. This is done by "intelligently" exploring this set, using an adequate complete algorithm, check e.g. [163].

Related to the previous approach, [295] presented a memetic algorithm, embedded with tabu search, for weighted constraint satisfaction problems (see next section for a more detailed discussion of this kind of problems) in which bucket elimination (BE) [202] is used as a mechanism for recombining solutions, providing the best possible child from the parental set. BE is an exact technique related to dynamic programming which based on variable elimination and is commonly used for solving constraint satisfaction problems. This algorithm, with another collaborative proposals, was applied to the resolution of the maximum density still life problem, a hard constraint optimization problem based on Conway's game of life.

Additionally, problem knowledge can be incorporated in the genotype to phenotype mapping present in many MAs, like when repairing an infeasible solution. This technique is used, for instance, in the MA designed by Chu and Beasley [127] for the multidimensional 0-1 knapsack problem. The use of complete techniques, again relaxed or truncated, can be here considered as well, check, e.g., [148] for an example of using backtracking to repair infeasible solutions.

Another place where an exact method can be particularly useful when used inside an MA is in the optimization of problems where different representations are considered. In these cases, an exact technique can be specifically useful in the codification-decodification phase. For instance, Puchinger and Raidl [742] represented another attempt to incorporate exact methods in metaheuristics. This work considered different heuristics algorithms for a real world glass cutting problem and a combined GA and B&B approach was proposed. The GA used an order-based representation that was decoded with a greedy heuristic. Incorporating B&B in the decoding for occasionally (with a certain probability) locally optimizing subpatterns turned out to increase the solution quality in a few cases.

Note also that applying always the ET in each generation of the MA (or initially on each individual in the initial population) is not always the best option (as shown in [858] for the application of LS on each generated new individual). For instance, if one considers LS as the technique to embed inside a MA, partial Lamarckianism [396], namely applying local search only to a fraction of individuals, can result in better performance. These individuals to which local search will be applied can be selected in many different ways [665]. Thus, LS can be applied to improve the individual with certain probability p_{LS}; in case of application, the improvement uses up a number of *LSevals* evaluations (or in the case of specific local search such as HC, until it stagnates, whatever comes first). It is easy to extrapolate these results from the use of LS to any ET embedded in a MA.

In general, the underlying idea of this kind of integration is to combine the intensifying capabilities of the embedded ET method, with the diversifying features of MA, i.e., the population will spread over the search space providing starting points for a deeper (probably local) exploration. As generations go by, promising regions will start to be spotted, and the search will concentrate on them. Ideally, this combination should be synergistic, providing better results that either the MA or the ET by themselves. Regarding this issue, one can find in the literature a number of proposals that explore the intensification/diversification balance within the memetic algorithm. Some works lean towards a more explorative combination, by using a blind recombination operator in the MA whereas other models incorporate an intense exploration of the dynastic potential of the solutions being recombined.

The other possibility for integrative combinations is to incorporate a metaheuristics into an exact algorithm. One example is the hybrid algorithm combining Genetic Algorithms and Integer Programming B&B approaches to solve MAX-SAT problems described in [287]. This hybrid algorithm gathered information during the run of a linear programming-based B&B algorithm, and used it to build the population of an EA population. The EA was eventually activated, and the best solution found was used to inject new nodes in the B&B search tree. The hybrid algorithm was run until the search tree was exhausted, and hence it is an exact approach. However, in some cases it expands more nodes than the B&B algorithm alone.

12.5 Collaborative Combinations

As mentioned in Section 12.3, the class of *collaborative combinations* includes hybrid algorithms which exchange information, but such that none of them is a subordinate of the other. Two subcases can be here considered in order to execute both algorithms:

- *Sequential execution*, in which one of the algorithms is completely executed before the other. Examples of this group are those in which one of the techniques can act as a kind of preprocessing for the other or those where the result of one algorithm can be used as data to initialize the other.
- *Parallel or intertwined execution*, where both techniques are executed simultaneously, either in parallel (i.e., running at the same time on different processors) or in an intertwined way by alternating between both algorithms.

As an example of sequential combinations we can cite the work of Klau et al. [469], in which a branch and cut algorithm is used analogously to the idea of dynastically optimal recombination mentioned in previous section, to combine the final population provided by a MA. This hybrid algorithm is applied to the prize-collecting Steiner tree problem. We will focus here on hybrid collaborative techniques in the second group. MAs and B&B techniques can be integrated by way of a *direct collaboration*, so that both techniques work alone in parallel (i.e., both processes perform independently) at the same level. Under this scheme, both processes will share the incumbent solution to the problem being solved. Whenever one of the algorithms finds a better approximation, it can update the solution. Two straightforward ways of obtaining a benefit of this parallel execution are [165]:

- The B&B algorithm can use the lower bound provided by the MA to purge its problem queue. Problems whose upper bound are smaller than the one obtained by the MA cannot improve the incumbent solution and can be safety removed.

- The B&B algorithm can provide information about more promising regions of the search space into the MA population. The aim of this process is to to guide the MA search towards these promising regions of the search space.

Several implementations of these schemes are possible. For example, Cotta *et al.* [165] proposed a collaborative approach such as the one described above for the TSP. Puchinger and Raidl [740] consider the parallel combination of a MA and a branch and cut algorithm for the multidimensional knapsack problem. The MA provides improved bounds to the branch and cut algorithm, and the latter provides both new best-so-far solutions and the corresponding dual variable values, to be used for repairing and local search. More recently, Gallardo *et al.* [294] defined a hybrid algorithm that starts by running a MA (with a randomly initialized population) in isolation, so that a first approximation to the solution is obtained. This initial solution is later used by a B&B algorithm to purge its problem queue. As it can be seen, no information from the B&B algorithm was used in this first execution of the MA. In a subsequent phase, the B&B algorithm starts its execution. New solutions found by the B&B are incorporated into the MA population (by replacing the worst individual). Whenever a new solution is found, the B&B phase is paused and the MA is run to stabilization. In addition, pending nodes in the B&B queue are incorporated into the MA population periodically. The intention of this transfer is to direct the MA to these regions of the search space, that represent the subset of the search space still unexplored by the Branch and Bound. In this way, the MA is used for finding probably good solutions in those regions. Upon finding an improved lower bound (or upon stabilization of the MA if no improvement is found), B&B is resumed. This process is repeated until the search tree is exhausted, or a time limit is reached. One interesting property of this hybrid algorithm is that it acts as an anytime algorithm, providing both a quasi-optimal solution, and an indication of the maximum distance to the optimum. In [293, 294] this implementation schema is used to tackle large instances of the multidimensional knapsack problem. Experimental results showed that the hybrid approach can provide high quality results, better than those obtained by the MA and B&B on their own.

An alternative implementation of the previous model consist on using beam search (BS)[46] instead of B&B. This is an incomplete derivative of the later and acts thus as an heuristic method. In essence, BS extends every partial solution from a set $\mathscr{B}$ (called the *beam*) in at most k_{ext} possible ways, generating a new beam. When all solutions in $\mathscr{B}$ have been extended, the algorithm reduces the new beam by selecting the best up to k_{bw} (called the *beam width*) solutions and proceeds. A very interesting feature of this heuristic is that it extends in parallel a set of different partial solutions in several possible ways. For this reason, it can be used to provide periodically diverse promising partial solutions to a population based search method such as a MA. A general description of the resulting hybrid algorithm is given in Algorithm 23.

The beam search part of the algorithm can be iterated for each level of the search tree that corresponds to the problem at hand. The hybrid algorithm starts by executing this process for an initial number of levels (parameter l_0 of the algorithm). Subsequently, both parts of the hybrid algorithm are alternatively executed until a

Algorithm 23. Beam Search + MA hybrid algorithm

1 **for** l_0 *levels* **do** run BS;
2 **repeat**
3   select *popsize* nodes from problem queue;
4   initialize MA population with selected nodes;
5   run MA;
6   **if** *MA solution better than BS solution* **then**
7 | let BS solution ← MA solution;
8   **endif**
9   **for** *l levels* **do** run BS ;
10 **until** *timeout or tree-exhausted* ;
11 **return** *BS solution*;

termination condition is reached. Similar to the first implementation, for every execution of the MA, its population is initialized using the nodes in the BS queue. As the size of the BS queue is usually larger than the MA population size, a criteria, such as selecting the best nodes according to some measure of quality or selecting a subset that provides high diversity, has to be used in order to select a subset from the queue. Nodes in the BS queue represent partial solutions in which some genes are fixed but others are indeterminate, so they must first be converted to full solutions in a problem dependent way. This must be considered when instantiating the general template for different combinatorial problems. This kind of collaborative integration of Beam Search and MAs has been used to tackle different combinatorial optimization problems. In [297] the hybrid algorithm was experimentally evaluated on the multidimensional 0-1 knapsack problem and on the shortest common supersequence problem, a NP-hard classical problem from the realm of string analysis. For both problems, it was shown the benefits of using the hybrid approach when compared to the constituents algorithms. Additionally, an analysis of the dynamics and sensitivity on different parameters of the algorithm was carried out. In [298], the hybrid algorithm was applied to the inference of phylogenetic trees, an important problem in Systematic Biology, that aims to represent the evolutionary history for a collection of organisms. That work focused in the ultrametric model for phylogenetic inference. A robust setting for the different parameters of the algorithm was determined, and the hybrid algorithm was experimentally shown to also be synergetic for this problem.

A related hybridization model has been defined in [299] for weighted constraint satisfaction problems (WCSP) [795]. A WCSP is a constraint satisfaction problem (CSP) in which preferences among solutions can be expressed. Formally, a WCSP can be defined by a tuple $(\mathscr{X}, \mathscr{D}, \mathscr{F})$, where $\mathscr{D} = \{D_1, \cdots, D_n\}$ is a set of *finite domains*, $\mathscr{X} = \{x_1, \cdots, x_n\}$ is a set of variables taking values from their finite domains and $\mathscr{F}$ is a set of *cost functions* (also called *soft constraints* or *weighted constraints*) used to declare preferences among possible solutions. Each $f \in \mathscr{F}$ is defined over a subset of variables, $var(f) \subseteq \mathscr{X}$, called its *scope*. The objective function F –to be minimized– is defined as the sum of all functions in $\mathscr{F}$, i.e., $F = \sum_{f \in \mathscr{F}} f$. WCSP were tackled using a algorithmic model based on the hybridization of MAs with

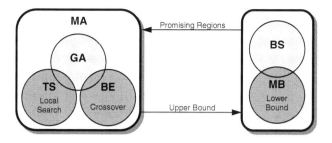

Fig. 12.2. Schematic description of the multilevel hybrid algorithm.

exact techniques at two levels: within the MA (as an embedded operator), and outside it (in a cooperative model). Figure 12.2 depicts the different components of the algorithm and their relationships. The first level of hybridization has already been described in Section 12.4, so we will describe here the second level, in which the MA cooperates with a beam search algorithm that further uses the technique of mini-buckets as a lower bound.

Algorithm 24. Hybrid algorithm for a WCSP
---
1 $sol \leftarrow \infty$;
2 $\mathscr{B} \leftarrow \{\,()\,\}$;
3 **for** $i \leftarrow 1$ **to** n **do**
4 $\quad \mathscr{B}' \leftarrow \{\}$;
5 $\quad$ **for** $s \in \mathscr{B}$ **do**
6 $\quad\quad$ **for** $a \in D_i$ **do**
7 $\quad\quad\quad \mathscr{B}' \leftarrow \mathscr{B}' \cup \{s \cdot (x_i = a)\}$;
8 $\quad\quad$ **endfor**
9 $\quad$ **endfor**
10 $\quad \mathscr{B} \leftarrow$ **select** best k_{bw} nodes from $\mathscr{B}'$;
11 $\quad$ **if** $i \geqslant k_{MA}$ **then**
12 $\quad\quad$ **initialize** MA population with best *popsize* nodes from $\mathscr{B}'$;
13 $\quad\quad$ **run** MA;
14 $\quad\quad sol \leftarrow$ **min** $(sol, \text{MA solution})$;
15 $\quad$ **endif**
16 **endfor**
17 **return** sol;

The proposed hybrid algorithm, that executes BS and the MA in an interleaved way, is depicted in Algorithm 24. Here, a (possibly partial) solution for a WCSP instance is represented by a vector of variables $s = (x_1, x_2, \ldots, x_i)$, $i \leqslant n$, where $s \cdot (x_i = a)$ stands for the extension of partial solution s by assigning value a to its i-th variable. The hybrid algorithm proceeds by constructing a search tree, so that its leaves are complete solutions to the problem and internal nodes at level i represent solutions that are partially specified up to the i-th variable. The algorithm traverses this tree heuristically in a breadth first way using a BS algorithm that only maintains the best k_{bw} nodes at each level of the tree. During each iteration of BS (lines 5-16), a variable is assigned for every solution in the beam (line 8). The interleaved execution of the MA starts only when partial solutions in the beam have at least k_{MA} variables (line 12). For each iteration of BS, the best *popsize* solutions in the beam are selected with the purpose of initializing the population of the MA (line 13). The solution provided after the execution of the MA is used to update the incumbent solution (*sol*), and this process is iterated until the search tree is exhausted.

The performance of this algorithm will depend on the quality of the heuristic function used to estimate partial solutions (line 11). In order to compute tight, yet computationally inexpensive, lower bounds for the remaining part of the solutions, the technique of mini-buckets (MB) can be used. As described by Kask and Dechter[445], the intermediate functions created by applying the MB scheme can be used as a general mechanism to compute heuristic functions that estimate the best cost of yet unassigned variables in partial solutions. This can be achieved by running MB as a preprocessing stage. The set of augmented buckets computed during this process can be used as estimations of the best cost extension to partial solutions (check [445] for details).

In [296, 299], such a multilevel algorithm was used to tackle the Maximum Density Still Life Problem, a hard constrained problem defined in the context of John Conway's game of life. The resulting algorithm was able to find optimal solutions for currently solved instances of the problem in considerable less time that state-of-the art approaches. Additionally, it was able to find new best known solutions for very large instances whose exact solutions are yet unknown.

12.6 Conclusions

Throughout this chapter we have surveyed existing work on MAs that incorporate at some level a complete technique. Several notes have to be done here. Notice firstly from a 'terminological' point of view that many evolutionary techniques hybridized with complete techniques can be considered memetic regardless of whether a classical trajectory-based local search algorithm is also used or not. For example, in an evolutionary algorithm that used an exact technique for recombination, the latter could be regarded as a generalized local-search operator working on set of solutions rather than on single solutions, and using a neighborhood composed of all solutions in the corresponding dynastic potential. This is also related to the so-called crossover hill-climbing idea defined in [536] for continuous optimization.

From a practical point of view, this kind of hybrid approaches must carefully control the computational complexity of the problems submitted to complete search. This draws again a connection to parameterized complexity by noting that this complexity is typically related to some structural parameter of the problem (e.g., a higher similarity of the parents during exact recombination reduces the size of the dynastic potential, thus making its exploration more amenable in principle). Even though no efficient (in the FPT sense) algorithmic resolution were available for the problem at hand, the combinatorial explosion could be kept within acceptable levels by checking these parameters and resorting to other approaches (truncated exact search, fast heuristic search, or even a blind procedure) if the possibility of a prohibitive computational cost cannot be excluded prior to a certain invocation of the complete method.

The any-time nature of MAs has to be considered as well. Whether a hybrid approach including complete techniques is itself complete or not, it is very important that it provides better and better solutions for any increasing computational budget allowed. This is not always possible within the context of complete techniques, e.g., a B&B algorithm using a best-first policy may exhaust its allotted time and/or memory without producing a single feasible solution. On the other hand, the very same B&B algorithm using a LIFO policy may quickly provide a solution but take a long time to improve it. MAs are however ideal for anytime search, and this can be exploited in a synergistic combination. Note for example that a parallel collaborative model using a MA and an exact technique may end up providing both an upper and a lower bound for the optimal solution, and the higher the computational budget available, the tighter these bounds will be.

Acknowledgements. This work is supported by Spanish MICINN under project NEMESIS (TIN2008-05941) and Junta de Andalucía under project TIC-6083.

Chapter 13
Multiobjective Memetic Algorithms

Andrzej Jaszkiewicz, Hisao Ishibuchi, and Qingfu Zhang

13.1 Introduction

Multiple conflicting points of view, which are often taken into account in real life applications, naturally result in a multiple objective optimization problem (MOP) [848]. In order to find the best compromise solution of a MOP, or a good approximation of it, Multiobjective Optimization (MOO) methods need some preference information from a decision maker. According to when and how the preference information is used in the solution procedure, MOO methods can be classified as either methods with a priori, a posteriori, or progressive (interactive) articulation of preferences [400].

In recent years, the demand for new applications and the increasing computing power have led to growing interest in computationally hard multiobjective problems, e.g. nonlinear or combinatorial optimization problems. These problems arise in many areas such as scheduling, timetabling, production facilities design, vehicle routing, telecommunication routing, investment planning and location. Problems of this kind are difficult to solve even in the single objective case. Encouraged by the success of metaheuristics in single-objective optimization (see e.g. [694]), much effort has been made in developing MOO metaheuristics.

Andrzej Jaszkiewicz
Poznan University of Technology, Institute of Computing Science, Piotrowo 2,
60-965 Poznan, Poland
e-mail: jaszkiewicz@cs.put.poznan.pl

Hisao Ishibuchi
Department of Computer Science and Intelligent Systems, Osaka Prefecture University,
1-1 Gakuen-cho, Nakaku, Sakai, Osaka 599-8531, Japan
e-mail: hisaoi@cs.osakafu-u.ac.jp

Qingfu Zhang
The School of Computer Science & Electronic Engineering University of Essex,
Colchester, CO4 3SQ, UK
e-mail: qzhang@essex.ac.uk

Traditional MOO methods usually assume that an underlying single-objective exact solver is available. This solver is used to solve a series of substitute single-objective optimization problems sequentially. The objective functions of these substitute optimization problems could be an aggregation function of the individual objectives of the MOP in question. Their optimal solution can be Pareto optimal solutions of the MOP under some conditions. However, for many hard MOPs, no efficient exact solvers are available. One can use a single-objective metaheuristic instead of an exact solver. A single run of the metaheuristic can generate a single approximate Pareto-optimal solution and therefore many runs are required to generate multiple solutions. This approach is simple but not very efficient.

Many dedicated multiobjective metaheuristics have recently been developed [131, 196, 418]. These methods aim at generating in a single run a set of solutions for approximating the whole Pareto optimal front. The set of solutions could be then presented to the decision maker (DM) to allow it to choose the best compromise solution in a posteriori or interactive manner [422].

A multiobjective metaheuristic is often a modified version of a single objective heuristic method. It is natural to expect that the best results may be achieved by adapting the most efficient single objective methods to the multiobjective problems. Memetic algorithms proved to be one of the most efficient metaheuristic paradigms for single objective optimization [618]. For this reason, many attempts have been made to extend memetic algorithms to multiobjective optimization.

The purpose of this review is to present and discuss basic concepts in multiobjective memetic algorithms and to characterize some state-of-the-art algorithms. In the next section, we introduce some basic definitions in MOO. In the third section, we discuss the main ideas in multiobjective memetic algorithms. The fourth section presents several typical multiobjective memetic algorithms. Some specific implementation issues are discussed in the fifth section. In the last section we discuss some further research topics in this area.

13.2 Basic Definition and Concepts

In this section, we introduce some basic concepts and aggregation functions in multiobjective optimization.

13.2.1 Basic Concepts

A *multiobjective optimization problem* (MOP) can be stated as follows:

$$\text{maximize } F(x) = (f_1(x), \ldots, f_m(x)) \tag{13.1}$$
$$\text{subject to} \quad x \in \Omega$$

where Ω is the *decision space*, $F : \Omega \to R^m$ consists of m real-valued objective functions. R^m is called the *objective space*. The *attainable objective set* is defined as the set $\{F(x)|x \in \Omega\}$.

13 Multiobjective Memetic Algorithms

Ω can be a subset of a base set S and often be described by several constraints $C_1,\ldots,C_k$. i.e.

$$\Omega = \{x \in S | x \text{ satisifies all the constraints } C_1,\ldots,C_k\}. \tag{13.2}$$

In this case, Ω is called the feasible solution space and any solution in Ω is a feasible (candidate) solution. A solution in S is infeasible if it is not in Ω, in other words, it violates at least one constraint.

If $x \in R^n$, all the objectives are continuous and Ω is described by

$$\Omega = \{x \in R^n | h_j(x) \leq 0, j = 1,\ldots,k\}, \tag{13.3}$$

where h_j are continuous functions, we call (1) a *continuous MOP*. If Ω is a finite or countably infinite set, then (1) is a *combinatorial MOP*.

Domination is widely used to compare different solutions in multiobjective optimization.

Definition 13.1. Let $u, v \in R^m$, u is said to *dominate* v if and only if $u_i \geq v_i$ for every $i \in \{1,\ldots,m\}$ and $u_j > v_j$ for at least one index $j \in \{1,\ldots,m\}$[1].

Domination defines a strict partial ordering in the objective space- not any two vectors are comparable based on domination. For example, $(1,0)$ and $(0,1)$ do not dominate each other.

Definition 13.2. Let $x, y \in \Omega$, x is said to *dominate* y if and only if $F(x)$ dominates $F(y)$.

Obviously, a rational decision maker prefers x to y if x dominates y.

Definition 13.3. $x^* \in \Omega$ is a *Pareto optimal solution* and $F(x^*)$ is a *Pareto optimal vector* to (13.1) if no other solution in Ω can dominate x^*. The set of all the Pareto optimal solutions is called the Pareto optimal set (PS) and the set of all the Pareto optimal vectors is the Pareto front (PF).

We should point out that the above definition refers to the global optimality. x is called locally Pareto optimal if it cannot be dominated by any solutions in a neighborhood of x.

In many real-life applications of multiobjective optimization, an approximation to the PF is required by a decision maker for selecting the final preferred solution. Most MOPs may have many or even infinite Pareto optimal vectors. It is very time-consuming, if not impossible, to obtain the complete PF. On the other hand, the decision maker may not be interested to have an unduly huge number of Pareto optimal vectors to deal with due to overflow of information. Therefore, many multi-objective optimization algorithms are to find a manageable number of approximate Pareto optimal solutions to approximate the Pareto set or Pareto front. Researchers and practitioners are often more interested in approximating the Pareto front than

---

[1] This definition of domination is for maximization. All the inequalities should be reversed if the goal is to minimize the objectives in (13.1).

the Pareto set because the objective space is of lower dimension and it is easy to visualize an approximate Pareto front. However, recent work has shown that approximation of the Pareto set is also very important.

Definition 13.4. Given a set of solutions P, $x \in P$ is called a nondominated solution in P if no solution in P can dominate x.

Many multiobjective metaheurstics are based on Pareto dominance. These methods often select nondominated solutions from a set of solutions.

Definition 13.5. $z^{id} = (z_1, \ldots, z_m)$ is called the ideal objective vector if z_i is the maximal function value of $f_i(x)$ over Ω.

Definition 13.6. $z^{nadir} = (z_1, \ldots, z_m)$ is call the nadir objective vector if

$$z_i = inf\{y_i | (y_1, \ldots, y_m) \in PF\} \tag{13.4}$$

Ideal objective vectors and nadir vectors in the objective space are the upper and lower bounds of the PF, which are of interest since they are useful for determining the range of the Pareto front and for normalizing the objectives so that all the objectives in the same range. A typical normalization is:

$$f_i(x) \leftarrow \frac{f_i(x) - z_i^{nad}}{z_i - z_i^{nad}} \tag{13.5}$$

It might be not practical to obtain the exact ideal and nadir vectors, one can substitute them by approximate ones.

13.2.2 Aggregation Functions

In traditional optimization, a widely-used strategy for dealing with a MOP is to aggregate all the individual objective functions and then optimize the aggregation function. In the following, we introduce three popular aggregation approaches.

13.2.3 Weighted Sum Approach

This approach considers a convex combination of the different objectives. Let $\lambda = (\lambda_1, \ldots, \lambda_m)^T$ be a weight vector, i. e., $\lambda_i \geqslant 0$ for all $i = 1, \ldots, m$ and $\sum_{i=1}^{m} \lambda_i = 1$. Then the aggregated function is

$$g^{ws}(x|\lambda) = \sum_{i=1}^{m} \lambda_i f_i(x). \tag{13.6}$$

where we use $g^{ws}(x|\lambda)$ to emphasize that λ is a coefficient vector in this objective function while x is the variables to be optimized. A maximal solution of a weighted sum function is Pareto optimal to (13.1) if all the weights are positive. Moreover, for any Pareto optimal solution x^* to a convex MOP, there exists a weight vector such that x^* is the maximal solution to (13.6). However, for a non-convex MOP, there may exist a Pareto optimal solution which is not a maximal solution to any weighted sum function.

13.2.4 Tchebycheff Approach

In this approach, the aggregation function to be minimized is in the form

$$g^{te}(x|\lambda, z^*) = \max_{1 \leqslant i \leqslant m} \{\lambda_i(-z_i^* - f_i(x))\} \tag{13.7}$$

where z^* is the ideal point or a point dominated by the ideal point. Each Tchebycheff aggregation function has at least one global minimum which is Pareto optimal to (13.1). Under some mild conditions for each Pareto optimal point x^*, there exists a weight vector λ such that x^* is an optimal solution of (13.7). Therefore, one is able to obtain different Pareto optimal solutions by altering the weight vector.

One weakness with this approach is that its aggregated function could be flat in some regions. To overcome it, the following aggregated function can be used:

$$g^{te}(x|\lambda, z^*) + \rho g^{ws}(x|\lambda) \tag{13.8}$$

Aggregation methods are still a very active research topic in traditional optimization. The readers interested in more detail about aggregation methods may wish to consult [233, 241, 599].

13.3 Adaptation of Memetic Algorithms for Multiobjective Optimization – Basic Concepts

Memetic algorithms have to evaluate or compare a set of solutions at each generation for determining their contribution to the next generation. In the single objective case, several different evaluation functions or mechanisms have been used and studied in solution evaluation, the objective function itself, however, is the most natural and widely used evaluation function [618]. In multiobjective optimization, no such a natural choice for the evaluation exists. The evaluation mechanism is one of the major issues in the design of multiobjective memetic algorithms. A good evaluation mechanism should guide the solutions generated to

- approach the Pareto front,
- and at the same time disperse over all (or some desired) regions of the Pareto front.

Two main classes of evaluation mechanisms have been proposed for multiobjective memetic algorithms, i.e., mechanisms based on the dominance relation and mechanisms based on aggregation functions. Of course, these two mechanisms could be hybridized together. Below we discuss these two evaluation mechanisms.

13.3.1 Dominance-Based Evaluation Mechanisms

As pointed out in Section 2, the dominance relation defines a partial order in the set of all feasible solutions. All the Pareto-optimal solutions are the best with respect to this order. Therefore, the use of dominance relation in evaluation mechanisms creates a selection pressure towards the Pareto front. Dominance relation alone leaves, however, many pairs of solutions incomparable. For this reason, dominance relation on its own may not be able to define a single best solution in a neighborhood or in a tournament. Thus, multiobjective memetic algorithms need additional evaluation mechanisms with dominance relation to distinguish different solutions.

Probably, the most popular dominance based evaluation mechanism is Pareto ranking, which was originally suggested by Goldberg [325] and has been widely used in multiobjective evolutionary algorithms (see e.g.[131, 196]). In this mechanism, the dominance relation is used to rank all the solutions in the current population. Different algorithms may use slightly different versions of Pareto ranking. Srinivas and Deb [844] used the most direct implementation of the Goldberg's idea in their Nondominated Sorting Genetic Algorithm (NSGA) [199]. It assigns rank 1 to all solutions nondominated in the current population. Then, the nondominated solutions are temporarily removed from the population and the next rank is assigned to the solutions nondominated in the remaining part of the population. The process is continued until all solutions in the population are ranked.

Dominance relation may also be used to guide local search-based memetic algorithms. For example, Knowles and Corne [472, 473] proposed a greedy local search method mainly based on dominance relation. Their idea is to accept a new neighborhood solution if it dominates the current solution. In population-based Pareto local search [21, 50, 705], the neighborhood of each solution of the current population is explored, and if no solution of the population weakly dominates a generated neighbor, the neighbor is added to the population.

An obvious advantage of dominance relation is its independence on any monotonic transformation of objective functions. Furthermore, particular dominance-based evaluation mechanisms are usually very simple and have no or few parameters. For example, Pareto local search is a fully parameter-free method.

Dominance-based evaluation mechanisms may have, however, some significant disadvantages. Although dominance relation assures the pressure towards the Pareto front, it does not necessarily assure the dispersion of the solutions over all regions of the Pareto front. Thus, the basic method often needs to be extended by introducing some additional dispersion mechanisms. For example, several researchers [276, 472, 844] suggested the use of fitness sharing to improve Pareto ranking. The idea is to penalize a solution if it is too close, either in the objective or in the decision space, to some other solutions in the current population. Note that many fitness sharing techniques use some kind of distance measures in the objective space. Hence, the techniques are not invariant of scaling and more general of monotonic transformation of objective functions.

Another disadvantage of dominance-based evaluation mechanisms is that the selection pressure decreases with the growing number of objectives. The larger the number of objectives is, the lower the chance that one of two solutions dominates the other. In particular, a population of a multiobjective memetic algorithm may easily contain mainly or only mutually non-dominated solutions which are incomparable based on the dominance relation.

Increasing the number of objectives may also deteriorate the efficiency of algorithms based purely on the dominance relation. For example, in the case of Pareto local search, the number of neighborhood solutions to be accepted may become very large, and the size of the population to be maintained may grow enormously.

Furthermore, efficient single-objective local search algorithms usually use a number of advanced, problem-specific speed-up techniques based on the properties of the objective function. Such techniques often cannot be directly adapted to dominance-based mechanisms. There is still very little work that applies speed-up techniques in local search algorithms based on dominance relations [542].

13.3.2 Aggregation Function-Based Evaluation Mechanisms

Evaluation of new solutions with the use of Aggregation functions is another typical evaluation mechanism. Aggregation functions have well-established theoretical properties as tools for generating Pareto-optimal solutions in traditional MOO (see section 2). Thanks to these properties of Aggregation functions, their use also induces a pressure towards the Pareto front. Of course, a single Aggregation function would guide a metaheuristic towards a single Pareto solution. This drawback could be, however, overcome by the use of multiple Aggregation functions defined by various weight vectors. For example, Serafini [805] used the mechanism of random walk to modify the weights randomly in each iteration. Ulungu et al. [897] and Zhang and Li [957] used a predefined set of well dispersed weight vectors. Czyzak and Jaszkiewicz [696] and Hansen [358] modified the weights deterministically in each iteration in order to obtain a form of repulsion between a population of solutions, Hajela and Lin [350] allowed the weights to evolve during the search. Ishibuchi and Murata [409] and Jaszkiewicz [419] generated weight vectors randomly in each iteration.

An important advantage of Aggregation functions-based evaluation mechanisms is the fact that by the use of various weight vectors they naturally assure dispersion of the search over all regions of the Pareto front. Thus, no additional dispersion mechanisms like the fitness sharing are needed. Another advantage of such mechanisms is that many speed-up techniques may easily be used in local search based on Aggregation functions.

A disadvantage of Aggregation functions-based evaluation mechanisms is their dependence on monotonic transformations of objective functions. A simple change of units in one objective may significantly deteriorate the algorithm performance. It is thus very important to assure that all the objectives take their values in comparable ranges. It may be achieved with the use of normalized objective values (see section 2). Some methods, e.g. Jaszkiewicz's MOGLS [420], perform automatic scaling of objectives.

According to the properties mentioned in section 2, weighted Tchebycheff and composite Aggregation functions have the advantage over linear Aggregation functions of being able to generate all Pareto optimal solutions. Note, however, that the properties concern only optimal solutions of the Aggregation functions. A suboptimal solution of a linear Aggregation function found by a metaheuristic may appear to be a nonsupported Pareto optimal solution. Some experiments indicated that the use of linear Aggregation functions may yield better results for some particular problems (see e.g. [359, 419]). Nevertheless, weighted Tchebycheff or composite Aggregation functions should still be considered as the first choice for Aggregation functions-based evaluation mechanisms.

13.3.3 Problem Landscapes in Multiobjective Optimization

Intuitively, a problem (fitness) landscape is a graph where solutions play the role of vertices and edges indicate the neighborhood relation or a distance measure in the decision space between solutions [581, 618, 762]. In the single-objective case, it is labeled on vertices with real values of the fitness function. In the multiobjective case, it is labeled with vectors of real values.

A simple conclusion of the No free lunch theorem [940] is that no optimization algorithm may work for all possible landscapes. The properties of landscapes may be analyzed e.g. by the distance between local optima [76], fitness-distance correlation [581] or scatter plots of fitness versus distance. Several authors observed that single-objective memetic algorithms perform very well for problems with the 'big valley' property. This property means that there is a correlation between quality of solutions and their distance, i.e. good solutions tend to be located close according to some distance measure in the decision space.

Landscape analysis of multiobjective problems has not achieved significant attention yet. Very few such studies have been performed [276, 357]. However, it is natural to expect that (approximately) Pareto-optimal solutions do not need to be close in the decision space. For example, Pareto-optimal solutions corresponding to optima of particular objectives may be very distant in the decision space if the objectives are independent or conflicting. On the other hand, some solutions close in the objective space may also be close in the decision space [357].

This observation puts some new light on the typical statement that "population-based methods are ideal candidates for solving multiobjective problems" (see e.g. [196], Preface). In fact, single-objective population-based methods are rather designed to converge towards the vicinity of good solutions, and some natural convergence mechanism, e.g. genetic drift [328], may be beneficial in the single-objective case. The same convergence may, however, cause a multiobjective population-based method converge to a sub-region of the Pareto front only. Thus, dispersion mechanisms that are 'side' elements of single-objective algorithms may become crucial in the multiobjective case. Indeed, some studies indicate that various population-based methods have problems with assuring proper dispersion of solutions even if the convergence to the Pareto front is very good [420].

Furthermore, the 'big valley' property and the convergence of the population well explain the effectiveness of recombination operators. The recombination constructs a new solution by combining properties of the parents, and so, creating a solution being close to the parents and other good solutions. This offspring solution may be then efficiently improved by local search. In fact, some very successful recombination operators such as respectful operators [581] are directly designed to preserve properties common to both parents.

In multiobjective cases, the population may contain some very distant solutions with few or no common properties. Recombination of such solutions does not need to produce good offspring and may deteriorate efficiency and effectiveness of the whole algorithm. Thus multiobjective memetic algorithms may require some specialized mechanisms for selection of promising parents for recombination [412].

13.3.4 Archive of Potentially Pareto-optimal Solutions

In the single-objective case, the outcome of the algorithm should be the best solution found, even if in some cases it is not contained in the final population. In the multiobjective case, an analogue of the single best solution is the set of potentially Pareto-optimal solutions, i.e. solutions that are not dominated by any other solutions generated by the algorithm. Some initial population-based multiobjective memetic algorithms did not take this fact into account and assumed that their outcome is the final population. This means that many potentially Pareto-optimal solutions could have been lost. Thus, it is natural to maintain an additional archive of potentially Pareto-optimal solutions. Please note, however, that the size of this archive may become enormously large and its maintenance may become the main factor influencing the efficiency of the algorithm. Thus some techniques for reduction of the archive size have been proposed.

13.3.5 Evaluation of Multiobjective Memetic Algorithms

With the increasing number of multiobjective memetic algorithms and other metaheuristics, the issue of their evaluation and comparison becomes of crucial importance. Although full evaluation of single objective metaheuristics is already a complicated task that involves many aspects like quality of results and computational efficiency, some difficulties are specific to the multiple objective case. In the single objective case, when two algorithms generate two solutions, their comparison is straightforward. Either one of the solutions is better or they are equally good on the single objective function. In the multiobjective case we are dealing with evaluation and comparison of sets of solutions from the point of view of multiple criteria. In some cases, two sets of potentially Pareto-optimal solutions may be compared based on the dominance relation only with the use of so-called set outperformance relations [421, 970]. For example if all solutions in one set are covered (are dominated or equal) by solutions from another set the latter should be considered better. These relations leave, however, many pairs of sets incomparable. Thus, a number of quality indicators have been proposed. The quality indicators, usually, assign a single real value to each set. It is natural to expect that the proper quality indicators should properly rank sets comparable with set outperformance relations. Several quality indicators like hypervolume or R-indicator have this property. For detailed analysis of this issue see e.g. [421, 969, 970].

13.4 Examples of Multiobjective Memetic Algorithms

13.4.1 MOGLS of Ishibuchi and Murata

Ishibuchi and Murata [409] proposed multiobjective genetic local search (MOGLS), which is the first well-known multiobjective memetic algorithm. Their MOGLS uses a weighted sum fitness function for parent selection and local search. Pareto dominance is used only for maintaining an archive population. The archive population is updated at every generation so that it includes all non-dominated solutions among examined ones during the current execution of MOGLS. At each generation, the weight vector is randomly updated when a pair of parents is selected from the current population by roulette wheel selection based on the weighted sum fitness function. An offspring is generated from the selected pair of parents. The current weight vector is used for local search from the generated offspring. When the next pair of parents is selected, the weight vector is randomly updated. In this manner, the next population is generated by iterating random weight update, parent selection, crossover, mutation and local search. Some non-dominated solutions in the archive population are randomly selected and added to the current population as elite individuals. MOGLS of Ishibuchi and Murata [409] has some good properties such as the use of archived non-dominated solutions as parents and the use of aggregation functions with multiple weight vectors. Its performance, however, is not so high because its implementation is too naive. Its performance can be easily improved by a number of simple tricks such as the increase in the selection pressure for parent selection, the choice of a good starting solution for local search with the current weight vector, and the specification of a good balance between genetic search and local search [409, 410].

13.4.2 M-PAES

Memetic Pareto Archived Evolution Strategy (M-PAES) method proposed by Knowles and Corne [472, 473] is a memetic algorithm based fully on the dominance relation. The method is composed of two sequential phases - local search phase and recombination phase. In the local search phase the local search is independently applied to each starting solution. The local search is based on the dominance relation. The new neighborhood solution is rejected if it is dominated by the current solution, and accepted if it dominates the current solution. If the two solutions are mutually nondominated, the two solutions are compared with a local archive of potentially Pareto-optimal solutions and the one from a less crowded region is accepted. This acceptance rule is an additional dispersion mechanism. In recombination phase, randomly selected solutions from the current population created in local search phase and solutions from global archive of potentially Pareto-optimal solutions are recombined. The acceptance criterion of the offspring again takes into account both dominance relation and location in the more or less crowded region of the global archive. Although the method uses some dispersion mechanism in both phases, the experiment in [420] indicated that the method may be strongly affected by genetic drift.

13.4.3 NSGA-II with LS

Deb and Goel [198] proposed an algorithm in which local search is used to improve results of a standard multiobjective evolutionary algorithm. The method combines dominance-based and aggregation functions-based guiding mechanisms. The method starts by using Nondominated Sorted Genetic Algorithm-II NSGA-II that uses recombination and some dominance-based elitist dispersion mechanism. The algorithm is fully based on the dominance relation. Pairs of solutions are selected from the current population by binary tournament selection based on the dominance relation and a crowding measure. In the second phase each solution generated by NSGA-II is a starting point for local search. The local search is based on weighted linear aggregation functions. The weight vector is set automatically depending on the location of the solution in comparison to other solutions. Intuitively, solutions located close to the best value on a given objective will have a large weight value corresponding to this objective. In other words, each solution is pushed in the direction in which it is already good. The authors report that the hybrid approach improves performance of NSGA-II on a number of engineering design problems.

Cheng et al. [125] also proposed a multiobjective memetic algorithm based on the NSGA II method [199]. In each generation of NSGA-II they apply local search to just one potentially Pareto-optimal solution. To choose this solution a weighted linear aggregation function is drawn at random. Then 2-tournament based on the current aggregation function is used to select the single solution to which local search is applied. The method is applied to the multiobjective job shop scheduling problem on which the method performs better than a benchmark non-memetic evolutionary algorithm.

Garret and Dasgupta [304] hybridized NSGA II with a variant of tabu search for the multiobjective quadratic assignment problems. They studied the influence of the length of tabu search runs and noticed that with increasing number of objectives it becomes more beneficial to perform more short runs.

13.4.4 MOGLS of Jaszkiewicz

Jaszkiewicz has proposed a multiobjective genetic local search (MOGLS) method based on the aggregation functions guiding mechanism [419]. Alike the MOGLS of Ishibuchi and Murata a random weight vector of the aggregation function is drawn in each iteration. In a single iteration, two solutions are recombined and local search, or, more generally, a specialized heuristic, is applied to the offspring. The local search optimizes the current aggregation function. The random selection of weight vectors assures dispersion over all regions of the Pareto front. In each iteration, the search is pushed in a different direction but always towards the Pareto front. The method uses a relatively large population of solutions and some effort is made to define its size automatically. The solutions for recombination are also drawn based on the current aggregation function. In the original version parents are drawn at random from among some (e.g. 20) solutions being the best on this function. Thus only solutions being very good from the point of view of the current aggregation function could be recombined. This technique gives a high chance of constructing a new solution that performs well on the same function. In the improved version called Pareto memetic algorithm [421], this selection was based on the tournament selection with many solutions taking part in the single tournament. This mechanism improves efficiency of the selection. The method has been applied to the multiobjective traveling salesperson problem (TSP) [419], multiobjective knapsack problem [420], and multiobjective set covering problem [421]. In [423] it has been combined with a very efficient Lin-Kernighan local search for single objective TSP. Since a weighted linear aggregation function was used in this case, it was possible to directly apply the Lin-Kernighan method for the standard single objective TSP.

13.4.5 RM-MEDA

RM-MEDA (Regularity Model-Based Multiobjective Estimation of Distribution Algorithm) [957] for continuous MOPs is a an example of utilizing problem-specific knowledge in designing multiobjective heuristics. Under certain smoothness assumptions, the *PS* of a continuous MOP defines a piecewise continuous $(m-1)$-dimensional manifold in the decision space. where m is the number of the objectives. Therefore, the *PS* of a continuous bi-objective optimization problem is a piecewise continuous curve in R^n while the *PS* of a continuous MOP with three objectives is a piecewise continuous surface. The idea behind RM-MEDA is to force its population to converge to a $(m-1)$ piecewise continuous $(m-1)$-dimensional manifold. At each generation, RM-MEDA firstly extracts statistical information from some selected good solutions and then estimates the distribution of Pareto optimal points by using a probability model whose centroid is a $(m-1)$-dimensional manifold. New solutions are generated by sampling from the model thus built. The major computational overhead in RM-MEDA lies in model building. It is very costly to build a very accurate model. The local principal component analysis, a low-cost statistical algorithm, has been used for modeling. The experimental results have demonstrated that RM-MEDA works well, particularly when the PS is not a linear manifold. Recently, RM-MEDA has been generalized to the case when the dimensionality of the PS is unknown [961].

13.4.6 MOEA/D

MOEA/D (multiobjective evolutionary algorithm based on decomposition) [955] is a simple and generic multiobjective metaheuristic. It uses a aggregation method to decompose the MOP into N single objective optimization subproblems and solves these subproblems simultaneously (where N is a control parameter set by users). In MOEA/D, N procedures are employed and different procedures are for solving different subproblems. A neighborhood relationship among all the subproblems (procedures) is defined based on the distances of their weight vectors. Neighboring subproblems should have similar fitness landscapes and optimal solutions. Therefore, neighboring procedures can speed up their searches by exchanging information. In a simple version of MOEA/D [955], each individual procedure keeps one solution in its memory, which could be the best solution found so far for its subproblems; it generates a new solution by performing genetic operators on several solutions from its neighboring procedures, and updates its memory if the new solution is better than old one for its subproblem. A procedure also passes its new generated solution on to some (or all) of its neighboring procedures, who will update their current solutions if the received solution is better. A major advantage of MOEA/D is that single objective local search can be used in each procedure in a natural way since its task is for optimizing a single objective subproblem. Several improvements on MOEA/D have been made recently. Li and Zhang suggested using two different neighborhood structures for balancing exploitation and exploration [516]. Zhang et al [961] proposed a scheme for dynamically allocating computational effort to different procedures in MOEA/D in order to reduce the overall cost and improve the algorithm performance, this implementation of MOEA/D is efficient and effective and has won the CEC'09 MOEA competition. Nebro and Durillo developed a thread-based parallel version of MOEA/D [652], which can be executed on multi-core computers. Palmers et al. proposed an implementation of MOEA/D in which each procedure recorded more than one solutions [699]. Ishibuchi et al. proposed using different aggregation functions at different search stages [413].

13.4.7 MGK Population Heuristic

Gandibleux at al. [301] proposed a hybrid population heuristic for combinatorial problems. They used a Pareto ranking-based evolutionary algorithm as the population heuristic. The algorithm starts by seeding the initial population with some very good solutions. The solutions may be supported by Pareto-optimal solutions found by either an exact or approximate method. Furthermore, local search is applied during the run of the method. The method has been applied to a permutation scheduling problem and to the knapsack problem.

13.4.8 Memetic Approach by Chen and Chen

Chen and Chen [124] combine a dominance-based local search with a Pareto ranking-based multiobjective evolutionary algorithm. They use a special kind of local search with several species exploiting different regions in the objective space. The method has been applied to the problem of flexible process sequencing.

13.4.9 SPEA2 with LS

Schuetze et al. [797] combined SPEA2 [968] algorithm with local search for continuous multiobjective problems. They developed Hill-Climber with Sidestep procedure based on the dominance relation than can move either towards or along the Pareto front. They reported significant improvements of the performance in comparison to the standard SPEA2 algorithm.

SPEA2 was also hybridized with a gradient-based local search for continuous problems by Harada eta al. [365]. They compared two approaches, GA with local search and GA then local search. They reported better performance of the latter for continuous problems.

13.4.10 Interactive Memetic Algorithm by Dias et al.

Dias et al. [212] proposed an interactive multiobjective memetic algorithm. The algorithm optimizes an aggregation function based on the preferences of the decision maker. The algorithm works like a standard single objective memetic algorithm, however, the whole set of potentially Pareto-optimal solutions may be presented to the decision maker. The algorithm uses hot start technique. When the decision maker changes his/her preferences the current population is optimized further with a new aggregation function. The algorithm has been applied to the dynamic location problem.

13.4.11 SMS-EMOA with Local Search

Koch et al. [475] hybridized SMS-EMOA [64] with a gradient-based local search for continuous problems. SMS-EMOA in an indicator-based evolutionary algorithm that optimizes the hypervolume of the dominated space. The authors used a multi-objective Newton method.

13.5 Implementation of Multiobjective Memetic Algorithms

When we implement a multiobjective memetic algorithm for a particular MOO problem, we have a large number of options in its design. This means that we have a number of implementation issues to be taken into account. Typical issues are as follows: choice of a population-based multiobjective global search algorithm, choice of a local search algorithm, timing of local search, selection of starting points for local search, and allocation of the available computation time to global search and local search. Each of these issues is briefly explained in the following:

1. The choice of a population-based multiobjective global search algorithm: This choice includes the related settings in the chosen global search algorithm (e.g., coding of solutions, genetic operators, parameter specifications, etc.). In early proposals of multiobjective memetic algorithms, evolutionary algorithms were mainly used for global search. This is because other population-based multiobjective algorithms were not popular in 1990s. Recently various multiobjective algorithms have been proposed based on different population-based global search mechanisms such as particle swarm optimization [134], ant colony optimization [214], and differential evolution [30, 769]. New types of multiobjective evolutionary algorithms have been also proposed based on estimation of distribution algorithms [957], indicator-based algorithms [64, 244, 967], and multiple aggregation functions [967]. As a result, we have a wide variety of options with respect to the global search part of multiobjective memetic algorithms.

2. The choice of a local search algorithm such as hill-climbing, simulated annealing and tabu search: This choice includes the related settings in the chosen local search algorithm (e.g., a generation mechanism of a neighboring solution, an acceptance criterion of neighbors, a termination condition of local search, etc.). The specification of an acceptance criterion is usually the same as the choice of a local search guiding mechanism. Problem-specific heuristics can be incorporated into local search, which usually improves the search ability of multiobjective memetic algorithms [413]. In the case of combinatorial optimization, generation mechanisms of neighbouring solutions in local search are usually similar to mutation operators in evolutionary algorithms. That is, new solutions are generated in a similar manner in local search and mutation whereas they have different acceptance criteria. On the other hand, different mechanisms are often used to generate new solutions in local search and mutation when multiobjective memetic algorithms are designed for continuous optimization. This is because the gradient information of objective functions is often used in local search to find better solutions whereas mutation usually modifies a part of the current solution randomly.

3. Timing of local search: Local search can be combined with a population-based multiobjective global search algorithm in various manners with respect to the timing of local search. Usually local search is invoked at every generation of a population-based multiobjective global search algorithm. In this case, local search starts from an offspring in global search. That is, global search can be viewed as providing local search with good starting points. Then the improved solutions by local search are used as parents in global search. That is, local search can be viewed as providing global search with good parents. In this manner, a population of solutions is improved by alternately using global search and local search. Local search is not necessarily to be used at every generation. For example, it can be used at every 10 generations or every 100 generations. One extreme case is the use of local search only before global search. In this case, local search can be viewed as generating a good initial population for global search. The basic idea behind this implementation is "better

solutions may be obtained by recombining good locally-optimal solutions". Another extreme case is the use of local search only after global search. In this case, global search can be viewed as generating good starting points for local search. In other words, local search can be viewed as being used for the final improvement of global search results. The basic idea behind this implementation is "the local search ability of population-based algorithms is not high" and "better solutions may exist in the vicinity of good solutions".

4. Choice of staring points for local search: When local search is applied, starting points should be chosen from the current (or offspring) population. One naive implementation is to apply local search to all solutions in the current population. Another implementation is to apply local search to each solution probabilistically. Of course, other mechanisms can be implemented to choose starting points for local search such as the choice of only a small number of very good solutions with respect to some local search guiding mechanisms and the application of local search only to non-dominated solutions.

5. Allocation of available computation time to global search and local search: In single-objective memetic algorithms, many more solutions are usually examined by local search than population-based global search. This is not a bad strategy because the goal of single-objective optimization is usually to find a single optimal solution. In the case of multiobjective optimization, however, it is not a good strategy to spend too much computation time on local search for a specific direction even if some Pareto optimal solutions can be found by local search. This is because the goal of multiobjective optimization is not to find some Pareto optimal solutions but to approximate the entire Pareto front. We need to search for Pareto optimal solutions in various directions. Thus we should not spend too much computation time on local search for a specific direction. As a result, it is very important to allocate available computation time to global search and local search [411]. Moreover the computation time for local search should be appropriately reallocated to various local search directions.

As we have already explained in a previous subsection, landscape analysis is very useful in the design of efficient multiobjective memetic algorithms. For example, if a multiobjective problem has many local optima where local search is trapped, it may be a better idea to shallowly examine only a few neighbors of many starting points rather than to deeply examine many neighbors of a few starting points. However, the landscape of real-world multiobjective problems is often unknown. In that case, it is important to understand characteristics of each component of multiobjective memetic algorithms. For example, if a local search algorithm with high search ability towards the Pareto front is available, it may be a good idea to use a population-based global search algorithm with high diversity maintenance ability. On the other hand, if we use a population-based global search algorithm with high convergence property towards a part of the Pareto front, the diversity improvement by local search is important. The point is to fully utilize the synergy effect of using both global search and local search.

13.6 Conclusions

Multiobjective memetic algorithms constitute a very promising class of multiobjective metaheuristics. In many experiments they proved their efficiency for both combinatorial and continuous MOO problems.

Despite the need for new efficient memetic methods and the need for further applications, a number of other important directions for further research could suggested:

- The use of landscapes analysis in the design of recombination operators or the whole methods. Despite of some promising preliminary results discussed above, we are far from full understanding of the influence of landscapes of MOO problems on the performance of multiobjective memetic algorithms.
- The use of Pareto local search in multiobjective memetic algorithms. PLS has recently proved [542] to be a powerful technique for some multiobjective combinatorial optimization problems, being able to compete with memetic algorithms based on standard local search. PLS, however, becomes prohibitively inefficient with increasing number of objectives. Thus, hybridization with some global search techniques seems to be a promising approach.
- The use of advanced local search techniques, e.g. candidate moves, in MOO. Such techniques may have crucial influence on the performance of the local search component, and thus on the performance of the whole multiobjective memetic algorithm.
- Hybridization of other population-based algorithms, e.g. ant colony optimization, particle swarm optimization, differential evolution, with local search in MOO. Such algorithms may provide alternative global search components often competitive to evolutionary algorithms.
- Handling of many objectives in multiobjective memetic algorithms. Since, in general, the size of the Pareto front and the time needed to approximate it grows fast with the increasing number of objectives, interactive approaches seem to be a promising direction. In this case, some partial preference information may be used to focus the search on the desired regions of the Pareto front.

Chapter 14
Memetic Algorithms in the Presence of Uncertainties

Yoel Tenne

14.1 Motivation

Memetic Algorithms have proven to be potent optimization frameworks which are capable of handling a wide range of problems. Stemming from the long-standing understating in the optimization community that no single algorithm can effectively accomplish global optimization [940], memetic algorithms combine global and local search components to balance exploration and exploitation [368, 765]: the global search explores the function landscape while the local search refines solutions. In literature the terms memetic algorithms [615, 673] and hybrid algorithms [325] refer to the same global–local framework just described. The merits of memetic algorithms have been demonstrated in numerous publications, [374, 375, 686, 688].

However, while optimization algorithms are often conceived and tested on synthetic mathematical problems, real-world applications can be significantly different. One such major difference is that real-world problems often induce uncertainty in the optimization problem and studies identify four common scenarios [425]:

1. a model approximates the objective function and provides the optimizer with predicted objective values having an unknown error
2. the variables can stochastically fluctuate and it is required to find a solution which is insensitive to these fluctuations
3. the responses from the objective function are corrupted by noise and
4. the problem (objective function, constraints) is dynamic, that is, varies with time.

As such, baseline memetic algorithms developed using synthetic problems can perform poorly in such uncertain settings and this has motivated research into new

Yoel Tenne
Department of Mechanical Engineering and Science-Faculty of Engineering,
Kyoto University, Japan
e-mail: yoel.tenne@ky3.ecs.kyoto-u.ac.jp

and dedicated memetic frameworks. As such the goal of this chapter is to survey representative studies on memetic algorithms in the four uncertainty classes. In the remainder of the chapter we consider without loss of generality the minimization problem

$$\begin{aligned}\min\ & f(x) \\ \text{s.t.}\ & g_i(x) \leqslant 0, i = 1 \ldots k\end{aligned} \quad (14.1)$$

as the baseline optimization problem.

The remainder of this chapter is as follows: Section 14.2 surveys Algorithms for optimization with uncertainty due to approximation, Section 14.3 deals with Algorithms for robust optimization, Section 14.4 surveys Algorithms for noisy optimization problems, Section 14.5 deals with dynamic optimization problems and lastly Section 14.6 concludes the chapter.

14.2 Uncertainty Due to Approximation

Current research in engineering and science often replaces real-world laboratory experiments with analysis-codes, that is, computationally-intensive simulations which model real-world physics with high accuracy [881]. The approach allows to reduce the cost and duration of the design process and is being widely used, for example in aerospace [307, 725] electrical engineering [484] and chemistry [603]. Such computer simulations are typically *computationally expensive*, that is, each simulation call requires minutes to hours of CPU time. This makes many optimization algorithms, and particularly computational intelligence ones (such as evolutionary algorithms, particle swarm optimizers and so on) inapplicable since they require many thousands of function evaluations making the optimization process prohibitively expensive.

There are two main approaches to combat this difficulty. First, parallelization allows to reduce the wall-clock time [192, 725]. While the approach can be efficient one potential obstacle is that for commercial analysis-codes there is typically a licence restricting the number of concurrent simulations which can be run.

A complementary approach is that of modelling. Based on the 'plug-in' concept in statistics, the idea is to create a computationally cheaper mathematical approximation of the expensive simulation and to use it instead during the optimization search. The optimization algorithm then obtains the (predicted) objective values from the model in a fraction of the time when compared to using the true (expensive) simulation [283, 603]. Representative model types include:

- Quadratics [646]: the simplest models which capture function curvature and have the general form

$$\mathscr{S}(x) = \frac{1}{2}x^{\mathrm{T}}Hx + x^{\mathrm{T}}g + c \quad (14.2)$$

where coefficients are typically determined by a least-squares fit.

- Radial Basis functions (RBFs) [85]: the model is defined as a linear combination of kernel basis functions

$$\mathscr{S}(x) = \sum_{i=1}^{k} \alpha_i \phi(\|x - x_i\|) \qquad (14.3)$$

where α_i is a scalar coefficient and x_i is an interpolation point. The coefficients are obtained from the Lagrangian interpolation condition

$$\Phi \alpha = f \qquad (14.4)$$

where Φ is the interpolation matrix ($\Phi_{i,j} = \phi(\|x_i - x_j\|)$) and f is the vector of responses ($f_i = f(x_i)$).

- Kriging [172]: a statistically-oriented approach which models the function as a combination of a global 'drift' function (typically a constant β) and a stochastic function $Z(x)$ providing local adjustments so the model becomes

$$\mathscr{S}(x) = \beta + Z(x). \qquad (14.5)$$

The stochastic function is a Gaussian process with a zero mean and variance σ. Model parameters are typically calibrated by maximum-likelihood to best fit the data [554].

- Artificial Neural Networks (ANN) [68]: a biologically-inspired approach which uses an array of inter-connected 'neurons' (processing units). The ANN is trained using available data and learns the input-output mapping.

While models alleviate the bottleneck of high computational cost they introduce uncertainty into the optimization problem: the optimizer now needs to operate based on the responses of the model but those are inherently inaccurate as the model is trained using a typically small sample (since evaluations are expensive). The extent of inaccuracy is unknown and depends on various factors such as the dimension and landscape complexity of the objective function and the sample size [544, 851].

Model inaccuracy implies that the optimizer is searching on a deformed landscape with uncertainty regarding its 'goodness'. If the model accuracy is poor then the optimizer may even converge to a false optimum (an optimum of the model which is not an optimum of the true expensive function) [426]. This implies that to be effective model-assisted frameworks must account for this *uncertainty due to approximation* and several approaches have been proposed.

In [307, 439] the authors proposed the Inexact Pre-Evaluation (IPE) framework which uses the expensive function in the first few generations (typically 2–3) and then uses the model almost exclusively while only a portion of the elites are evaluated with the expensive function and are used to update the model. The approach was later incorporated into a hierarchical distributed algorithm [803] which uses 'layers' of optimization, for example, at each layer an EA uses an analysis code of different fidelity. Promising individuals would then migrate from the computationally cheap low-fidelity layer to the expensive high-fidelity layer to obtain a more accurate fitness and vice versa. The idea was later expanded such that each layer may use different solvers, for example an EA and a gradient-based resulting in a memetic like framework [440]. By using the high-accuracy simulation and a gradient search the framework can diminish the effect of the low fidelity simulations.

In [426] the authors proposed the Controlled Evaluations (CE) framework which monitors the model accuracy using cross-validation: a cache of previously evaluated vectors is split into two disjoint sets and a model is trained using one set and tested on the complementary set. Model accuracy is then measured by the mean squared error (MSE)

$$MSE = \frac{1}{k} \sum_{i=1}^{k} (\mathscr{S}(x) - f(x))^2 \qquad (14.6)$$

for a test set of k vectors. The authors examined both individual-based control (at each generation evaluating a few vectors with the expensive function) and generational-based control (every few generations evaluating all individuals with the expensive function). A fuzzy logic rule adapted the frequency of expensive evaluations, that is, it increased the number of expensive evaluations when the MSE is too large and vice versa. A related memetic approach was proposed in [305] where for an expensive multiobjective optimization problem. The EA was used for a certain number of generations and then an ANN was trained to predict objective responses. The framework then used a gradient local search to refine solutions while monitoring the goodness of the ANN using (14.6).

The trust-region (TR) framework is another option for managing optimization with approximation uncertainty and has a long standing history in nonlinear programming (and unrelated to expensive black-box optimization). The idea is to perform a sequence of restricted steps around the optimum instead of a one-shot global optimization of the model. Starting from an initial guess $x^{(0)}$ then at each iteration $i = 0, 1, \ldots$ a model is trained and the framework performs a trial step, that is, it seeks an optimum of the model constrained to the trust-region ($\mathscr{T}$) where

$$\mathscr{T} = \{x : \|x - x^{(i)}\|_p \leq \Delta\}, \quad p = 1 \text{ or } 2, \qquad (14.7)$$

where Δ is the TR radius. This defines the constrained optimization problem

$$\begin{aligned} \min \quad & \mathscr{S}(x) \\ \text{s.t.} \quad & x \in \mathscr{T} \end{aligned} \qquad (14.8)$$

which gives a minimizer x_m. Next, the framework examines the success of the trial step with the merit value

$$\rho = \frac{f(x^{(i)}) - f(x_m)}{\mathscr{S}(x^{(i)}) - \mathscr{S}(x_m)}, \qquad (14.9)$$

where $\rho > 0$ indicates the trial was successful, that is, the predicted optimum indeed improves on the current iterate ($\rho = 1$ indicates a perfect agreement between the model prediction and the true function). Based on the value of ρ the framework then updates the iterate and the TR, for example:

- if $\rho > 0$: centre the TR at x_m (so $x^{(i+1)} = x_m$) and increase Δ.
- otherwise decrease Δ.

A merit of the TR framework is that it guarantees asymptotic convergence to an optimum of the true (expensive) objective function [141, 771] which has motivated using it in memetic settings.

Reference [78] seems to be among the first to propose a TR-based memetic framework. It used a variant of the pattern search algorithm as a global search which gradually restricted the search to zoom in on an optimum. In case no improvement was made over the current iterate the authors proposed invoking a gradient-free local search to refine solutions.

Later [681, 682] proposed memetic frameworks combining an EA as a global search where at each generation every non-duplicated vector in the population was refined using a TR local search with local RBF models. The extent of the memetic refinement was limited to k iterations (prescribed a-priori by the user). If the local search found an improved (true) solution after k iterations then another round was performed but otherwise it terminated and the resultant solution replaced its original in the population in a Lamarckian updating scheme.

In [878, 879] the authors proposed a TR memetic framework which uses quadratic models and clustering. An EA performs global exploration and it directly evaluates the expensive objective function. Every several generations the framework would cluster the population using the k-means algorithm [543] to identify if the population is converging around previously found optima. The idea is to improve the search by identifying basins of attractions (by clustering) and invoking the local search only from solutions considered to lie in yet unexplored basins [891]. The local search is based on the DFO algorithm which is a gradient-free TR local search algorithm [140, 141].

To further improve search efficiency and leverage on the power of models several studies have proposed using models both in the global and local search phases. For example, [963] extended the framework from [681]: an EA searches over a global Kriging model and a number of solutions were then refined using a TR local search with RBF models. After the local search the refined solutions replace the originals in the population in a Lamarckian update scheme. A related study [962] proposed a framework which uses a global Kriging model but with multiple local searches (possibly performed in parallel) where each is performed based on a different model type. The idea is that occasionally an inaccurate model can actually yield a fast improvement in the search [685] and so performing multiple searches and selecting the best solution among them can improve the search effectiveness (the study used a quadratic model and an RBF one during the local search). Continuing the multiple models approach, [522] has recently proposed a framework relying on ensembles of models as well as smoothing models. The framework uses an ensemble of different local models where the individual predictions by each model are aggregated into a single response based on the models' accuracy. The framework also employs a smoothing-model (low-order polynomial) to reduce the number of optima and simplify the landscape. During the search the framework chooses between the optimum predicted by the smoothing model and the ensemble. The authors have also presented a multiobjective variant of the framework.

Another development was that of model-adaptive frameworks [876, 879, 880]. The approach is motivated by the tenet that an optimal model is problem dependant but often there is insufficient a-priori information to select the optimal type [476, 557]. As such, a model-adaptive framework aims to autonomously select the best model from a family of candidates. To achieve this the framework leverages on a rigorous statistical model selection theory: it assesses the goodness of a model based on its maximum likelihood which is a statistical measure indicating how well a model fits the data [526, 718]. When comparing different candidate models the one having the highest likelihood is chosen as the best predictor of the data.

Leveraging on these ideas, [876] proposed a model-adaptive memetic framework which uses a DFO-like local search with Kriging models and selected at each iteration an optimal local model type. A follow-up study [880] then extended model-adaption to select an optimal global model as well. The proposed framework used an RBF neural network as a global model and selected an optimal RBF kernel for it out of the four candidate kernel functions based on the MSE criterion (14.6). Next, an EA would search for an optimum of the model and then a TR local search would improve the predicted optimum. The local search followed the classical TR procedure described earlier but with the addition of monitoring the number of points in the TR. If the trial step was unsuccessful and there were too few points in the TR a new point would be added to improve the model. Also, the framework selected an optimal model during the local search iterations. The global–local process would repeat until the optimization budget was exhausted. Algorithm 25 gives a pseudocode of the framework. Three variants of Ratle's algorithm [757] were used each with a different RBF model (multiquadric, linear and inverse multiquadric) where the model type was fixed throughout the search. The proposed framework showed statistically significant performance advantage over the three variants indicating the merit of model adaption. Lastly, the framework and Ratle's algorithm were also used in an airfoil shape optimization (an 11 dimensional problem) and again showed a statistically-significant performance advantage. Overall, performance analysis showed that adapting the model improves the optimization search.

14.3 Uncertainty Due to Robustness

In many real-world applications a system needs to operate under a range of conditions and not a single fixed one. For example, an engine should maintain efficiency over a range of operating speeds or an aircraft fleet assignment should maintain punctuality while accounting for a range of weather conditions and so on. In these cases and similar ones elements of the problem are not crisp but can stochastically assume any value within a known range. In such settings the optimization goal is typically not to find the best global optimum but rather a robust solution which yields a 'good' objective response and which is relatively insensitive to fluctuations.

14 Memetic Algorithms in the Presence of Uncertainties

Algorithm 25. A Global–Local Model-Adaptive Memetic Framework [880]

1 generate initial sample;
2 **repeat**
3   global search phase: select model type by maximum likelihood;
4   train global model;
5   locate model optimum with EA;
6   select starting point for local search;
7   local search phase: **repeat**
8   select model type by maximum likelihood;
9   train local model;
10   perform trial step;
11   update TR based on step, improve model if necessary;
12   **until** *k iterations or convergence* ;
13 **until** *until evaluations budget exhausted* ;

Robust optimization problems can be classified according to which elements of the problem vary:

- objective function (for example, noise in instruments measuring the objective values).
- variables (for example, manufacturing inaccuracies).
- operating conditions (for example, the ambient temperature in which a system operates).

As a side note, a solution which can be adapted to yield a high-quality response is termed flexible [837]. In contrast, a robust solution requires no adaption.

Given the stochastic nature of the variations, statistical decision theory [203, 610] suggests three main criteria for selecting robust solutions (for simplicity we consider an unconstrained minimization problem). The robust solution should provide a bound on the worst case performance, implying (in minimization) a min-max formulation, that is

$$\min\max f(x). \tag{14.10}$$

The robust solution should minimize the expected objective value, mathematically

$$\min F(x) \tag{14.11}$$

where

$$F(x) = \int_{-\infty}^{+\infty} f(x+\delta)p(\delta) \tag{14.12}$$

and x is the baseline design vector (nominal settings), δ is a fluctuation and $p(\delta)$ is its probability density function. In practice both the distribution $p(\delta)$ and the effect of fluctuations on the objective response (or uncertainty propagation [229]) are unknown and so algorithms use Monte Carlo sampling [526] to generate the empirical unbiased estimate

$$\hat{F}(x) \simeq \frac{1}{N} \sum_{i=1}^{N} f(x+\delta). \qquad (14.13)$$

The robust solution should minimize both the expected objective response and its variance since (14.12), (14.13) can still yield a small expected value even when there are large positive and negative responses cancelling each other. This scenario also considers the objective variance

$$Var(f(x)) = \int_{-\infty}^{\infty} (f(x+\delta) - F(x))^2 \, p(\delta). \qquad (14.14)$$

The problem formulation is then

$$\begin{aligned} &\min F(x) \\ &\min Var(f(x)) \end{aligned} \qquad (14.15)$$

which is a bi-objective optimization problem. As before, when the exact information is unavailable algorithms use the empirical unbiased estimate of the variance

$$\widehat{Var}(F) = \frac{1}{k-1} \sum_{i=1}^{k} (f - \hat{F})^2 \qquad (14.16)$$

In [869] the authors proposed a memetic algorithm for robust optimization of digital filters where the uncertainty in performance is due to material imperfections. The problem formulation involves both three parameters (which can assume a range of values) and 12 design variables defining the filter geometry (termed control factors). The goal of the optimization was to find a filter with a robust frequency response. Following the Taguchi method [870], the authors used a full-factorial design for the three parameters (defining 'low' and 'high' settings for each parameter and evaluating all 8 combinations). The authors then defined a sound-to-noise ratio (SNR) as a measure of robustness and maximized it. The optimizer was a memetic algorithm which combined a real-coded EA and the the variable neighbourhood search algorithm [604], which searches in increasingly larger local neighbourhoods around the current iterate.

In [811] the authors considered the problem of optimizing a robust aircraft control system using a memetic algorithms. The problem was formulated as a quadratic minimization problem where the goal was to find a set of matrix elements which optimize a prescribed system robustness measure. The memetic algorithm combined an EA with a hill-climbing local search.

In a multiobjective formulation [835] proposed a memetic algorithm for robust optimization while considering both the expected value and variance of the objective function. The study applied the algorithm to robust optimization of airfoils where the goal was to identify an airfoil shape yielding a low drag (aerodynamic friction) over a range of aircraft velocities. The proposed algorithms used a variant of the NSGA-II algorithm [196] to approximate the Pareto front and then invoked a gradient-based local search to refine solutions. For each solution the local search minimized one function at a time while treating the other as a constraint, and the resulting vector was used as a starting point for subsequent steps, repeating the procedure for the two objective functions.

In another multiobjective study [691] used the Design-for-Six-Sigma (DFSS) approach which considers both the mean and variance of the objective and proposed using a particle swarm optimizer (PSO) to obtain the Pareto front of the mean–variance objectives. A follow-up study [690] then extended the idea to a memetic algorithm combining an EA as a global search algorithm and then using a finite-differences quasi-Newton local search to further refine the solutions, an approach termed memetic algorithm for robust solution search. The local search was applied to a certain percentage of the population chosen at random but without considering the variance of the fitness, that is, a single objective refinement of the solutions. The authors also applied an ageing operator which adjusted the expected mean fitness based on the duration an individual has survived.

Considering multiobjective optimization and robustness [324] proposed a multi-objective EA for robust and constrained optimization. The algorithm uses a micro-GA (that is, having a very small population) as a form of a local search to obtain the worst-case performance of candidate solutions. It also uses a tabu-like approach which restricts and guides the EA and periodic re-evaluation of cached solutions to reduce uncertainty regarding their fitness.

In [521] the authors addressed the problem of robust optimization when no a-priori information is known about the uncertainties. Commonly, algorithms assume some a-priori statistical distribution for the unknown uncertainties (for example Gaussian) but this can be unfounded. The authors proposed the *inverse robust evolutionary design methodology* which combines an EA with a constrained local search (performed by an SQP solver). The idea is to replace the classical problem (termed *forward optimization*) with *inverse optimization* which locates a target solution satisfying some prescribed criteria:

$$\min f(x) - T \\ \text{s.t.} \quad x_l \leqslant x \leqslant x_u \qquad (14.17)$$

where T is the target output performance. The authors proposed a single objective variant which considers only the robustness function (the maximum uncertainty a design variable handles before violating the worst-case performance), bi-objective (robustness function and objective function) and tri-objective which also considers the opportunity fitness.

In [98] the authors proposed a memetic algorithm for robust airline scheduling where the goal was to obtain a fleet assignment which accounts for flight re-timing and aircraft rerouting. Using a multiobjective approach the study considered two objectives: schedule reliability and schedule flexibility. The proposed algorithm used a tailored representation (the adjacency representation often used in traveling salesman problems) and multi-memes (multiple local searchs) to improve effectiveness. The algorithm used three variants of local search, each considering the schedule reliability, schedule flexibility or both. The study also used a host of additional features such as archiving and biased sampling (to encourage exploration).

Also in scheduling problems, [837] proposed a memetic algorithm for the stochastic capacitated vehicle routing problem (CVRP). The baseline CVRP is that of determining the sequence in which a fleet of vehicles visits spatially distributed customers such that some cost measure (time, distance) is minimized. In the stochastic CVRP the customer demands and travel costs are no longer crisp which motivates a robust approach. The proposed algorithm samples the objective function around a set of solutions and selects (based on the problem formulation) either the expected (mean) response or the worst-case (max). The algorithm refines solutions using a local search combined with tabu search.

Following the worst case performance approach to robust optimization [684] proposed a memetic algorithm designed for expensive objective functions. The algorithm builds upon the earlier genetic algorithm with robust solution searching schemes (GARSS) [895] in which a random perturbation was added to a chromosome before evaluation. In its single evaluation variant each chromosome was perturbed once while in the multiple evaluations variant it was perturbed repeatedly and the final fitness was taken either as the mean or worst of the perturbed set. Empirical tests show that the multiple evaluations variant was more reliable than the single evaluation one but obviously required more function evaluations which makes the algorithm inapplicable to expensive problems. As such, the authors proposed an algorithm which combines a max-min optimization strategy with a TR model-assisted approach and a Baldwinian updating scheme. The algorithm starts with an initial sample (random or by design of experiments) and uses the baseline GARSS algorithm with the worst fitness of the perturbed set taken as the chromosome fitness. The GARSS is run for several generations while evaluating the true (expensive) function and all vectors and associated fitness are cached. Next, each individual in the population is refined with a TR local search where the goal of the latter is to find the worst case performance. To reduce function evaluations the local search used RBF models which were trained using cached vectors adjacent to the TR centre and the TR procedure follows that described in 14.2. The goal of the local search was to find the worst case performance for each population member by solving the max-min problem

$$\min f(x+x_c)$$
$$\text{s.t.} \quad x \in \Omega$$
(14.18)

where x is the vector of perturbations, x_c is the baseline candidate and Ω is the feasible range of perturbations. The search was performed using an SQP solver and the TR iterations terminated after a prescribed k expensive function evaluations. If the TR local-search found a lower objective value then it replaced the fitness of the original population members (that is, before the local search was invoked) in a Baldwinian learning scheme (the chromosome was not changed). Algorithm 26 gives a pseudocode of the framework (adapted from [684]). Performance analysis was based on a robust airfoil shape optimization problem with a parametrization

Algorithm 26. Trust-region Enabled Max-Min Surrogate-Assisted EA [684]

```
1  initialize database;
2  repeat
3      for each individual i in population do
4          if status is database building then
5              evaluate individual with true (expensive) function and cache;
6          endif
7          else improve individual with TR–SQP search
8              initialize TR;
9              repeat
10                 train local RBF model using neighbours from database;
11                 establish domain where uncertain variables vary Ω;
12                 find point of worst-fitness in TR using RBF models (trial step);
13                 evaluate predicted point with expensive function and cache it;
14                 update TR based on success of trial step;
15             until TR termination condition met ;
16             set individual's fitness to worst-case value;
17
18     endfor
19     Apply standard EA operators to create a new population;
20 until EA termination condition met ;
```

resulting in a 24-dimensional problem. The authors first obtained an airfoil shape without considering any perturbation (a classical non-robust optimization) as a reference shape. Next, they applied the framework to robust optimization in the presence of manufacturing errors ($\pm 5\%$ error bounds on design variables). Analysis showed the performance of the robust airfoil is indeed more stable than that of the non-robust one. The authors also optimized the airfoil for perturbations in operating conditions (cruise velocity). Similarly to the previous case, the robust airfoil performance was more stable over the entire range of cruise speeds while the performance of the non-robust airfoil degrades quickly outside the nominal operating point. Overall, the framework was able to generate robust designs on a limited computational budget.

14.4 Uncertainty Due to Noise

In many real-world applications repeatedly evaluating the same vector returns slightly different objective values, a scenario termed *noisy* optimization. Such fluctuations in the response imply uncertainty regarding the true function value. Noisy functions are encountered in two main scenarios:

1. The response is obtained by measuring some real-world process and noise is either inherent in the process or in the measurement instruments. For example, reading electrical signals from an electric motor [104].

2. The objective function depends on some random process. For example, when optimizing the topology of neural networks the same vectors (candidate topologies) can produce different responses due to random initialization of network weights [949].

The dominant (and sometimes implicit) assumptions in noisy optimization problems are that the noise is random (so it cannot be filtered out a-priori) and that its amplitude is much smaller than the underlying objective response (so it only moderately deforms the landscape). Many studies also assume that the noise is Gaussian.

Since the observed responses are corrupted by noise some additional sampling mechanism needs to be introduced to estimate the true objective value. These mechanisms come in two main flavours:

1. Explicit Averaging: a better estimate of the true response can be obtained by using multiple samples. In temporal sampling the same vector is re-sampled n times and under the assumption of random Gaussian noise this allows to improve (reduce) the estimated response variance by $\sqrt{n}$ [808]. A complementary approach is that of spatial sampling where the samples are taken from neighbouring points around the current individual [788].
2. Implicit Averaging: simply increasing the population size provides more samples of the objective function and implicitly combats noise. The population size can be either fixed (set a-priori to a high value) or adapted during the search.

With the first category (noise due to external processes), in [104] the authors tackled the problem of optimizing the control system of an electric motor. They used online optimization, that is, where each candidate control settings were tested in real-time and the resulting performance was fed back into the algorithm. The measurements of the motor were inherently noisy and to combat noise the study proposed a memetic algorithm which monitored the population diversity to control the degree of mutation: high diversity invoked more local searches while low diversity invoked a higher mutation rate. Also, the algorithm selected between two types of local search (Hooke-Jeeves pattern search [391] and Nelder-Mead simplex [653]) to refine vectors.

Also in this category, [462] proposed a memetic algorithm combining a real-coded EA with the Bacteria Foraging local search. The latter is inspired by the swim pattern of the E. coli bacteria in the presence of favourable/hostile environment (rich/poor with nutrients). The idea is to perform tentative moves (similar to the bacteria's swim pattern) and adapt the step size based on the success/failure of these moves. The authors applied the memetic algorithm to optimization of a Proportional/Integral/Derivative (PID) controller for an automatic voltage controller subject to a sine wave noise.

In [601] the authors proposed a memetic algorithm based on differential evolution where the scale factor was adjusted with a local search. The algorithm also employed a noise analysis component to determine whether to replace a parent with an offspring. Specifically, it compared the samples of fitness (for each) to determine whether the means were sufficiently distinct (so a comparison is meaningful). If so, the better solution was retained but otherwise the algorithm sampled more points and repeated the comparison.

In [43] the authors considered the noisy pattern recognition problem of inexact graph matching, that is, determining whether two images match when one is corrupted by noise. They proposed a memetic algorithm in a combinatorial framework where each graph is represented by a chromosome of its vertices. The GA uses tournament selection and a new position based cross-over but no mutation. A tailored local search explored the neighbourhood of a solution and if it succeeded in locating a better individual then the latter replaced the original in a Lamarckian update scheme. The operators were designed to be insensitive to vertex location to provide better immunity to noise.

Also in this class, [695] studied the problem of matching an input image to one from an available data set. The difficulty being that the input image may be partially obscured, deformed and so on which results in a noisy optimization problem. They used a specialized encoding to represent both the input image and the database images by segmenting them into lines and connecting angles. They proposed a memetic algorithm which combined a real-coded EA (one point cross-over, uniform mutation) and a hill-climbing local search. For each database image the algorithm matched each segment to the those of the input image while ignoring small differences (to combat minor image deformations).

Related to the second category of noise due to a random process, [171] proposed an EA which uses a self-organizing map (SOM) [477] as a local search operator. The algorithm was designed to solve the vehicle routing problem (VRP) with emphasis on noisy data. The SOM was used to allow immunity to noise and to fluctuations in customer demands. The authors have also proposed several dedicated operators which work in conjunction with the SOM to improve the search.

In [656, 660] the authors tackled the problem of training a neural network used for controlling resource discovery in peer-to-peer (P2P) networks. They considered a multi-layer perceptron (MLP) network with a topology of 22 input neurons and 10 hidden-layer neurons plus a bias channel resulting in 298 weights to optimize. Since the network needs to operate under a variety of query conditions this results in a noisy objective function. The authors proposed the adaptive global–local memetic algorithm (AGLMA) which combined a real-coded EA with self-adaptation and two local searches: the stochastic simulated-annealing (SA) [468] and the deterministic Hooke-Jeeves. To combat noise the algorithm adjusted the objective response by explicit averaging. The proposed algorithm used a measure of population diversity

$$\psi = 1 - \frac{\hat{F}_{\text{avg}} - \hat{F}_{\text{best}}}{\hat{F}_{\text{worst}} - \hat{F}_{\text{best}}} \qquad (14.19)$$

where the measures are the average, best and worst fitness values in the population at the end of a generation. It follows that $\psi \simeq 1$ indicates high diversity and $\psi \simeq 0$ low diversity. The algorithm used this diversity measure to determine when to invoke each local search by the heuristic rules

$$\psi \begin{cases} \in [0.1, 0.5] & \text{invoke simmulated annealing} \\ < 0.2 & \text{invoke Hooke-Jeeves} \end{cases} \quad (14.20)$$

The idea is to use an explorative search (the SA) when the population diversity is decreasing (low ψ values). The Hooke-Jeeves local search was applied to the best individual and does not have the same explorative qualities but is more localized. It follows that for $\psi \in [0.1, 0.2]$ both local searchs are applied. The algorithm also leveraged on implicit resampling by adjusting the population size based on the diversity measure using the rule

$$S_{\text{pop}} = S_{\text{pop}}^f + S_{\text{pop}}^v \cdot (1 - \psi) \quad (14.21)$$

where S_{pop}^f, S_{pop}^v are a prescribed lower and upper bounds on the population size, respectively. Algorithm 27 gives a pseudo-code of the framework. As mentioned, the authors applied the algorithm to the topology optimization of a P2P network and benchmarked it against the Checkers Algorithm (CA), the Adaptive Checkers Algorithm (ACA) and a baseline real-coded GA while the optimization budget was 1.5e6 function evaluations. The proposed algorithm (AGLMA) performed best, closely followed by ACA and lastly the CA and baseline GA. Although the AGLMA converged more slowly than the CA it obtained a better final solution. The paper explains that the ACA can be viewed as an AGLMA without the memetic (that is, local search) component which explains its slightly degraded performance and highlights the merit of the memetic approach. Further, the AGLMA and ACA also effectively filtered noise which was evident from the convergence analysis (given in the paper) when compared to the CA and baseline GA. Overall, the AGLMA framework was able to handle this high-dimensional and noisy optimization problem.

14.5 Uncertainty Due to Time-Dependency

In the three categories covered so far (expensive evaluations, robustness, noise) the underlying optimization problem was fixed. However, in many real-world applications the problem is time-dependant so (14.1) becomes

$$\begin{aligned} & \min \ f(x, t) \\ & \text{s.t.} \ \ g_i(x, t) \leq 0, \quad i = 1 \ldots k \\ & \quad t = 1, 2, \ldots \end{aligned} \quad (14.22)$$

that is, the objective landscape, constraints and hence the problem optima may vary with time. Such problems arise in diverse applications such as scheduling [525] and control [755].

Algorithm 27. Adaptive Global–Local Memetic Algorithm [660]

1 sample weights W and self-adaptive parameters h;
2 evaluate fitness of initial population with explicit averaging;
3 calculate merit value: $\psi \leftarrow 1 - \frac{\hat{F}_{avg} - \hat{F}_{best}}{\hat{F}_{worst} - \hat{F}_{best}}$;
4 **while** *budget conditions and* $\psi > 0.01$ **do**
5   **for** *all individuals i* **do**
6   **for** *all variables j* **do**
7   update weights and self-adaptive parameters;
8   **endfor**
9   **endfor**
10   evaluate fitness of population by explicit averaging;
11   sort population (parents+offspring) based on fitness;
12   **if** $0.1 \leqslant \psi \leqslant 0.5$ **then**
13   execute simulated annealing on *2nd* best individual;
14   **if** $\psi < 0.2$ **then**
15   execute Hooke-Jeeves on best individual;
16   **endif**
17   **if** *simmulated annealing successful* **then**
18   execute Hooke-Jeeves on individual improved by SA;
19   **endif**
20   **endif**
21   calculate $S_{pop} \leftarrow S_{pop}^{f} + S_{pop}^{v} \cdot (1 - \psi)$;
22   select S_{pop} best individuals as the next generation;
23   calculate merit value: $\psi \leftarrow 1 - \frac{\hat{F}_{avg} - \hat{F}_{best}}{\hat{F}_{worst} - \hat{F}_{best}}$
24 **endw**

The time-dependant nature of such problems introduces several specific algorithmic considerations:

1. Since the optimization algorithm effectively needs to solve not one but a series of problems it should not drive the population of candidate solutions to fast convergence but should rather maintain diversity to allow the population to adapt to the changing landscapes.
2. Between subsequent time steps changes to the problem formulation are often small so it may be beneficial to search in the vicinity of the recent optimum (optimum tracking).

Due to their unique nature dynamic problems are often tested with a tailored suite of problems termed the Moving Peaks [82, 613, 906] which define a time-varying multimodal landscape where peaks deform and translate. There are also specific performance measures for dynamic problems where the commonly used one being the mean offline performance

$$f_{off} = \frac{1}{T} \sum_{t=1}^{T} f^*(t) \qquad (14.23)$$

where $f^*(t)$ is the best objective value found at time step t [927].

In [904, 905, 906] the authors proposed a memetic algorithm combining a binary EA with the variable local search (VLS) operator to track optima in dynamic problems. The EA invoked the operator when the averaged best performance of the population dropped below a prescribed threshold. Once a change in the landscape was detected the VLS operator enabled a local search around individuals from the pre-change population, an approach motivated by the assumption that changes are gradual (as mentioned above). The extent of the search was variable and calibrated based on the observed degree of change. When the VLS operator was invoked the evolutionary operators of recombination and mutation were temporarily suspended and the EA generated new vectors by adding or subtracting (with equal probability) from the population a random binary vector (whose range of values was limited to define a small search neighbourhood). After a single application of the VLS the EA reverted back to standard recombination and mutation and observed the performance of the population elites over a period of several generations. If the mean performance did not reach its previous (pre-change) value then the range of the VLS operator was increased and the process was repeated.

In [915] the authors proposed a memetic algorithm based on a particle swarm optimizer (PSO) and a hill-climbing local search. The algorithm combined several techniques to improve its performance in dynamic problems:

1. when updating a particle's position the algorithm considered the best solution found by the particle and its neighbours (termed local-PSO) to avoid rapid convergence
2. particles were refined by a local search which stochastically perturbed an elite vector to perform a neighbourhood search
3. particles were positioned on a virtual 'ring' and communicate only with their ring-wise neighbours (irrespective of the Euclidean distance in the search space) as an additional measure to avoid rapid convergence and lastly
4. to increase diversity the worst solutions were extracted and allowed to evolve in a sub-swarm independently from the main swarm.

In [635] the authors proposed a memetic algorithm which combined the Extremal Optimization algorithm (EO) [77] and a deterministic local search. The former (EO) starts from a baseline solution and perturbs it to generate a population and then probabilistically eliminates the worse member. As such, it aims not for fast convergence but for gradual adaption, which has motivated the authors to apply it to dynamic problems. In a follow-up study [633] the authors proposed another variant which at each generation refined one population member with a local search using the Hooke-Jeeves algorithm. Another follow-up study [634] evaluated both the EO with the Hooke-Jeeves variant and with an improved local search which scanned along each coordinate with an initial step and adjusted the step size depending on the search progress.

In [232] the authors tackled dynamic and highly constrained problems and proposed a memetic algorithm based on the scatter search framework [321] which combines a global search (diversification) and a local search (intensification). The global search generated solutions similarly to an evolutionary recombination operator and where an offspring could replace only a parent. Solutions were also generated in a Nelder-Mead simplex-like move which explored along promising directions. Next, solutions were chosen for refinement based on competitive ranking (considering both their fitness and diversity) and were refined with one of several local optimizers (the authors considered variants of SQP and hill-climbing). The algorithm handled constraints by a static penalty method.

Recently [482] proposed a memetic algorithm for dynamic multiobjective problems. The idea is to accelerate the convergence of a multiobjective EA (or similar algorithms) by predicting the change in the Pareto set based on the observed pattern in past time steps under the assumption that the Pareto set does not change erratically but follows an identifiable pattern. The approach used a *predictive gradient* (g) which approximated the shift in the population between consecutive time-steps. The idea was then to shift individuals in the population using the rule

$$x_{new} = x + \mu g. \qquad (14.24)$$

The predictive gradient was calculated based on changes in the centroid of the non-dominated solutions. The algorithm monitored landscape changes by comparing the fitness of a subset of individuals and so a mismatch between consecutive time-steps indicated a landscape change. This then triggered a population update where a predetermined number of individuals were randomly selected and updated with the predictive gradient. The approach was implemented within a multiobjective evolutionary gradient search algorithm.

In [913] the authors proposed a memetic algorithm for dynamic optimization which used a binary representation where at each generation the elite was refined by a local search and added several tailored enhancements. First, it used two hill-climbing variants for the local search:

1. greedy crossover hill climbing (GCHC): used the current elite and another parent (chosen by roulette wheel selection) and generated an offspring by uniform crossover and
2. steepest mutation hill climbing (SMHC): the elite individual was mutated by randomly flipping its bits.

In both variants the offspring replaced the elite if it was better. Another feature was that the algorithm adapted the probability of applying each variant based on their success in previous steps (starting from an equal probability of 0.5 for both). The success of a step was measured by

$$\eta = \frac{|f_{imp} - f_{ini}|}{f_{ini}} \qquad (14.25)$$

where f_{imp}, f_{ini} were the improved and initial objective values, respectively, and the probability of applying each variant was updated by

$$p(t+1) = p(t) + \Delta \cdot \eta(t) \qquad (14.26)$$

where Δ was prescribed by the user. Lastly, the algorithm safeguarded the population diversity using two procedures:

1. adaptive dual mapping (ADM): before starting a local search the method evaluated the bit-complementary of the initial solution and used the better of the two as the resultant initial vector and
2. triggered random immigrants (TRI): when the population diversity was deemed low a portion of the population was replaced by randomly generated individuals while the population diversity was measured by

$$\xi = \frac{\sum_{i=1}^{s} d(x^\star, x_i)}{s} \qquad (14.27)$$

where s is the population size and $d(x^\star, x_i)$ is the Euclidean distance between the current elite and the ith individual in the population.

Algorithm 28 gives a pseudo-code of the framework. The authors evaluated the proposed framework using tests derived from stationary problems (the 100-bit binary coded variants of the OneMax, Plateau, RoyalRoad and Deceptive). The authors used memetic variants with the GCHC, SMHC, AHC operators described above, a

---

Algorithm 28. Memetic Algorithm for Dynamic Problems [913]

1 initialize population and evaluate individuals;
2 calculate algorithm parameters;
3 select elite for local search;
4 **if** *ADM is used* **then**
5   create a dual of the elite and evaluate;
6   if dual is better set as new elite;
7 **endif**
8 perform AHC with elite;
9 **repeat**
10   apply standard EA operators(selection,recombination,mutation) to create offspring;
11   evaluate offspring and select individuals for next generation;
12   select elite for local search;
13   calculate algorithm parameters;
14   **if** *ADM is used* **then**
15   create a dual of the elite and evaluate;
16   if dual is better set as new elite;
17   **endif**
18   perform AHC with elite;
19   **if** *TRI is used* **then**
20   **if** $\xi < \theta_0$ **then**
21   replace a prescribed number of worst individuals in new generation with random immigrants;
22   **endif**
23   **endif**
24 **until** *stop condition is met* ;

baseline GA, a baseline GA with population restart when a change is detected, a GA with random immigrants, a GA with elitism-based immigrants and the population-based incremental algorithm (PBIL). Performance analysis indicated that:

1. the diversity-based procedures improved performance in dynamic problems
2. the ADM approach performed better when there were significant changes in the environment while the TRI performs better in corrensponcence to small changes
3. the optimal local search was problem dependant and there was no clear winner and lastly
4. the AHC approach used multiple local searches which provided more robustness.

Overall, results indicated that the combination of the AHC as a local search with ADM and TRI provided an effective memetic framework for dynamic problems.

14.6 Conclusion

Optimization problems arising in real-world applications can differ significantly from synthetic mathematical test problems and one such major difference is uncertainty induced by approximation, robustness, noise or time-dependency. While computational intelligence algorithms have been applied to such problems, memetic algorithms offer enhanced capabilities which significantly improve search efficacy under such challenging settings, as surveyed in this chapter. The complexity of real-world problems can be expected to grow, for example, to problems with multiple uncertainties (expensive and robust or noisy and dynamic). In such settings memetic algorithms will likely further establish their standing as a potent framework for optimization in the presence of uncertainties.

Acknowledgements. This research is kindly supported by the Japan Society for Promotion of Science.

Part III
Applications

Chapter 15
Memetic Algorithms in Engineering and Design

Andrea Caponio and Ferrante Neri

15.1 Introduction

When dealing with real-world applications, one often faces non-linear and non-differentiable optimization problems which do not allow the employment of exact methods. In addition, as highlighted in [104], popular local search methods (e.g. Hooke-Jeeves, Nelder Mead and Rosenbrock) can be ill-suited when the real-world problem is characterized by a complex and highly multi-modal fitness landscape since they tend to converge to local optima. In these situations, population based meta-heuristics can be a reasonable choice, since they have a good potential in detecting high quality solutions. For these reasons, meta-heuristics, such as Genetic Algorithms (GAs), Evolution Strategy (ES), Particle Swarm Optimization (PSO), Ant Colony Optimization (ACO), and Differential Evolution (DE), have been extensively applied in engineering and design problems.

On the other hand, population-based meta-heuristics do not guarantee detection of the global optimum and they might either prematurely converge to solutions with a poor performance or stagnate without successfully improving upon the current best solutions. In order to overcome these problems and as a consequence of the No Free Lunch Theorem [940], engineers realized that real-world problems can be efficiently solved by means of an ad-hoc combination of algorithms. This fact led to an employment in recent years of Memetic Algorithms (MAs). As a matter of fact, MAs, if properly designed and implemented, can be a valid alternative to classical meta-heuristics in engineering and design. In some cases, MAs can lead to results which are orders of magnitude more accurate and efficient than other popular optimizers.

Andrea Caponio
Technical University of Bari, Via E. Orabona 5, 70121 Bari, Italy
e-mail: caponio@deemail.poliba.it

Ferrante Neri
University of Jyväskylä, P.O. Box 35 (Agora), 40014, University of Jyväskylä, Finland
e-mail: ferrante.neri@jyu.fi

This chapter aims to summarize the main results in the topic of MAs successfully applied to engineering and design. Although not exhaustive, the proposed survey is supposed to give some indications about the main trends and some suggestions about the future of MAs in engineering.

This chapter is structured in the following way. Section 15.2 presents a survey on applications of MAs for real-world problems. In particular, Subsection 15.2.1 focuses on single-objective optimization problems while Subsection 15.2.2 deals with multi-objective optimization problems. Regarding single-objective, a survey on MAs in image processing is given in Subsection 15.2.1.1, in telecommunications in Subsection 15.2.1.2, in electrical and electronic engineering in Subsection 15.2.1.3, and in other fields in Subsection 15.2.1.4. Regarding multi-objective optimization, a survey on MAs in hardware design is presented in Subsection 15.2.2.1, in electrical and electronic engineering in Subsection 15.2.2.2, and in image processing and telecommunications in Subsection 15.2.2.3. Section 15.3 presents a case of study: an ad-hoc MA applied to a specific control engineering problem. Finally, Section 15.4 gives the conclusions of this work and attempts to foresee the future trends in the field.

15.2 Applications of MAs in Engineering Problems

In engineering and applied science, many decision making problems need to meet several objectives: minimize risk, maximize reliability, minimize errors or deviations from desired levels, minimize costs, and so on. The solution to these problems can be found through a single-objective or a multi-objective method: each of these approaches presents different advantages and drawbacks. Whichever of them we follow, when designing MAs, other important questions to answer are which local searchers should be employed and how they should be effectively hybridized within the evolutionary framework and in relation to each other, as highlighted in [489]. Of particular interest are the guidelines that lead to the execution and the coordination of the local searcher(s) in MAs. Some algorithms bluntly apply them to each point generated by the evolutionary framework, resulting in a very thorough search for the optimum, which, on the other hand, can also be extremely slow to converge. Some other algorithms follow instead one or more rules to choose when to launch a local searcher, possibly which one to run, and which individuals should it try to improve on: this logic leads to a less exhaustive, but generally quite faster, optimization.

In this section, we analyze various algorithmic solutions and strategies in Memetic Computing for facing engineering problems. In Subsection 15.2.1, we focus on MAs for single-objective optimization while in Subsection 15.2.2 we focus on MAs for multi-objective engineering problems.

15.2.1 Engineering Applications in Single-Objective Optimization

In single-objective optimization for real-world problems, we search for a solution corresponding to the minimum (or to the maximum) value of a single objective function. Although in real-world situations most of the problems are actually multi-objective, the problems can still be considered single-objective by constructing a fitness function that usually comprehends several different objectives into one. This means establishing an a-priori ranking of importance of the various objectives and implicitly accepting a compromise among them. For example, the scalarized approach, see [599], is a diffuse technique to aggregate various objectives: a weight factor is linked to each objective on the basis of its importance and the weighted sum is optimized. This is a very practical method that leads to a faster optimization process, while implicitly accepting that a ranking of the importance of each objective with respect to the others is already known, thus excluding some solutions that might still be interesting.

In the past years, several different single-objective meta-heuristics were developed and successfully applied to real-world problems. For instance, GAs [325], ES [354], PSO [458], and DE [787], were already widely used in real-world situations, showing extremely good performance. Furthermore, single-objective optimization becomes a mandatory choice when the time available to find a solution is limited, which often happens when dealing with real-time applications. For these reasons, single-objective MAs have been more popular than multi-objective MAs in the past and the greater part of the algorithms proposed in the literature are meant for single-objective optimization.

15.2.1.1 Memetic Algorithms in Image Processing

Many problems in image processing and analysis can be treated as optimization issues: feature extraction and recognition, filtering, image registration, and reconstruction are all situations in which, among a huge set of alternatives, we have to find the one that best solves the problem at hand.

In [888], the Memetic Differential Evolution (MDE), a hybridization of DE with the Hooke-Jeeves Algorithm (HJA) and a Stocastic Local Searcher (SLS), is proposed to design digital visual filters for flaw detection on a roll of paper produced in an industrial process. The two local search algorithms are coordinated by a rule that estimates the fitness diversity among the individuals in the current population. An improvement to this algorithm, namely Enhanced Memetic Differential Evolution (EMDE), that hybridizes the DE framework with Simulated Annealing (SA), SLS and HJA, is proposed in [889]. Particularly interesting is the rule used to coordinate the local search: every 1000 DE fitness evaluations, a measure of the fitness diversity and of the fitness values distribution within the population is computed. Then, according to a probabilistic scheme, one or more local search algorithms are run on selected individuals. In this way, according to the progress of the optimization process, the local search algorithms and the individuals selected are likely to give the best results. The performance offered by EMDE outrun those given by GA, ES, SA, DE and MDE.

In [258] a single solution population MA for the correction of illumination inhomogeneities in images is presented. In this case, the local search algorithm makes use of the gradient of the objective function. The algorithm is compared with ES, and the results show that a memetic approach is promising indeed for the problem under study.

Article [51] deals with discrete tomography reconstruction (DT), a highly multi-modal problem which cannot be properly solved through standard hill-climber algorithms. On the other hand, standard GAs are also not adequate for the DT reconstruction, since they are not originally designed to work with binary matrices. A new evolutionary approach, with crossover and mutation operators designed to handle binary images, is then proposed. In addition, a stochastic hill climb method is applied to each new solution, so that during each stage of the search, all individuals represent a local optimum in the search space. This MA offers good results for several different reconstruction problems, but the thoroughness of the local search algorithm considerably slows down the optimization process, limiting its applicability to images of size 50×50 or less.

In [211], new crossover and mutation operators are designed, and a switch operator and a compactness constraint are applied to the same problem. The resulting algorithm is much more greedy than the one in [51], and is able to process 100×100 binary images in reasonable time, but compatibility between the solutions found and the inputs is not assured.

In [789], Santamaria et al. investigate the effectiveness of MAs for the construction of a 3D model of forensic objects through image registration. Several MAs, based on CHC (which stands for Cross generational elitist selection, Heterogeneous Recombination and cataclysmic mutation), DE and Scatter Search, are compared. The Powell's method, the Solis & Wets method and the crossover-based local search method are used as local search methods. These local search algorithms are integrated into the evolutionary framework by means of two different laws: in one case the local search is applied to random selected individuals, in the other case it is applied to all those offsprings which outperform their own parents. Moreover, this study highlights the importance of a proper memetic design in order to obtain high quality performance in the image registration problem.

Article [564] deals with the problem of image registration for inspection of printed circuit boards arbitrarily placed on a conveyor belt. The GA framework is hybridized with a hill-climb procedure which is applied on every individual which manages to remain the fittest for a predefined number of iterations.

In [498], Kumar et al. apply MAs to feature selection in face recognition, showing that their approach considerably outperforms the most famous *Eigenface* method.

Ali and Topchy, in [9], use a memetic approach to solve the Video Chain Optimization problem. Three different MAs are obtained by hybridizing the GA with three different local searchers: the Next-Ascent Stochastic Hill-Climbing, the NMA and the Estimation Distribution algorithm. The goal of the optimization process is to find the optimum combination of parameter settings, implementation alternatives, and interconnection schemes of several image processing algorithms, in order to deliver the best final picture quality.

In [960] a combination of the ACO and GA with simplex is presented for the problem of setting up a learning model for the "tuned" mask in texture classification: the initial candidate masks are generated by means of the GA with simplex and the

ACO is then used to search the optimal mask. New solutions are created by GA operators.

Article [670] applies the MA proposed in [355], a GA enclosing a SA-like selection scheme, to train a morphological neural network used for image reconstruction problems. The proposed method outperforms the standard training techniques in terms of quality of the reconstructed images.

15.2.1.2 Memetic Algorithms in Telecommunications

Many situations which have to be solved through optimization procedures can also be found in telecommunications. Article [810] deals with signal processing and the problem of blind signal separation, i.e. how to separate a signal from the noise that affects it. The MA described combines a standard GA with a neighborhood local search which is applied to all the new individuals generated by GA. The results encourage the use of MAs for this kind of problem.

In [814], a MA is used to solve the Routing and Wavelength Assignment problem, an NP-complete graph-theoretical problem related to optical networks. The proposed algorithm hybridizes two different heuristics, developed for this specific case, and a GA with application-specific mutation and crossover operators. The probabilities that each of these operators are applied to an individual follow a credit-assignment rule. A more recent study for the same problem is shown in [262], where two MAs are proposed. The first one, using fixed probabilities to apply recombination or mutation, runs the local search on each new solution, pursuing a steady state logic for the survival selection. The second MA proposed is a distributed version of the first one on a network of optimization processes, and allows the exchange of individuals regularly by means of an epidemic algorithm.

In [747], a MA is developed to assign cells to switches in cellular mobile networks: each new individual, generated through recombination or mutation, undergoes a tabu search algorithm. In [748], a multi-population memetic approach is presented for the same problem. Article [749] combines a multi-population compact GA with the tabu search which is applied to each newly generated individual: the proposed MA is able to find a feasible solution and to outperform two comparison optimization algorithms.

Article [441] deals with location area management, another important problem in mobile networks: after introducing an evolutionary approach and a multi-population GA, the paper proposes a MA in which the local search is used to generate the initial population and as the mutation operator.

Paper [785] proposes two hybrid approaches combining a Hopfield Neural Network, used as local search, and GAs, to solve the terminal assignment problem, which involves determining minimum cost links to form a communications network. The first algorithm uses a binary-coded GA following an elitist strategy to transmit the highest fitness individual to the next generation. Each new individual undergoes the local search and the result of the neural algorithm replaces it in the new population. The second algorithm is an integer coded version of the first one.

In [464], Kim at al. propose a novel encoding in a MA to solve the channel assignment problem in frequency division multiple-access wireless communications systems. At first, the GA is applied, and if it fails to significantly improve on the solutions for a pre-defined number of generations, the local searcher is executed on a random individual; similarly, after the local searcher is executed without any improvement for a fixed number of iterations, the GA is invoked again. Crossover and mutation operators are designed in relation to the encoding proposed for the problem under examination.

A GA-based MA for dynamic design of wireless networks is described in [257], while in [786], a MA is used to develop an efficient centralized clustering algorithm for wireless sensor networks: the proposed algorithm mixes a GA with a local searcher which is performed on each new individual.

Neri et al. implemented in [656] the Adaptive Global Local Memetic Algorithm to train a neural network used to solve the resource discovery problem in Peer to Peer networks. Training of neural networks in this context is challenging due to the large number of weights and the (great) amount of noise in the dynamic testing environment. The local searchers used in this algorithm are SA and HJA, and the coordination is done through a parameter, namely ψ, which measures the population diversity and is specially designed for flat fitness landscapes. ψ is also used to control the size of the population which is adaptively adjusted during the optimization process.

In [180], a hybridization of DE and SA, namely Annealed DE (AnDE) is used to solve the spread spectrum radar poly-phase code design problem: the AnDE is fundamentally a DE in which the worst offspring can survive according to a decreasing probability rule inspired by SA.

Article [208] presents a MA integrating ACO and SA to design reliable communication networks: specifically, the SA obtains a seed network topology to initialize the pheromone trails, while the ACO searches for the best network solution using the trails which are continuously updated during the search.

In [480] a GA in which the mutation operator is replaced by the Cut Saturation Algorithm is applied to the problem of optimal backbone design of communication networks.

Article [899] proposes four memetic approaches for frequency modulation sound parameter identification: the GA and the Queen-Bee (QB) algorithm are combined with the random optimization method, while the PSO and DE algorithm are combined with the NMA. Results show that the memetic versions of GA, QB, PSO, and DE outperform their counterparts.

Alabau et al. present a MA for the problem of radio frequency assignment in [7, 8]. In this study the authors exploit an integer coded GA, with two crossover and two mutation operators developed for the problem under study; specifically the first mutation operator uses a greedy algorithm to decide which gene to change in order to obtain the best possible result, while the second mutation operator is based on the tabu search algorithm. Furthermore, the initial population is also generated by means of a greedy algorithm.

In [901], three MAs, differing in the way the heuristic search is applied, are compared for traffic engineering in an Internet Protocol version 6 (IPv6) domain by means of routing optimization.

Article [886] proposes a serialization of GA and SA applied to broadband matching network design for antennas, while in [517] the effectiveness of the subsequent application of GA and a direct search method is investigated for the synthesis of shape-beam array antennas.

In [174] the frequency assignment problem for a GSM network is faced with a MA combining a DE with a penalty assignment strategy for unfeasible solutions, and a local searcher, designed for the problem under study, which is applied to newly generated individuals. Results show that the proposed modifications considerably improve on performance of standard DE for this kind of problems.

15.2.1.3 Memetic Algorithms in Electrical and Electronic Engineering

Evolutionary techniques have been widely employed in electric and electronic engineering in order to solve optimization problems. Lately, MAs have also been applied in the field. Leskinen et al. study the performance of two kinds of MAs on the Electrical Impedance Tomography (EIT) problem in [513]. This paper proposes a comparison of five EAs, two of which are novel MAs employing a self-adaptive DE scheme: in one of them the local search is performed on the scale factor used by DE during the optimization, while in the other it is performed on the generated individual. Results show that the MAs are more promising when the geometrical configuration makes the problem harder to solve, i.e. for more difficult optimization problems.

In [900] a hybrid GA is used for the large Unit Commitment Problem (UCP) in electric power systems, a very complex mixed combinatorial and continuous constrained optimization problem. The proposed algorithm hybridizes a binary coded GA with a modified Lamarckian local searcher.

In [874] a MA based on a GA framework is proposed for performing very large scale integrated-circuit (VLSI) automatic design. The genetic operators are used only for exploration purposes, while exploitation of the promising regions is performed by the local search algorithms. Novel crossover and mutation schemes are proposed for the VLSI design problem. The local search algorithms are applied only to promising points, i.e. points whose fitness is performing above a predetermined threshold value.

Carrano et al., in [108], solve the problem of power distribution system design under load evolution uncertainties with an immune inspired MA. The algorithm presented is a Clonal Selection Algorithm hybridized with a local search algorithm explicitly designed for networks, namely the Network Local Search. This local search is used to improve each local optimum previously detected during the search.

Article [219] presents a MA based on Evolutionary Programming (EP) and SA for the tuning of the proportional-derivative (PD) and proportional-integral-derivative (PID) multi-loop controllers for a two-degree-of-freedom robot manipulator. After each generation of EP, the SA is run on all individuals in the new population so that only the local optima take part in the search. Similarly, in [812] a MA made up of an integer-coded GA and a hill-climb algorithm is used to tune a PID controller for a servo-motor system.

Caponio et al., in [106], propose the Super Fit Memetic Differential Evolution (SFMDE) which hybridizes two different evolutionary approaches and two different local searchers. At first, a PSO algorithm is run to generate some solutions with a high performance. These solutions are then integrated within a population of an evolutionary framework. This evolutionary framework employs the structure of a DE and employs two additional local search algorithms: a Rosenbrock algorithm and the NMA. Both local search algorithms are highly exploitative in comparison with the DE framework, but the Rosenbrock algorithm, being more "thorough", is more capable to finalize the optimization, while the NMA is more keen to further improve some fairly promising solutions. To coordinate the local searchers, a parameter called χ is calculated at the end of each DE generation: the value of χ measures the population diversity and the particular fitness value of the individual displaying the best performance with respect to the others. On the basis of this metric, the algorithm adaptively increases its exploitation pressure or attempts at exploring new search directions. The viability of the SFMDE is proved through some test problems and two real-world problems: the design of a proportional-integral (PI) speed controller of a direct current electric motor, and the design of digital filters for defect detection in paper production (see [888, 889]). A similar real-world problem is addressed in [104] and summarized in greater details in Section 15.3.

In [105] the performance obtained by three mate-heuristics (DE, GA, PSO) and three MAs (MDE, FAMA, SFMDE [104, 105, 888]) is compared in order to optimally design a permanent magnet synchronous motor (PMSM) control system, realized with a Proportional-Integral (PI) scheme. This study shows that a DE-based MA can be successful for this kind of problem; in particular, SFMDE offers the best average performance on the problem examined.

Article [349] compares the results obtained by a MA, a MA with population management, and a real valued GA, in the design of a supplementary controller for high-voltage direct current links to damp oscillations in a power system. Results show that both MAs offer better solutions to the problem than the GA, but the MA with population management has better convergence characteristics.

Hazrati et al. use a MA for pricing and allocation of spinning reserve and energy in restructured power systems in [379]. The target of the optimization is to maximize the market benefits and to minimize the payments to energy and reserves. The proposed algorithm uses the SA to improve, after each generation, the best individual found by a GA framework.

In [945], a MA based on tabu search is proposed for the optimal coordination of power relays: the objective takes into account sensitivity, selectivity, reliability and speed of intervention. Results show that the proposed algorithm is fast and easily finds the optimal solution.

Articles [194, 746] implement a MA for loss reduction in power distribution systems under variable demands: the proposed algorithm optimizes the power distribution network in order to have less switch operations, which generally cause losses.

15 Memetic Algorithms in Engineering and Design   249

The MA is a GA, with a novel chromosome representation and crossover operators, hybridized with a local searcher, a variation of the branch-exchange procedure specific for this application, which is applied on the best solution every 50 generations.

In [921], a GA-based MA is presented for finding the optimal network structure and switching configuration in service restoration in power distribution networks. The proposed algorithm combines a two-stage GA previously applied to this kind of problem, with a local search procedure, a greedy algorithm, and an efficient maximum flow algorithm. The local search, a branch exchange algorithm, is run on all the feasible solutions after each GA iteration.

Crutchley and Zwolinski present in [173] a MA for direct current operating point analysis of non-linear circuits. In this case, a DE framework is supported by a Newton-Raphson solver which has the role of finalizing on the search and kicks in when the DE ceases to considerably improve the best solution.

In [175], a combination of GA and SA is used to compute the optimal scheduling of generator maintenance in power systems. According to a steady state logic, a new individual is always inserted in the next population when it outperforms its best parent; if this does not happen, then the probabilistic acceptance approach of the simple SA is used to decide whether or not the new solution should be included in the population.

Hidalgo et al. propose in [385] a hybrid approach for multi-FPGA (multiple Field Programmable Gate Array) system design: after a predetermined number of compact GA iterations, a local searcher tries to improve upon the best solution by randomly changing its genes.

Article [919] deals with the problem of fault diagnosis in a power transformer: in order to pursue this aim, a probabilistic neural network tuned by means of a combination of PSO and Back Propagation (BP) is used. The two algorithms are serialized so that when the PSO stops improving on the best solution, the BP algorithm is activated in order to find the global optimum.

In [23], a GA and a least square curve fitting method are combined to identify the parameters of some peculiar transistors (NMOS in this case). Also in this case the local searcher is run at the end of the optimization process, i.e. when the GA does not manage to improve on the best solution. Results show that this memetic approach outperforms a simple GA and other standard techniques used for this kind of problem. Article [943] also deals with the parameter identification of electronic devices (MOSFET).The voltage parameter identification is performed by means of a MA in which a hill-climb algorithm assists a GA in generating the first population and in performing mutation operation.

Tian et al. refers in [885] to the problem of circuit maximum power estimation. For this aim, a GA employing two problem-specific components, namely input sharing and bit climbing, has been designed.

Liu et al. optimize the design and the sizing of the power train components of hybrid electric vehicles by means of GAs combined with Sequential Quadratic Programming (SQP) in [529]. In this case the GA is run at first, and when the search slows down the SQP method is applied to 20% of individuals randomly selected among the population, and to the best individual.

In articles [101, 102], a combination of float coded GA and trust region algorithm is proposed for parameters identification of strain and dynamic hysteresis model for magnetostrictive actuators. In [340] a GA is alternated with an approximation based local searcher for the problem of optimal electromagnetic design: the SQP is used in this algorithm, and the velocity of the search is increased by using an approximated model for the local search procedure. The SQP is run cyclically after a predefined number of GA iterations.

In [403], a MA for electromagnetic topology optimization is proposed. A 2-dimensional encoding technique is introduced, along with the corresponding crossover and mutation operator. The GA is used as the main evolutionary algorithm, aided in its search by a novel on/off sensitivity method launched according to a probabilistic rule. The MA was applied to three real-world problems, proving to be a very promising optimization method in the field of electromagnetism.

Article [944] applies a GA/SA hybrid algorithm for the parameters identification of the flux linkage model for switched reluctance motors. Simulated and experimental results prove the accuracy of the model tuned by the proposed technique.

In [459] the acceptance criterion of the SA is used for chromosome selection in a binary GA. The resulting optimization algorithm is used to decide where to place measurement devices for power system state estimation.

Bui and Moon describe in [87] a MA mixing GA and a weak variation of the Fiduccia-Mattheyses algorithm, applied to each individual after crossover and mutation: this algorithm is used for partitioning electronic circuit hyper-graphs into two disjoint graphs of minimum ratio cut. The application of the proposed approach to several benchmark circuit graphs demonstrates its validity.

15.2.1.4 Other Engineering Applications of Memetic Algorithms

MAs were also applied in other fields of engineering, or for problems that do not specifically fit into the categories cited before. Article [854] proposes the use of two MAs to train a neural network for non linear system identification. The first MA is a hybrid between GA and BP, the second is a hybrid between DE and the same BP. In both cases BP is applied to each new individual generated by the evolutionary framework. The authors eventually show that the DE-BP algorithm outperforms the GA-BP and the other reference algorithms in this specific application.

In [261] a particular SA algorithm is developed for the Global Positioning System (GPS) surveying network problem. Since SA is a local searcher and has no evolutionary components, the authors speak of a *memetic SA* because they replace the canonical SA perturbation steps with an internal local search step.

Tagawa et al., in [868, 869], introduce a MA for the optimum design of surface acoustic wave filters: the Variable Neighborhood Search algorithm is applied as local searcher to each new solution and a distance-based mutation is proposed to keep diversity among the population.

Article [724] presents a memetic approach to the problem of smooth map identification in electronic control units for internal combustion engines. A simple local searcher is implemented as mutation operator in a GA framework, thus obtaining a memetic GA.

In [251] a MA incorporating two local optimization operators in a micro GA were used to solve a structural optimization problem. One local searcher is a direct search technique derived from the HJA; this algorithm is used at each generation to improve upon the offspring obtained by applying genetic operators to the population. When this algorithm gets stuck, the second local search algorithm, a hill climber, is applied to get the search out of this impasse. This approach is then applied to design a minimum weight 18-bar truss structure subject to node forces.

Article [877] presents a MA assisted by an adaptive topology Radial Basis Function (RBF) network and variable local models for airfoil shape optimization: after the evolutionary algorithm has been run, its solution is processed by a trust region approach.

Kim et al. present in [463] a MA which hybridizes a clustered GA with a neural network, local search, and random search for parameter identification of rolling element bearings. SQP is adopted as local search algorithm, and a novel random search technique is developed in order to find unexplored regions of the search space.

In [499], a hybrid between GA and tabu search is proposed to minimize production costs of thermal units: at each iteration, the tabu search is used to improve promising solutions, and the results show that the MA is fast and reliable for the problem considered.

Ong et al. propose a surrogated assisted MA for aerodynamic shape design in [679]. Alternating exact and approximated evaluation for aerodynamic performances of wing profiles, the proposed algorithm evolves the population by means of standard operators and applies to all new design points a local search strategy which implements a trust-region framework to interleave the exact and approximate models.

Article [38] provides a comparison of several evolutionary approaches to the problem of optimization of causal infinite impulse response filters with applications to perfect reconstruction quadrature mirror filter banks. Four approaches for this problem are studied. At first, a constrained genetic algorithm searches a promising valley in the fitness landscape, and then the suboptimal filter parameters obtained are further optimized using four different methods: a GA-based "creep code", a gradient-based constrained SQP method, a Quasi-Newton method, and a non-gradient-based downhill Simplex method.

Burke et al. propose a memetic approach for the thermal generator maintenance scheduling problem in [95, 96]. More precisely, in [96], hybridizations of GA with tabu search, a basic hill-climber and a SA are compared for the problem under study, while in [95], the GA combined with tabu search is further modified to produce a multi-stage approach.

In [566], two different strategies are applied and compared for the problem of seismic image analysis. One of the MAs proposed applies the local searcher, a waveform steepest ascent, to each member of the population at every generation of GA. The second approach runs the local search to each individual after a predefined number of GA generations.

Tao, in [875], applies a MA to train a fuzzy neural network controller for a truck backer-upper. GA is chosen as the main evolutionary framework, and at each iteration, some individuals are processed by a BP algorithm while the remaining ones undergo standard genetic crossover and mutation.

In article [176], GAs are combined with a quasi-Newton method to solve the non-linear equation of a helicopter trim model. Zhang et al. in [958] try to solve the problem of inverse acceleration in robots with degrees-of-freedom less than six. The proposed approach makes use of a hybridization of a GA framework with a random search algorithm to avoid the calculation of the inverse Jacobian matrix and the second order influence coefficient matrix.

Article [972] applies a simple MA, made up of a GA and a local searcher applied to each newly generated individual, to the problem of spatial-temporal electroencephalogram dipole estimation, which is an ill-posed not fully determined inverse problem.

In article [693], a MA coupling an EA and a gradient search is designed for optimization of structures under dynamical load. In addition, an artificial neural network was used to control the parameters of the gradient-based algorithm.

15.2.2 Engineering Applications in Multi-Objective Optimization

While single-objective optimization techniques quickly provide a final unique solution, multi-objective algorithms give the chance to fully comprehend and model a problem, describing more thoroughly the connections between objectives and inputs. Multi-objective optimization eventually leads to a set of compromised solutions, known as the Pareto-optimal solution front, each of which minimizes (or maximizes) at least one objective, without simultaneously increasing (or decreasing) one or more of the others. The multi-objective approach is more thorough and usually requires more time, and, besides, once the final set is available, a decision making process is needed to select the most suitable solution. A comparative analysis between single-objective and multi-objective optimization can be found in [132].

Two important problems to solve in designing Multi-Objective Memetic Algorithms (MOMAs) are how to define a local search in a multi-objective environment, and how to optimally balance the global and the local search when dealing with simultaneous competing objectives [411]. In the following, special attention will be paid to how these problems were faced in real-world situations.

15.2.2.1 Memetic Algorithms in Hardware Design

During the design of some specific tools or hardware devices, one must satisfy many conflicting needs and one may be interested in understanding their mutual interactions. For this reason MOMAs have been widely applied to optimally plan hardware. One interesting application can be found in paper [196], where a MOMA is implemented to solve various problems of mechanical shape optimization. The study stresses the importance of a fast and good convergence and on the need to reach a final set of solutions well spread across the Pareto front. The proposed algorithm is realized by mixing a binary encoded NSGA-II [200] with a single objective local searcher, performing the weighted sum approach, which is applied to each non-dominated solution.

In [470], a multi-tiered MOMA for design of quantum cascade lasers is proposed. The evolutionary algorithm used by Kleeman et al. is the General Multi-Objective Parallel algorithm, while the local searcher used, applied after predefined generations throughout the entire process, is a multi-tiered neighborhood search, i.e. a neighborhood search algorithm which changes different alleles according to the number of generations done. The new non-dominated points returned by the local searcher are then reinserted in the population. Several strategies to apply the local search are implemented, and the results obtained are compared.

Article [835] describes the use of MOMAs in aerodynamic shape optimization through computational fluid dynamics. Song integrates within a NSGA-II framework a fitness sharing method in the design space, in addition to the fitness sharing in the objective space. The local searcher used is a single objective SA that tries to cyclically improve on each objective while treating the others as constraints. The SA is run on a certain number of points in the Pareto set: the more successful the previous local search step was, the more points will be selected.

Wang et al. in [917, 918, 919, 920] apply different MOMAs to the optimization of structures under load uncertainties. In all cases the proposed algorithm is a hybridization of multi-objective GA with HJA, but while in [917] the HJA is used as a standard local searcher applied to each solution generated by mutation operator, in [918, 919, 920] it is integrated as a worst-case scenario technique of anti-optimization, leaving to the evolutionary framework the duty to solve the multi-objective optimization. The algorithm presented in [917], is also applied in [871] to the automatic design of a compliant grip-and-move manipulator by topology and shape optimization.

15.2.2.2 Memetic Algorithms in Electric and Electronic Engineering

Electric and electronic engineering problems have also been intensely studied through multi-objective optimization techniques. Article [178] applies a MOMA to the automated synthesis of analog circuits in order to optimize circuit topologies and parameters. The evolutionary framework is implemented ad hoc and application specific crossover and mutation operators are used. The classification procedure is done through the crowded comparison operator introduced in the NSGA-II [200], and SA is applied to each new solution generated by the evolutionary framework and to non-dominated individuals after each ranking process.

In [103] the Cross-Dominance Multi-Objective MA (CDMOMA) is proposed and applied to design the control system of a direct current electric motor. The CDMOMA is composed of a NSGA-II [200] framework and two local searchers: the novel Multi-Objective Rosenbrock Algorithm (MORA) and the Pareto Domination Multi-Objective Simulated Annealing (PDMOSA) proposed in [860]. To coordinate the evolutionary framework and the two local searchers, the algorithm employs the so-called cross dominance concept after each generation. This novel concept consists of the calculation of a metric; this metric, namely λ, represents the mutual dominance between two sets of candidate solutions. This metric is then used, with the aid of a probabilistic scheme, to coordinate the MORA and the PDMOSA within the evolutionary framework. In the logic of the designer, the PDMOSA helps find non-dominated solutions in unexplored areas of the decision space, while the MORA tries to improve the individuals that already have a high quality by exploring their neighborhood. After showing the validity of the CDMOMA with several benchmark functions, the authors apply it to optimally tune a DC motor speed control system.

Mori and Yoshida, in [611], present an efficient power distribution network expansion planning method in the presence of uncertainties. The article presents a novel MOMA based on the Controlled-NSGA-II [197] combined with a local searcher run on the non-dominated points after each generation of the evolutionary framework. Results given prove the efficiency of this method for the problem under study.

In [4] a MOMA is applied to aircraft control system design. A multi-objective GA, working in the decision variable space, is supported by a local search that fine tunes the population directly in the objective space. The results of the local search process are then re-mapped into the decision space by means of an artificial neural network which is trained during the global search process.

Katsumata and Terano in [449] design a MOMA improving Bayesian optimization algorithm with tabu search and Pareto ranking. The proposed algorithm is then applied to an electric equipment configuration problem in a power plant.

In [177], a bi-objective MA is proposed for the optimal design of resonator filters of arbitrary topology. A local search algorithm assists the EA for fitness improvement of candidate circuits, refining their parameters in order to prevent good topologies with non-optimized parameters values from being prematurely discarded. Local search is run on each elite individual after the classification process, and on the topologies of new circuits after the crossover/mutation procedure.

15.2.2.3 Memetic Algorithms in Image Processing and Telecommunications

Some interesting examples of MOMAs in real-world optimization regard image processing and telecommunications problems. In article [754], the authors deal with the problem of intelligent feature extraction of isolated handwritten symbols by means of a multi-objective optimization algorithm: after coding the problem to treat it as a two-objective optimization, Radtke et al. propose a multi-objective GA hybridized with an annealed based heuristic. The MOMA follows a Pareto ranking approach, and the local searcher is applied to each individual generated by the genetic operators; at the end of each generation, the most promising individuals are stored in an archive. Results show that implementing a local searcher considerably improves the convergence speed of a stand-alone GA.

In [474], two different telecommunication problems are studied by means of several Multi-Objective Evolutionary Algorithms and one MOMA, namely the M-PAES, proposed in [472]. Numerical results show how a multi-objective approach can be very successful for this kind of problem.

Martins et al. propose a multi-objective memetic approach to the design of wireless sensor networks in [558]. The proposed algorithm is composed of a global and a local strategy. The global strategy has the role of designing the entire network of sensors, while the local one is used to repair the neighborhood of a failing node in the network.

In [530, 531], a MOMA based on a GA, is applied to the design problem of a capacitated multi-point network. During the search the generation of the new population is done by mixing four methods: an elitism reservation strategy, the shifting Prüfer vector, genetic crossover and mutation and the complete random method. Each of these strategies creates a subpopulation, and these are then merged. Comparisons between the proposed approach, the single objective GA with weighted sum approach and the vector evaluated GA (VEGA), show that the proposed MOMA finds most non-dominated solutions and offers the best performance.

15.3 A Study Case: The Fast Adaptive Memetic Algorithm

In this section, we will further discuss the study presented in [104], in which a Fast Adaptive Memetic Algorithm (FAMA) is used to design on-line and off-line the optimal control system for a Permanent Magnet Synchronous Motor (PMSM). The FAMA is an interesting example of MA applied to real-world problems. The FAMA is composed of an ES evolutionary framework with dynamic population size and two different local search algorithms, the HJA and the NMA. The local search is coordinated by means of an adaptive rule based on the concept of fitness diversity.

15.3.1 An Insight into the Problem

The performance offered by an electric motor is strictly connected to the quality of its control. Although many control structures are available, a common and convenient alternative is the Proportionate Integrator (PI)-based control. This control structure allows, despite its simplicity and low cost, high-performance if properly designed. Thus, an efficient algorithmic solution for tuning PI controllers is a very relevant topic in an industrial environment. In a nutshell, the control system of an electric motor is a device which guarantees that the motor does not encounter malfunctioning when a dynamic operation is performed. In other words, a control system is supposed to guarantee that the motor reacts quickly and accurately to an external event. For example if while a motor is working and an additional torque is suddenly applied (this is a typical scenario in industries), a good control system should ensure that the motor counterbalances the extra torque without damages to the structure. It is important to remark that with damage we do not mean only major damages which immediately compromise the functioning of the motor but also micro-damages which may significantly shorten the life of the devices.

Fig. 15.1 shows the block diagram of a vector controlled PMSM drive studied in [104].

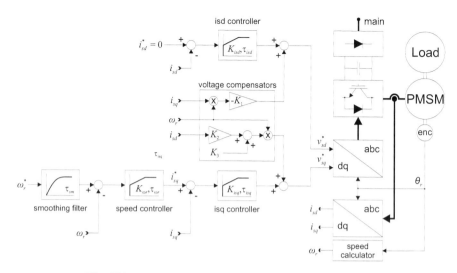

Fig. 15.1. Block Diagram of a vector-controlled PMSM drive

In [104], the main features of a good control system have been conceptualized as the capability of the motor to provide a quick and accurate response to speed command, load disturbance, and measurement noise. Thus, the PI tuning can be seen as a multi-objective optimization problem. More specifically, to evaluate the quality of each solution a training test, made of 8 speed and load torque steps, was designed. Each individual was used in this training test and its performance was given by the fitness in 15.1

$$f = \sum_{i=1}^{4} \left(a_i \cdot \sum_{j=1}^{n_{step}} f_{i,j} \right) \tag{15.1}$$

where j indicates the number of the generic speed step, i indicates the number of the performance index, and a_i is the positive normalization factor of the respective performance index $f_{i,j}$. Specifically, $f_{1,j}$ measures the speed error in the settling phase, $f_{2,j}$ is the overshoot index, $f_{3,j}$ measures the rise time, and $f_{4,j}$ takes account of the undesired d-axis-current oscillations, which increase losses and vibrations in the motor and drive.

It is interesting to notice that, since during the on-line optimization (the fitness function is not calculated by a computer but measured from an actually functioning motor) an unstable solution can be tested, to overcome the danger of possibly stressing the hardware, each performance index is constantly monitored during each experiment so that when a dangerous situation is recognized, the motor is stopped and a penalty factor is applied to the objective value.

15.3.2 Fast Adaptive Memetic Algorithm

The FAMA is a MA based on an ES framework. Initially a set of points is pseudo-randomly generated in the search space. Then, at the end of each iteration, the index ξ is calculated according to equation 15.2:

$$\xi = \begin{cases} \left|\frac{f_{best}-f_{avg}}{f_{best}}\right| & \text{if } \left|\frac{f_{best}-f_{avg}}{f_{best}}\right| \leqslant 1 \\ 1 & \text{if } \left|\frac{f_{best}-f_{avg}}{f_{best}}\right| > 1 \end{cases} \quad (15.2)$$

where f_{best} and f_{avg} are respectively the best and average fitness at the last iteration. Parameter ξ measures the diversity and, indirectly, the current state of convergence of the algorithm: the condition $\xi = 1$ means that there is a high diversity (in terms of fitness) among the individuals of the population and that the solutions are not exploited enough, while when $\xi \to 0$ the convergence is getting closer and since it could be premature, a higher search pressure is needed. According to this logic, the coefficient ξ is used to adaptively set several parameters of the optimization algorithm:

- The size of the population is set according to this rule:

$$S_{pop} = S_{pop}^{f} + S_{pop}^{v}(1-\xi) \quad (15.3)$$

where S_{pop}^{f} is the minimum size of the population deterministically fixed and S_{pop}^{v} is the maximum size of the variable population. When $\xi = 1$ the population contains high diversity and a small number of solutions need to be exploited, if $\xi \to 0$ the population is going to converge and a bigger population size is required to increase the exploration.
- The probability of mutation is set in the following way:

$$p_m = 0.4(1-\xi) \quad (15.4)$$

Furthermore, the value of ξ is also used to decide which local searcher should be run and when: defining η as the number of the current generation, when $(\xi < 0.1)$ AND $(\eta > 8)$ the algorithm is likely to converge soon and the HJA is applied to the best performing individual to refine the final stages of the search. On the contrary, if $(0.05 < \xi < 0.1)$ AND $(\eta > 4)$, the NMA is applied on 11 individuals, i.e. the dimension of the search space $+1$, pseudo-randomly selected in the population, in order to find promising search directions. These two local search algorithms are both direct methods and can be applied to the given objective function which, being non-linear and not-differentiable, and without an explicit analytical expression (the fitness is generated by an experiment and its measures), could not have been tackled with any analytic approach. Furthermore, HJA and NMA show different and complementary behaviors: while the HJA is highly deterministic converging to the closest local optimum, the NMA retains some stochastic features, since its outcome depends on the initial sampling and the solutions are periodically sampled at random (during the shrinking phase).

The FAMA is stopped either when the number of generation η reaches a pre-arranged number, or when the coefficient ξ gets smaller than a predetermined value.

15.3.2.1 Experimental Results

The FAMA was compared with a pure GA and a simplex algorithm for the off-line optimization, and with a pure GA only for the on-line optimization. With off-line optimization we mean that the fitness function is calculated by means of a simulation model of the control system simulated within a computer. With on-line optimization we mean that the fitness is measured by means of experiments on an actual motor and an actual control system. The necessity of repeating the optimization twice allows an initial identification of the interesting region of the decision space which contains the optimum. The optimization must then be replicated by means of the actual devices because the model, although accurate, cannot fully simulate the real-world. As a matter of fact, similar motors of different producers can have very different responses in stress conditions. In addition, even apparently identical motors characterized by the same nameplate might have some different behaviors. Even measurement devices unavoidably influence (although in a minor way) the motor performance. For these reasons, it is important to design a specific control system tailored to the features of the available devices.

Figures 15.2 and 15.3 compare the performance trends obtained in the off-line and in the on-line case respectively. It is worth noticing that experimental results are, as expected, considerably different than the simulation results. This is due to non-linearities and uncertainties of the system, which were impossible to accurately model. In both cases, the results obtained by FAMA are strictly better than the initial commissioning and than the results offered by the other optimization techniques.

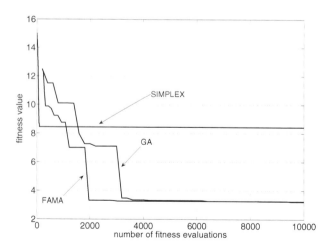

Fig. 15.2. Performance trend of three optimization methods (Simulation result)

15 Memetic Algorithms in Engineering and Design

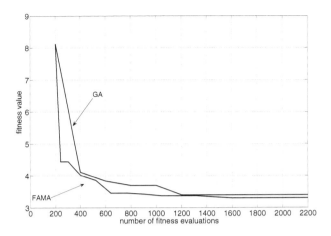

Fig. 15.3. Performance trend of two optimization methods (Experimental result)

The Fast Adaptive Memetic Algorithm presented in [104] is a good example of a MA applied to real-world optimization. Facing a non-differentiable problem which could not be solved with analytical techniques, the FAMA was designed keeping an eye on the peculiarities of the specific context under study. Nonetheless it includes some guidelines which are useful in similar conditions, i.e. when the fitness landscape is highly multi-modal and contains high gradient areas. Finally, FAMA demonstrates that the optimization performance is increased not only by the integration of a local search algorithm within an evolutionary framework, but also by a smart coordination strategy between the algorithmic components.

15.4 Conclusions

Many real-world problems are too complex to be solved by means of standard analytical techniques. In theses situations, direct search methods have become more and more popular. Specifically MAs, joining the exploration characteristics of Evolutionary Algorithms with the exploitative abilities of local searchers, have found a continuously increasing success in engineering problems.

When designing MAs, special attention must be paid to the peculiarities of the specific optimization problem to deal with. Putting together an evolutionary framework with one or more local searchers could not be enough to get good results, and a strategy to combine and harmonize the different components of a MA should be designed.

This chapter offered a panoramic view of several fields in which MAs were successfully applied so far. Researchers could see how different situations have been faced. The most interesting cases were analyzed in more depth and a specific situation, the self commissioning of electric drives for a permanent magnet synchronous motor, was described in detail.

Future trends, in accordance to the No Free Lunch Theorem, will be oriented towards the design of domain-specific MAs for addressing each engineering problem. On the other hand, this trend might lead to the design of overwhelmingly complex optimization algorithms which can require an extensive parameter tuning if minor modifications are made to the original problem (e.g. variation of working conditions). For this reason, in our view a keyword in future MA design in engineering will be "algorithmic robustness". Finally, in our opinion, it will be important that future MAs have a relatively simple structure and are fairly easy to modify and control.

Thus, we think that engineers and computer scientists will attempt to find a compromise between high performance and algorithmic flexibility. A crucial role will be played by the adaptation rules and their capability of being employed in various optimization problems, thus attempting to push towards "the outer limit" of the No Free Lunch Theorem. By giving up a marginal part of the algorithmic performance, future MAs will attempt to solve not only a very specific case but a restricted set of problems having common features. A suitable trade-off will be in our opinion a future topic of discussion. The final aim would be the implantation of "fully intelligent algorithms" which can automatically detect the suitable algorithmic components or might even be able to design the algorithms during the run time on basis of the fitness landscape response without any human decision. Although some interesting work has been already done, completely avoiding human decision within the algorithmic design phase is still very far from achievable

Acknowledgements. This research is supported by the Academy of Finland, Akatemiatutkija 130600, Algorithmic Design Issues in Memetic Computing.

Chapter 16
Memetic Algorithms in Bioinformatics

Regina Berretta, Carlos Cotta, and Pablo Moscato

16.1 Introduction

Bioinformatics is an exciting research field for memetic algorithms (MAs). Its core activity is the integration of techniques from Computer Science, Mathematics and Statistics to address challenging computational problems related with the analysis of large volumes of data. Due to its huge relevance as a means to understand biology in the 21st Century, this field has attracted the attention of many pioneers in MAs, including the authors of this chapter.

During the past two decades, the field of molecular biology and the new high-throughput technologies associated with it has spawned a number of interesting problems. These problems can, in many cases, be posed as optimization problems which are combinatorial, non-linear, and often have aspects of both. Some examples arise in the analysis of large scale genetic datasets (e.g. gene expression using microarrays, massive datasets of single nucleotide polymorphisms derived from genome-wide association studies, etc.).

The field of bioinformatics is characterized by a constant evolution in computational methods for clustering and feature selection, analysis of phylogenetic trees

---

Regina Berretta
Centre for Bioinformatics, Biomarker Discovery and Information-Based Medicine,
School of Electrical Engineering and Computer Science, The University of Newcastle,
University Drive, Callaghan, NSW, 2308, Australia
e-mail: Regina.Berretta@newcastle.edu.au

Carlos Cotta
Escuela Técnica Superior de Ingenieria Informática, Universidad de Málaga,
Campus de Teatinos, 29071 - Málaga, Spain
e-mail: ccottap@lcc.uma.es

Pablo Moscato
Centre for Bioinformatics, Biomarker Discovery and Information-Based Medicine and
Hunter Medical Research Institute, School of Electrical Engineering and Computer Science,
The University of Newcastle, University Drive, Callaghan, NSW, 2308, Australia
e-mail: Pablo.Moscato@newcastle.edu.au

(inference and reconstruction), image processing, protein analysis (structure prediction, sequence alignment), drug therapy design, among many others others. As we said before, many aspects of these problems are combinatorial in nature, involving the selection or the arrangement of discrete objects. Many of these combinatorial problems are NP-optimization problems, thus biologists are generally interested in finding the optimal solution of a given problem, but if that is impossible to obtain, they also rely for their investigations in high-quality solutions, provided by some metaheuristic technique. In this sense, MAs are a good strategy as they can provide solutions quickly, but then if they are coupled to an exact solver (thus forming a complete MA – check chapter 12), they can also prove the optimality of the final solution.

In general, researchers employ exact methods developed by themselves, and highly crafted for the problem at hand, or rely on Integer Programming reformulations of their problems. References in Mathematical Programming, Integer Programming for problems in computational biology can be found in works by Lancia [501] and Althaus *et al.* [15]. A hands-on approach to modeling using commercial packages can be found in [338] and [278]. Our experience with students, coming from different academic backgrounds, also suggest that the book by Williams [936], and the reviews of Greenberg, Hart and Lancia [332] and Festa [259], are not only useful but they have the added value of being very motivational for those interested in crossing fields and to jump into this new area. However, it is clear that since the size of the datasets associated to these challenges problems is, in general, is massive, in many cases it is necessary to develop efficient metaheuristics to deal with the large instances of these problems. As usual, research on metaheuristics is important as it can provide good upper bounding schemes to guide exact search procedures.

This chapter provides an review of MAs that have been developed to address some of the problems mentioned above. For an eagle's view of the contents, in Table 16.1 the reader can find a list of references grouped by application. For the sake of completeness we have also included in this table some applications in the wider area of biomedicine, where applications of memetic algorithms are also manifold. In particular, it is worth mentioning the deployment of MAs for optimizing cancer treatment, both in radiotherapy [347, 348] and chemotherapy [519, 520, 894]. Precisely related to this later issue of drug scheduling we can cite the work of Neri *et al.* for HIV multidrug therapy [658]. Imaging applications in tomography and imaging are also numerous [99, 144, 210, 211, 789] (please check [716] for a review of metaheuristic methods applied to microwave imaging). In the following sections we will focus on the purely bioinformatic tasks defined in the table though.

16.2 Microarray Data Analysis

With the introduction of DNA microarray technologies, it is now possible measure the expression of thousands of genes simultaneously. However, this obviously comes at a price as even a single microarray experiment leads to the need to deal with large datasets. This has posed a challenge primarily for statistics, as researchers

Table 16.1. An overview of MA applications in Bioinformatics

Area	Subarea	Reference
Microarray analysis	clustering	[406, 592, 698, 840, 841]
	gene ordering	[167, 576, 631]
	feature selection	[339, 402, 953, 964, 965, 966]
Phylogenetics	inference and reconstruction	[153, 155, 157, 298, 767, 937]
	consensus tree	[723]
Protein analysis	structure prediction	[53, 148, 150, 495, 496, 677, 790, 959]
	structure comparison	[107, 488]
Molecular design	ligand docking	[373, 612]
	PCR product primer design	[947]
Sequence analysis	DNA sequencing	[218]
	multiple sequence alignment	[883]
	supersequence problem	[151, 297]
Systems biology	cell models	[773]
	gene regulatory network	[465, 466, 671, 842, 843, 893]
Biomedicine	3D reconstruction of forensic objects	[789]
	Radiotherapy	[347, 348]
	Drug therapy design	[519, 520, 658, 894]
	Tomography	[99, 144, 210, 211]

now need to deal with the "large n, small m" problem (where n denotes the number of measurements on a single sample and m is the total number of samples). Statisticians obviously prefer to deal with the reverse situation, with more samples than measurements. When multi-variate methods are required, researchers resort to obtaining "molecular signatures", searching for a more coherent, reliable and robust set of molecular changes [668]. They count on Computer Science (allied of course with statistical methods) for the development of sophisticated algorithms to analyze such data.

The approaches for the analysis of microarray datasets can be primary classified as *unsupervised* and *supervised methods*. At this description level, we can understand that these microarray datasets are basically two-dimensional arrays of values (the measurements) and that a re-assignment of labels to the samples (and, analogously, to the measurements) helps to uncover some structure within the data.

Clustering algorithms are the most common example of unsupervised methods to find these structures. Another unsupervised method, which can be seen as a particular type of clustering algorithm is called *gene ordering*. In this case the overall objective is to find a permutation of either the rows or columns of this two-dimensional array such that those having the same patterns of global expression are relatively close in the permutation. An example of supervised method is feature selection, in which the aim is selecting a subset of features (genes in this case) such that a main goal is optimized, for example, classification accuracy.

We now give a brief description of some MAs that have been proposed to address the clustering and feature selection problems in microarrays.

16.2.1 Clustering

From the description we have given before, it is clear that clustering encompasses a wide number of different problems, as the word "scheduling" in Production Planning and Operations Research encompasses different specific problems. Merz and Zell's proposal [592] for the clustering problem in microarray data analysis is based on a model in which the task is to define an assignment of objects into clusters, such that the sum of squared distances to the centroid of the cluster is minimized. They proposed a MA which uses the K-Means algorithm as a local search technique. They use uniform crossover and they also propose a new one denominated replacement recombination operator. They compare the MA with a multi-start *k*-means local search using five different microarray datasets.

Speer *et al.* used in [840, 841] a Minimum Spanning Tree (MST) to represent the data, where each node is a gene and each edge between nodes i and j represent the dissimilarity between genes i and j, thus modeling the clustering problem as tree partitioning problem, i.e., deleting a set of edges to find the clusters. They proposed a MA based on the framework presented by Merz and Zell in [592]. They use two fitness functions, the sum-of-squared-error criteria (the same used in [592]) and the Davies-Bouldin-Index [186], which minimizes the intra-cluster and maximizes the inter-cluster distances. Using four microarray datasets, they compared the MA with two other popular clustering algorithms, the average linkage algorithm [242] and the Best2Partition [950], which is also based on a MST-representation of the data.

Palacios *et al.* [698] present the results of different population based metaheuristics (genetic algorithms, MAs and estimation of distribution algorithms) to obtain biclusters from microarray datasets. According to the authors, the advantage of finding biclusters in microarray datasets (instead of traditional clusters) stems from the ability to find a group of genes that are similar in a specific subset of samples. To analyze the performance of each algorithm, they used a yeast expression dataset comprising 17 samples on 2,900 probes.

Gene Ordering is another unsupervised method that can be interpreted as a special type of clustering algorithm. The objective is, given a gene expression dataset, to rearrange the genes, such that genes with similar expression patterns stay close to each other. MAs to tackle this problem have been proposed in [167, 576, 631]. In [167], Cotta *et al.* represent a solution as a binary tree, using hierarchical clustering as a start point. The crossover operator is similar to the one used in [155], using subtrees from the parents to create an offspring. Flipping subtrees are used as the model for the mutation operator. Two local searches are applied, the first one works by inverting branches of subtrees and the second one employs a pairwise interchange local search. They test the MA in instances with up to 500 genes. Mendes *et al.* [576] uses the same MA, but with the objective to evaluate the impact of parallel processing in the performance of the MA and ability to apply it in larger instances (up to 1,000 genes). More recently, in [631] these MAs are improved significantly, with the inclusion of new local searches which employ Tabu Search. The MA is tested not only in microarray instances (containing more than 6,000 genes), but as well in images, where the objective is unscramble the rows of an image when the image has all its rows permuted at random. The images are excellent as benchmark instances and help to evaluate gene ordering and different clustering algorithms, making it easier to understand the quality of the results. The MA proposed by Moscato *et al.* [631] has been successfully applied in different microarray studies [63, 170, 330, 397, 577, 768].

16.2.2 Feature Selection

Feature selection methods are used primarily in bioinformatics to reduce the dimensionality of a dataset to help to discriminate between classes of samples under study. We note that the definition of a feature is rather general, it can be a gene expression (as in microarray datasets), a single nucleotide polymorphism (SNP) (as in genome-wide association studies), protein abundances (as in ELISA kit panels), among many others sources of biological information. Feature Selection methods can be classified as *filter* or *wrapper methods*. In filter methods, the features selected are evaluated based only on the characteristic of the data and in the wrapper methods, a classification algorithm is embedded in the method, giving constant feedback regarding the quality of the set of features selected.

Zhu *et al.* [965] present a MA for feature selection problems with the objective to improve classification performance. Each individual in the population is composed of a set of selected features (X) and a set of excluded features (Y). The local search procedure move features between sets X and Y based on some filter ranking methods, such as ReliefF, Gain Ratio and Chi-Square. They evaluated the performance of their approach using four UCI datasets (UC Irvine Machine Learning Repository[1]) and four microarray datasets, showing improvements in the classification accuracy.

In [953], Zhu and Ong present a similar MA, but now using a Markov blanket approach in the local search procedure. In [964], the same authors present a comparison study between the MAs presented in [965] and [953]. They evaluated the results on synthetic and real microarray datasets. Both MAs perform well in regards to classification accuracy, but the one that uses Markov blanket approach gives smaller feature sets. Finally, in [966], they present a memetic framework that combines the previous approaches with a hybridization of wrapper and filter feature selections methods. The computational tests were done in fourteen microarray

---
[1] http://archive.ics.uci.edu/ml/

data sets containing 1,000 to 24,481 genes. They have also tested their methods for hyperspectral imagery classification. The classification accuracy was good and the number of features selected varies depending on the local search used.

Other MAs for feature selection problems were proposed in [339, 402]. However, as stated by Zhu *et al.* [966], due to the inefficient local search methods a large amount of redundant computation is incurred on evaluating the fitness of feature subsets. This is an issue worth considering in detail when designing an MA as we rely on the power of local search, associated with good data structures, to speed-up the process. This is an area of great interest and we hope more sophisticated MAs will be developed during this decade.

16.3 Phylogenetics

The aim of plylogenetics is to study the evolutionary relationship between species, which can be represented by a phylogenetic tree. The inference of phylogenetic trees, known as Phylogeny Problem, is a very challenging task and is certainly important in molecular biology. It has connections with other problem domains in bioinformatics like *multiple sequence alignment, protein structure prediction,* among others [153]. The aim of the Phylogeny Problem is to find the tree (or in certain cases the network), that best represents the evolutionary history of a set of species. Several criteria have been defined in order to measure the quality of a certain tree given certain input data (typically, molecular data corresponding to a collection of different organisms or taxa); these can be broadly grouped into sequence-based methods (such as maximum parsimony and maximum likelihood) and distance-based methods (e.g., minimal ultrametric trees). Unfortunately, NP-hardness has been shown for phylogenetic inference under most of these models [190, 191, 277, 942]). Due to the complexity of the problem, the research focuses in the development of powerful metaheuristics, like MAs [153, 155, 157, 298, 767, 937].

Cotta and Moscato proposed several MAs for hierarchical clustering from distance matrices under a minimum-weight ultrametric tree model (i.e., finding an ultrametric tree of minimal overall weight, such that its associated distance matrix bounds the observed distances from above). The first approaches [155] were based on the use of evolutionary algorithms endowed with heuristic decoders, which could be viewed as greedy hill-climbers for genotype-to-phenotype mapping. Although these provided much better results than other simpler decoder-based approaches and tree-based evolutionary algorithms, their computational cost was also large. Later [157] an orthodox memetic approach was presented based on the use of a tree representation and a local search operator based on tree rotations.

A scatter search method using path relinking was subsequently presented by Cotta [153]. Scatter Search (SS) [314, 320, 500] is a powerful metaheuristic which can be considered as a particular type of MA that often relies more on deterministic strategies rather than randomization. In this work, the author used a ultrametric model and a minimum weight criterion as in previous works [155, 157]. The SS algorithm was evaluated using five real biological data sets from an online repository –the TreeBase site[2]– and was shown to compare favorably to an evolutionary algorithm and a MA. Related to this, Gallardo *et al.* [298] propose an hybrid algorithm that combines Branch and Bound (BnB) and MA in an interleaved way. The idea is to have both techniques sharing information between them. They used the same five biological data sets from as [153] and showed improved results.

Williams and Smith [937] use maximum parsimony as the optimization criteria, which means that the tree with the least evolutionary events is the best. They propose a MA, which uses diverse and elitist populations (similar with the ones used in scatter search methods). More precisely, their approach is based on maintaining a collection of Rec-I-DCM3 trees (Recursive-Iterative DCM3, a powerful heuristic for designing maximum parsimony trees [777]) which cooperate within a selectorecombinative evolutionary algorithm. They evaluate their method using biological datasets with up to 4,114 sequences, obtaining better results than parsimony ratchet [669] and TNT (Tree Analysis using New Technology[3]). Richer *et al.* [767] also uses maximum parsimony as the optimization criteria. They propose a MA that uses progressive neighborhood as local search (similar with VNS - variable neighborhood search [364]). They used twelve instances from TreeBase, and obtained results that were generally equal or better than TNT.

A problem related to phylogenetic inference is that of finding consensus trees, namely finding a tree that summarizes the information comprised in a collection of trees (e.g., finding a unique tree that faithfully amalgamates the outcome of different phylogenetic inference methods). A seminal approach to this problem using evolutionary methods can be found in [152] on the basis of the TreeRank distance measure [916] between trees. Pirkwieser and Raidl [723] tackled this problem using VNS, evolutionary algorithms (EAs), MAs (using EAs endowed with local search on different tree-based neighborhood structures), and multi-level hybrids based on the intertwined execution of VNS and EA/MA which ultimately produced the best results.

16.4 Protein Structure Analysis and Molecular Design

Problems involving analysis of protein structure are fundamental in bioinformatics. We refer to Oakley *et al.* [677] who present a review of problems involving analysis of protein structure (including structure prediction, structure comparison, aggregation of structures, etc.).

The protein structure prediction (PSP) problem aims to find the 3D structure with minimum energy (based in a specific energy model) given the primary sequence of the protein (i.e., the linear sequence of amino acids composing the protein). Krasnogor *et al.* [495] analyzed three main factors affecting the efficacy of evolutionary algorithms for PSP: the encoding scheme, the way illegal shapes are considered by the search, and the energy (fitness) function used. In [148] the protein structure

---

[2] www.treebase.org
[3] http://www.zmuc.dk/public/phylogeny/tnt/

prediction problem on the hydrophobic-polar (HP) model was considered. The HP model [213] is based on classifying each amino acid into two classes: hydrophobic or non-polar (H), and hydrophilic or polar (P), according to their interaction with water molecules. In this case the binary sequence of H/P amino acids is embedded in a cubic lattice subject to non-overlapping constraints, with the aim of maximizing the number of H-H contacts, namely the number of H-H pairs that are adjacent in the lattice. The MA featured the inclusion of a backtracking operator in order to repair infeasible protein configurations. A similar approach was used in [150] in the context of the HPNX energy model, an extension of the HP model in which polar amino acids are split into three classes: positively charged (P), negatively charged (N), and neutral (X). Krasnogor *et al.* [496] presented a multimemetic algorithm for protein structure prediction using four different models (HP in square and triangle lattice, and functional model proteins in the square and diamond lattice). Bazzoli and Tettamanzi [53] also considered the HP cubic lattice model. They presented a MA using a self-adaptive strategy, where the local search is applied with a probability guided by a function similar to the one used in simulated annealing, with the aim to either control exploitation or diversification. According with the authors, the MA was strongly based on the MA proposed by Krasnogor and Smith [491], where the authors compared self-adaptation against other local-search approaches for the traveling salesman problem. Santos and Santos [790] presents a MA for the protein structure problem using 2D triangular HP lattice model, whose main feature was the use of caching in order to reuse computation and speed-up fitness evaluation. The study of Zhao [959] addressed HP models as well. They described several metaheuristics such as MAs, tabu search, ant colony optimization, self-organizing map-based computing approaches and chain growth algorithm PERM, highlighting their advantages and disadvantages.

Protein structure comparison or protein alignment is another important problem in the area of protein structure analysis problem. In this case the goal is to identify structural similarities between proteins. Some MAs developed to deal with this problem can be found in [107, 488, 568, 911]. Carr *et al.* [107] considered the maximum contact map overlap problem. They presented a multimemetic algorithm where a family of local searches is used: selection of the particular local search to be applied depends on the instance, stage of the search or which individual is using it. The MA proposed is a combination of the genetic algorithm proposed by Lancia *et al.* [502] and six different local searches. Their computational results have showed that the results obtained by their method are compatible with the state of art in this problem. Also, Krasnogor [488] proposed a self-generating MA to obtain structural alignment between pair of proteins using the Maximum Contact Map Overlap (MaxCMO) problem as a model. MaxCMO is an alignment of two proteins that maximizes the structural similarity. They tested the approach in four different data sets, of which one was composed of randomly generated proteins and the other three data sets with real world proteins.

A bioinformatics area closely related to protein structure analysis is that of molecular design, which actually can be regarded as a superset of the former. Indeed, conformational analysis, namely determining the low-energy configurations a molecule can adopt is a natural generalization of the protein structure prediction problem (for example, Zacharias *et al.* [954] presented a MA based on a genetic algorithm endowed with simulated annealing to determine the ground state geometry of molecular systems). In general, molecular design is a very hard problem, and numerous evolutionary approaches have been proposed in the literature to deal with problems in this area, e.g., [128, 935].

Ligand docking, i.e., the identification of putative ligands based on the geometry of the latter and that of a receptor site, is a problem within the area of molecular design with paramount interest for structure-based drug discovery. MA approaches to this problem have been proposed by Hart *et al.* [373, 612] using an evolutionary algorithm endowed with the Solis-Wets method for local search (see Chapter 12), aimed to minimize the free energy potential of the docking. This MA is used in the AutoDock[4] software package. MAs have also been used for PCR (Polymerase chain reaction) product primer design [947], taking into account constraints such as primer length, GC content, melting temperature, etc.

16.5 Sequence Analysis

Sequence analysis is arguably one of the lowest-level tasks in bioinformatics, albeit it remains a very important one due to its role in generating the input data for further biological problems. Within this general subarea we can cite problems such as DNA sequencing and the alignment of genomic/proteomic sequences.

DNA sequencing amounts to determining the correct order of nucleotides in a certain DNA sequencing. This order must be ascertained by assembling short fragments of DNA obtained from the fragmentation by chemical or mechanical means of a larger sequence. These fragments are typically randomly distributed across the sequence and partially overlap, thus leading to a permutational problem with strong similarities to that of finding a minimum weight Hamiltonian path. In [218] a spatially-structured evolutionary algorithm endowed with a so-called problem-aware local search (PALS) procedure is presented for this purpose.

Another important problem in sequence analysis is that of aligning sequences of nucleotides or amino acids. This problem actually bears some relationship with sequencing, since the determination of the best overlap among DNA fragments requires finding the best pairwise alignment. The applications of sequence alignment are not limited to this case though; thus, they are very important in phylogenetic studies to cite a relevant example. This alignment problem is easily solvable in polynomial time for two sequences using a dynamic programming approach, but its complexity quickly grows for when a multiple sequence alignment is sought. Not surprisingly, evolutionary methods have been commonly applied to this problem – see [813] for a survey. Some of these evolutionary approaches can be actually regarded as memetic. For example, the evolutionary Clustal/improver presented in [883] incorporates a seeding mechanism (using the outcome of the Clustal[5] software

---
[4] http://autodock.scripps.edu/
[5] http://www.clustal.org/

package) for creating a high quality initial population, and an improvement strategy based on the removal of matched gap columns which can be regarded as a simple form of local search.

Closely related to sequence alignment, the problem of finding the shortest common supersequence (SCS) for a collection of biological sequences stands as another important task. A supersequence of a given sequence is a possibly longer sequence in which all the symbols of the former can be found in the same order (although not necessarily consecutively). Finding the SCS for a given collection of sequences is a NP-hard problem that has been commonly dealt with in metaheuristics [70, 83, 149] including MAs. Thus, Cotta [151] considered a MA defined on the basis of an evolutionary algorithm endowed with a repairing mechanism (based on a greedy heuristic) and a local search operator based on the iterative removal of symbols in the tentative supersequence. Later, Gallardo *et al.* [297] presented a multi-level MA that combined the previous algorithm with a beam search algorithm (see Chapter 12), executed in an intertwined way. This MA was shown to provide much better results than the combined algorithm as stand-alone techniques.

16.6 Systems Biology

Systems biology [13] is a prominent interdisciplinary area of bioscientific research focusing on the holistic study of cellular systems from the perspective of (and using tools from) complex systems and dynamical systems theory. This encompasses the analysis and modeling of cell systems, including the study of networks of genomic/proteomic/metabolomic interactions. The latter are very amenable to the use of network-theoretical results and graph-based algorithmic tools, among which MAs excel. Thus, Spieth *et al.* consider a memetic approach to gene regulatory network modelling using linear weight matrices [924] and S-systems [914]. They use a binary genetic algorithm to evolve the topology of the network, and an evolution strategy to do local search on the parameters of the model representing the network. They consider a so-called *feedback MA* in which the outcome of the local search is used to filter gene dependencies whose strength is below a certain threshold. This can be regarded as a Lamarckian learning procedure, as opposed to the Baldwinian learning of the simpler MA [842] without feedback. An analogous approach is followed by Norman and Iba [671]: they consider time series data of gene expression and use a differential evolution endowed with hill climbing to determine the structure of the network and the kinetic parameters; an information-based criterion is used for fitness evaluation. It is also worth mentioning the work of Kimura *et al.* [465] in which a genetic local search method is used to solve the inference problem in the context of S-systems. In a later work [466], they consider a cooperative approach based on multiple subpopulations and problem decomposition and use golden section search in order to do local improvement. Tsai and Wang [893] consider a differential evolution hybridized with local search for S-system inference too.

A wider perspective on cell models is provided by [773]. They consider the use of P-systems [738], a computing model included in the ampler paradigm of membrane computing [739]. These computational models are inspired by cellular processes, and can be roughly described a system of so-called *membrane structures*, namely permeable (and potentially nested) containers that comprise collections of symbols and grammar-like rules for their evolution. By an appropriate definition of the rules and a wise arrangement of membranes it is possible to carry out an arbitrary computation. The biological inspiration of these systems make them specifically suited for cell modelling and simulation though. Romero-Campero *et al.* use a two-level genetic algorithm to evolve the structure of a P-system: the upper level is devoted to searching in the space of rules, and the lower level performs numerical adjustment of the kinetic parameters determining the probability of application of each rule.

Acknowledgements. C. Cotta is supported by Spanish MICINN under project NEMESIS (TIN2008-05941) and Junta de Andalucía under project TIC-6083.

Part IV
Epilogue

Chapter 17
Memetic Algorithms: The Untold Story

Pablo Moscato

17.1 Motivation, or Something Like That

I believe this is, in some sense, the end. It is, however, only the end of one journey. We are not abandoning our quest as a new destiny is on the horizon. As happens with the end of any cycle, new opportunities and a large number of challenges arise. Certainly, there is much more to be done for memetic algorithms than what we have collectively achieved until now. I truly believe this is the end of one journey, the *end of the beginning* of memetic algorithms. And now, although I am very happy that we have finally established the field, I am equally concerned about becoming *"the establishment"*. Writing and editing a handbook, for somebody who dislikes academic dogmas and appreciates the continuous criticism his own work, the task becomes intrinsically challenging. And now, in a non-rhetorical way I ask, like a child in a car: *"Are we there yet?"* and *"What's next?"*

Paradoxically, this is not the first time I ask myself about the future of MAs, or indeed, if we are at an inflexion point in their seemingly continuous evolution. To use an old cliché, *"it seems like yesterday"* that, when I was returning to La Plata from Pasadena in October 1989, I thought about which strategy I should follow to develop that incipient research field. I was bringing back to my country half a cubic meter of papers and preprints, and all my computer files on the 80MB hard drive of my gorgeous Macintosh SE/30. It was only a week before that flight, and just a few days before Loma Prieta's earthquake to be more precise, that I had finally managed to compile all I knew about MAs at the time, on a technical report that many told me I would never be able to publish (and, to be honest, I never really intended to publish yet I wanted to be public, a great contradiction at the time). I recall that during my flight back to Argentina I thought my country will give me the peace of mind and a more relaxed research atmosphere on which to develop this field. How wrong I was.

Pablo Moscato
Centre for Bioinformatics, Biomarker Discovery and Information-Based Medicine and Hunter Medical Research Institute, School of Electrical Engineering and Computer Science, The University of Newcastle, University Drive, Callaghan, NSW, 2308, Australia
e-mail: Pablo.Moscato@newcastle.edu.au

Against all odds MAs, initially only supported from very distant positions (La Plata, Argentina, and Edinburgh, Scotland), evolved from the collaboration of researchers of two countries that had been recently enemies at war, a few technical reports, a conference paper in 1992, and a couple of papers in journals, it all led to what the field is today. Twenty years have passed, and perhaps it is now the right time, while the memories are still fresh, to tell a few good lessons learned before *"all those moments will be lost in time, like tears in the rain."* (did I mention that I may be using old clichés to disguise a lack of creative inspiration?).

The reader may ask, and with reason, *"what is the point of a nostalgic-induced racconto of events in this quite technical book?"* I have no real defence. There are no good reasons, I am afraid. My first colleague in these endeavours of MAs, Mike Norman, once told me: *"Pablo, you know, Life is a cheeseburger!"*, and when I asked why he completed the thought: *"It has no sense."* But, jokes apart, both our publisher and our readers expect more than a random collection of anecdotal evidence of that. I think there are some interesting stories to tell, linked to the mathematics we knew at the time, and the constraints we faced, that explain many of our decisions. I also think that I have the opportunity to point towards where the future of MAs may be, by letting readers understand the path so far, sometimes hidden by the cold ink of our manuscripts. I like the irony that our Handbook may look at the future of MAs by understanding the past.

What can I offer in return for a few biographical notes? After all, at forty-five, I have not yet won a Grand Slam tournament and I have no previous fashion modelling, singing, or a political career to be writing biographical notes about. I do, however, feel that I have some good stories to tell, including a few mistakes I have made that could help newcomers and our students to avoid committing, and that I can highlight a few good choices that immodesty compel, and dishonesty permit, to account as virtues of intuition or talent instead of just being attributed to mere sheer strikes of luck.

With no further ado, blaming no more motivation than three years of peer-pressure, the several *"you must write all these things down"* I have received from some colleagues, and, perhaps, a short dose of *"keeping-the-record-straight"* on some technical issues and historical remarks, the recollection of these life events may give a privileged insider's view of this fascinating scientific and social phenomena, a narrative from which at least I hope to harvest a smile.

17.2 In the Beginning, There Was no Evolutionary Computation

'How did it all start?' I could say that, at least for me, the work that finally lead to the initial development of MAs really started way before 1988, and does not really involve any major influence of evolutionary computation.

In September 1985, two events happened close in time that in a certain way changed my life. I recall them happening on the same week. I was a fourth year Physics student at University of La Plata, Argentina, when I attended a seminar of somebody who was announced as being: *"Argentina's most famous physicist of all time"*. Either that exact phrase or another tone of equal fanfare was used to introduce him to the audience. After all, his research credentials were of the highest standards; the Virasoro algebra was named after him (he introduced it into string theory in 1970), and many particle physicists in the audience were eagerly waiting for his seminar. Now in Rome, Virasoro, together with Marc Mezard and Giorgio Parisi, were changing focus, actively working on *spin glasses* [66, 231, 467, 594, 712]. That seminar was an *"smörgåsbord"* of information for us geographically challenged undergraduates in a Third World country deprived of Internet access, with no emailing systems, and with a library full of missing journals, volumes and issues. Virasoro presented a *"physicist view"* of a number of things that later will take my full-time dedication including: computer science fundamentals, design of algorithms, theory of computation, Turing Machines, NP-Completeness and combinatorial optimization, Monte Carlo methods, the travelling salesman problem, Simulated Annealing, Hopfield's artificial neural network approach for optimization, and *ultrametricity* [593, 711, 756]. I recall that I wrote down every single word he said that I could not understand, and that was a big set of words indeed. I rushed to the library to try to find the papers he recommended, with a notoriously consistent pattern of bad luck most of the time.

Virasoro was in La Plata for a couple of days. We (the students) thought he had an imposing presence. He really looked like a reincarnated Michelangelo to me, but we dared to ask him to give us photocopies of his latest preprints. I recall several other visits, perhaps in 1986 and 1987, and he was always very helpful and I will always be grateful for that inspiring seminar. He was also very inspiring with stories of how amazingly Parisi was defeating all odds at making hardware and software for the APE computers and the good science coming from those great maverick Italians. Every time he would come he would share his work and other manuscripts he was reading. Each year he would return bringing precious preprints that for me were eye openers of the *"New Physics"*. In that unorthodox way I had my first contact with what could really be called *'Computer Science'*.

Right after Virasoro's seminar, I travelled to Mar del Plata to attend the Festival of the Union Internationale du Cinéma (UNICA), the amateur cinema association. Watching movies from 9AM to midnight was my preferred way for a mid-semester break at the time. What a geek, many would say, but I would say that I met fantastic people at the Festival, completely out of my scientific world. For the purpose of this narrative, however, I should mention only one anecdote. The director of the Festival's Jury was Krzysztof Kieslowski, who had previously shot *Camara Buff* (*"Amator"*, in its original name, 1979) and was screened off competition at the event. Most of his other films (e.g. *Workers*, 1971; *Blind Chance*, 1981; *No End*, 1985), had trouble with the tight censorship existing in Poland at that time. An Argentinean journalist on a press conference, immediately after the screening of *"Amator"*, aimed to challenge Kieslowski with a "reality check". Insolently, yet naively, the journalist dared to fire a "killer" question, something on the lines of: *"Aren't you frustrated as an artist that the general public can not see your films?"*, she said. Kieslowski answer was fast as a lightning strike: *"It is clear that you do not live in Poland"*, he answered. *"In our country we have shoes whose soles fall off, new cars with engines that do not start, why shouldn't we produce films that nobody can see?"* A few years later some of those films were released, then in 1989 he shot *The Decalogue*, and the rest is more or less history, quickly becoming one of the best cinema directors of all times. And I inherit a tiny bit of his courage, against all advice, four years later, to convince myself to write a long manuscript, potentially unpublishable, that I truly enjoyed writing analogous to his movies, hardly anybody would be able to read.

Looking back, I took from that unforgettable week, and the unforgettable scientists and artists I met in between 1985 and 1988, a common lesson: *"take risks, work in your craft with love and passion (otherwise do not do it, it is not worth the effort), take more risks, shoot the movie you want to see, write the paper you want to read, and when you are close to 45 years of age, take bigger risks"*. Both Kieslowski and Virasoro had my age when I first met them in 1985. I am making a reminder note for myself now.

When I returned to La Plata, I undertook a more systematic search at the Physics library without much hope or success; nothing was really very useful on the topics of NP-completeness and computer science, almost nothing about "artificial neural networks" could be found. At the time there was in La Plata one bachelor degree in *"Calculista Científico"*, which, if you completed it, left you in the hard position of deciding if you would be literally translated to English in your CV as *"Scientific Estimator"* or *"Scientific Calculator"*. Although it was a relatively good degree it did not open many doors academically speaking. Rumors existed at the time that these "Scientific Calculators" were sometimes employed by researchers as associates but were not allowed to be co-authors of the papers as *"they have just written the computer programs"*.

However, I did find a great book which gave me an introduction to Computer Science holding hands with a problem [508]. This is something I recommend for the following reason. I recall that one of our colourful Physics academics at La Plata, Victor Kuz, who taught a course on Fluid Mechanics (something that he indeed liked to teach) gave unexpected excellent advice. He may not even remember this. One morning, just before starting his lecture, he said on arrival:

> "Poor you! You will graduate soon, and when you are my age you will probably be one more physicist in a world which will already have one million physicists. I wonder what you will be working on!"

And then he took a piece of chalk and started to give his lecture while our heads were still spinning around that thought. In another opportunity he started the lecture saying something like this:

> "As in life, you must have only one true love, choose one problem and make steady efforts to know everything about it for the rest of your life."

I think the original quote was probably richer in detail, including the advice of having several previous romantic relationships followed by only one true love that you should then marry, but the novel idea, the part that involves scientific research, is original in some circles even today. He may have also said *"challenging problem"*, but that may be me projecting and not him actually saying it, I am not sure. Two great lessons from Victor.

In contrast, many years later, my wife and I were attending a conference in Operations Research. At the event, a relatively renowned expert in combinatorial optimization was giving the following advice to graduate students: *"choose relatively obscure problems, avoid those challenging ones, publish a lot on these obscure problems as the competition will be less on them, attempt all simple problem variations and write more papers, etc."* In essence, build a CV and not a career in science. For us both, it was a scandal. It looked for me as a chapter of: *"How to lose creativity and self-esteem in 10 days"*. My advice would be quite the opposite: choose a challenge early on in your career. Follow your conscience. Choose a challenging problem domain on which everything is being attempted, so you will learn all techniques, take risks. The challenges and the problem will guide you. I can tell from my own experience. Victor Kuz, you were right! The problem I chose was my mentor, my ultimate PhD advisor !

With the book I found [508], I chose a problem as early as 1986, the Traveling Salesman Problem (TSP). And while I was finishing the fifth and last year of my undergraduate Physics degree, I was studying the TSP book while at the same time was reading everything that I could find in the areas of simulated annealing (SA), artificial neural networks, and statistical mechanics of disordered systems. The TSP was my mentor and that book was my guide. I was truly reading everything I could find on this problem. With my Timex Sinclair 2068 (48K RAM + 24K ROM, my first computer) I wrote a program in BASIC using SA according to the physicists rules of thumb of the time, including different annealing schedules, as well as randomized versions of the constructive heuristics. Quickly it became clear to me that the so-called good properties of SA were more the product of insufficient experimental testing against the existing Computer Science heuristics. It was not until further empirical testing with programs I wrote in Pascal in 1987 and early 1988 which run on a Digital MicroVAX and an IBM PC XT 286, that I truly convinced myself that SA was not delivering to its high promises (no matter the high praise of the physics community). I became, now officially, a professional sceptic, secretly challenging the dogma.

It is clear, at least for me, that scepticism is a key attribute in research, but you do not develop a career by just being a sceptic. You must be able to invent something. In 1986, I was seeking how to make these new ideas from Physics work well against standard methods in Computer Science. My interest on Hopfield neural networks for the TSP soon moved to a less enthusiastic view of the field after reading about other approaches from Operations Research and Management Science, including Lagrangian Relaxation for integer programming formulations of NP-optimization problems. I also explored the *Boltzmann machine* paradigm [1, 2] to address these instances and wrote a couple of programs for the TSP. While I was still not convinced of these approaches, I got the first idea trying to sort out the problem of an undesirable parameter of Hopfield networks, the so-called 'temperature'. After all, similar issues were lurking in SA and Boltzmann machines. The beginning of a long battle, as I consider that any "parameter" you have in an algorithm often shows a design flaw you have not been able to address by other means. In this case, too high a temperature, a Hopfield network would not converge to a configuration that would satisfy the restrictions imposed by the problem constraints and would not produce feasible solutions; too cold a temperature and it will do it, at the cost of not "exploring" other alternatives, thus preventing the network to visit a "region of configuration space" where low lying local minima probably exist. *What if the solution to this problem would be* to use more than one *Hopfield network, each one working at a different temperature and each one coding for feasible solutions of the problem?*

This idea was probably inspired from Lapedes and Farber's approach that used two interacting neural networks in which one network "teaches" another one to perform a content addressable memory task [505]. I thought that this idea was promising, and that perhaps "emergent" phenomena could be achieved by having a collection of neural networks all addressing the same optimization problem.

I noticed that my Timex Sinclair 2068 could handle only two of these Hopfield networks I was running simulations with (due to its memory limitations), so I started to experiment on different ways by which I could couple the two neural networks to obtain better solutions. The basic idea was that when one of the networks would find a better solution than the other (which means a smaller tour) it will switch to a colder temperature ("fix" this configuration) while the other network will have an increment of temperature, allowing it to continue the exploration of configuration space. I recall I needed to install an oscillating mechanism allowing the coupling to vary, so that the higher temperature network at the time could have an independence of action during some time period. It looked a good idea on the drawing board, but would it really work in practice? The answer, as usually in this field, is both a *'Yes'* and a *'No'*.

My "pretty good" idea had a kind of fatal flaw; the combined coupling system of two networks was also creating a number of extra local minima, which were detrimental to the search process of my population of two networks. Overall, it seemed that the variance of the final results were greater than those obtained with a single network. Sometimes I would obtain dramatically better results, but in many cases the networks were trapped in low-quality local optima, sometimes not even a feasible solution of the original TSP problem instance. And, overall, the whole enterprise seemed to me, in comparison with my experiments with SA, a colossal waste of CPU time. I knew that, after all, I was simulating on a computer a *"physical computation"* device, a process that could be eventually embedded in custom hardware. However, I had the impression that if I needed something for local optimization by a *"software agent"* (a Hopfield neural network in this case), that "something" should be much faster. Efficient local search algorithms were key, but I got the first insight that there might be a way of recombining partial solutions to explore the configuration space. The "exploratory" long jump could be guided by the solutions already obtained.

17.3 Caltech and the Red Door Cafe

In November 1987, just around the time I was defending my degree thesis in La Plata, I won a Rotary Foundation Scholarship to do one year of full-time graduate studies overseas. That was the first time I would be paid to do research! *"Crime pays!"*, my friends used to tease me. The title of my proposal was: *"Collective computational properties of neural networks"*, with an intention to expand that research into utilizing several networks concurrently solving the same problem. Since there were many talented researchers with interests in neural networks, Complex Systems and Physics in Caltech, I wrote a letter to Steve Koonin, Chairman of the Faculty, in which I basically described what I had been doing (including my work on the t-expansion, a non-perturbative analytic method for the calculation of ground-state energies for Hamiltonian systems in lattice gauge theories, as well as my interests in the statistical mechanics of disordered systems). I also asked Osvaldo Civitarese, one of my lecturers in La Plata, to write a recommendation letter for me (Osvaldo had just returned from Caltech, where he used the same desk that Hans Bethe had used in an office next to W. Fowler and he was very supportive of me going there). Koonin might have passed the letters around and Geoffrey C. Fox sent me a telegram (yes, a telegram, I am that old) that basically said something as brief as this: *"I agree to be your supervisor. Come here."*. I rushed to the library to see who he was; I found that he had worked with Richard Feynman, Robert L. Walker and with Stephen Wolfram. Not bad for a landing!

Caltech then accepted me and also provided a Special Scholarship that combined with Rotary's support would pay for a whole year of studies and living expenses in Pasadena. The ticket finally arrived twenty-three hours before the flight's departure time, and only a week before the beginning of classes, and finally I was on my first international trip!

Caltech was great, and probably still is, at networking its people to produce innovative ideas. Already at the graduate student welcoming party I met fellow arriving students that I immediately started respecting as great professionals, and truly valued friends: Jose Tierno (coming from Montevideo, Uruguay), Enrico Santi (from Padova, Italy), Edoardo Amaldi (from EPFL, Lausanne), David MacKay (from Cambridge, UK), etc., and with some of them I have maintained close contact through all these years. Curiously, a few days later I discovered that Edoardo had been working at Virasoro's group in Rome for a short period and that he also had Fox as advisor. Amaldi was originally coming from EPFL Lausanne (he is now at Politecnico di Milano). It was Fox's idea to put me in the same 9 square meters office with Edoardo and another Italian, Roberto Battiti (who, as far as I remember spent most of the year working in computing optical flows [52], shape-from-shading, and speeding-up backpropagation learning with more sophisticated conjugate-gradient algorithms). With three big desks, three computers, shelves and a window (yes, we had sunlight !), I recall we had to tell each other when we were going to stand up or move a chair. It was a bit packed, like a trio of masters of Italian football "catenaccio" (Amaldi, Moscato and Battiti) but it was also mutually inspiring and, measuring by Caltech standards, a lot of fun.

Edoardo immediately resonated with me on my scepticism of SA for combinatorial optimization. He had the view that we should be using Tabu Search instead (*"you should use Tabu Search for everything!"*, he used to say literally or by using some other form of "subtle" Italian advice). He was quite right in his excitement, to some extent. A large extent, I should add. Tabu Search was also an incipient methodology at the time, which Edoardo was aware of due to Fred Glover's visits to EPFL, Glover's seminars and Edoardo's own experimentation with it on neural networks with Ising bonds (a joint work with S. Nicolis, while in Virasoro's lab in Rome) [17]. Tabu Search thus enters into the world of MAs via a similar mechanism; it came mainly due the dissapointment we had with SA, but it took until 1990-92 to use it more systematically in population-based approaches for optimization. It did prove very useful [616] and even two decades later we still use this combination [631].

As a student, I discovered that most of the courses in Physics at Caltech included some sort of reiteration of subjects that I had already covered in my last year in La Plata, sometimes using the same textbooks (as a B.Sc. in Physics was a five-year degree, several of the courses contained material that highly overlapped my previous courses taken two years before). Since my research plan for the Fellowship was highly interdisciplinary, I took courses that were valid for Physics but offered and aligned with other disciplines. I will discuss two of them: *Theory of Neural Networks* and two trimester courses in *Computational Physics* (which were mainly related to Concurrent Computation, and were based on the work by Fox and other members of his group who were pioneering the use of hypercubes for scientific computing). The former course was coordinated by Yaser Abu-Mostafa (still at Caltech) and J. Stephen Judd (now at University of Pennsylvania and Princeton) and it was three trimesters long. David, Edoardo, Jose and myself took it. For me that course was an eye-opener and I enjoyed it a lot. We covered issues on neural networks that related to Information Theory, computational complexity, generalization, learning theory and the Vapnik-Chervonenkis dimension [702, 703] (an absolutely novel concept at the time). I believe that those courses had an influence in Edoardo, Jose, David and me; in my case they were part of my final "informal re-entry program" into Computer Science.

In the other course, early in October 1988, Fox gave a lecture about the application of SA for the TSP on hypercube computers that made me think that the approach was really wrong, doomed, even. Basically, he described a strategy called *"domain decomposition"*; each processor of the hypercube computer was receiving only a subset of the cities, which led to a lot of problems inherent to this algorithmic design decision. I recalled that I thought that it should be better to follow an approach like the one I used for my neural nets, so that each of the processors can have the whole problem instance (the inter-cities distance matrix). I asked Fox to have a meeting to discuss my idea. To test if there was an improvement, I could employ the same basic SA I was very familiar with (and witness if true progress was being made). I had plans for a competitive interaction between optimization agents, but a mechanism for sharing information (which we later called "cooperation") was missing. He suggested: *"Perhaps what you are looking for is something that some people called 'crossover'. A bright young guy that I have hired for a couple of months gave a talk about that, why don't you talk with him"*. That "guy" was Michael G. Norman, who was coming from the Edinburgh University (the group that later became Edinburgh Parallel Computing Centre (EPCC)), and had given a talk about a new idea called *"Genetic Algorithms"*.

Three years later, when I gave a seminar at the EPCC in Dec. 1991, my opening line was: *"Memetic algorithms were born from previous work and also from two deceptions, the one that I had with Simulated Annealing, and the one that Mike Norman had with Genetic Algorithms."* Mike was at Caltech to set up a new type of hardware and software facility for Fox's group, the Meiko Computing Surface, a beauty that had 32 T800 transputers. Mike had worked on implementations of parallel SA, following Geman and Geman [306]) on the much larger Computing Surface at the University of Edinburgh. He had also looked at the problem of optimizing the topology by which the processors of the Computing Surface were connected, using a Genetic Algorithm and the *Order Crossover* operator, previously introduced by Lawrence Davis for TSP. His paper was rejected for publication and never published, a theme we will return to below.

The GA parallelizes in a different way from SA because the population itself can be split across multiple processors. However, the population evaluation (ranking etc.) of a traditional Genetic Algorithm (without any sort of hybridization) that determines the "survival of the fittest" is a global operation, which is inefficient to implement on a multi-processor machine because it introduces communication overhead and synchronization delays. To avoid this, influenced by a seminar from Hans Mühlenbein who had visited Edinburgh in 1988, Mike started to introduce a concept of locality into his GA. He ran a standard GA on an individual processor (thereby defining a sub-population local to that processor), and then sent random members to other random sub-populations which were accepted into those populations with a probability depending on their fitness relative to the receiving population. Mike noticed that this approach also reduced the phenomenon of premature convergence, where slight variations on a "quite good" solution came to dominate a population (basically different "sub-species" can evolve in each subpopulation). It also matched well my formed belief on how to search a "quasi ultrametric" structure in configuration space, so it looked as the right thing to do.

The Meiko was late arriving so Mike, in a friendly rivalry with Steve Otto (with whom he shared a house), had spent a few weeks implementing an Order Crossover based Genetic Algorithm for the Traveling Salesman Problem in C, first on an early IBM RS/6000 that nobody wanted to use, and then on a i386-based Sun Workstation (the ill-fated Roadrunner) that also nobody wanted to use. The GA was no better than the SA, but for different reasons. The Order Crossover operator added huge amounts of what we perceived as "noise", i.e. the introduction of a small number of long edges that, with high probability, caused the newly generated solutions to be rejected from the population. This ultimately led to severe premature convergence, since only solutions that were very similar could be crossed over without introducing long edges.

When we met and started discussions it was clear that Fox had actually put us on the route to solving both of our problems. The Order Crossover operator could provide the mechanism for sharing information, and the SA could quickly remove the "noise" generated by the recombination procedure. We then discussed ways of elegantly leveraging the locality of the Computing Surface, analogous to the locality of competition in the natural world. But even during our first discussion we convinced ourselves that the locality for competitive interactions could be made different than the one for "cooperation" (i.e. the recombination algorithm). This would prevent the recombination of solutions that could be very similar (an issue which is key in MAs even today). In this way we designed the topology on a toroidal grid that was the first population structure introduced in our MAs.

After our first meeting, Mike challenged me to see if I could put some of our ideas into *"a decent pseudocode that I can understand and implement"*, a task that I rushed to complete that same night, aiming to address any problematic corner that may be remaining. It was not a long description, perhaps only seven pages long, but in the morning I left the pseudocode on his desk. He called me back in the afternoon when he arrived to work (Mike's office had no windows and he worked nights) and after a few explanations, more discussions and iterations on the same day he decided he would code it by amending the codebase already written for TSP GAs. In less than a week our first MA was running on the Sun Roadrunner. It was subsequently ported to the Meiko Computing Surface when it arrived, and also to the Intel Hypercube.

He basically did what he was paid for and also wrote the code of our first MA at the same time. It would have taken me a couple of years I guess. I was still in technological shock with my rudiments of Pascal and MS DOS, in a new world of UNIX and C, to achieve anything like that. We were very impressed with our own creation; this hybrid was clearly much better than the sum of parts and only borrowed the OX operator from GAs. The rest was a truly original new design. The first MA was then a solution for the problem of efficiently running a parallel optimization heuristic on an MIMD machine, an instance of a new paradigm for stochastic search. Several of the design decisions of our first MA (the ring topology, four interconnections for each agent, different neighbourhoods) were based on the first hardware that was used to run it, a Meiko Computing Surface. The core codebase still compiles and runs (although the graphics libraries for the UI which displayed tours and population fitness no longer exist).

From that day on, Mike and I would meet at the Red Door Café, *"the only real place in Caltech"*, as Mike used to say and I used to agree. A real student's café, with bits of borrowed or used furniture that did not match, but it did a decent espresso (Mike needs this) and cheesecake. It was not the high tech place of today, perhaps Mike will find it a scandal, now it has a .pdf file of the menu available from a website, with no cheesecake! It was a good and relatively quiet place to discuss and to get away from the neat spaces. It was conveniently located half-way from where we both had our offices. During these meetings we discussed changes of our basic MA (not many, I would say) but also my first computational experience on the TSP instances that I had worked on for the past two years. We were very soon convinced we had something "big" in our hands, and we crafted on a table what the title of the technical report should be. It should contain the words *'Cooperative'*, *'competitive', 'complex', 'combinatorial', 'search'* and Mike insisted on *"agents"*, as our approach was driven by the complex interaction of *'software agents'* as we envisioned that they can have different algorithms running on each of the agents [615]. We went together to a seminar that Steve Otto was giving, in which he explained their method (basically the one that Fox was describing in his lectures) and it gave us more conviction that this approach was, I would not say wrong, but intrinsically uninteresting from our point of view. We thought: *"What is the point of using one hundred processors to achieve a less than linear speed-up of a heuristic that searches a space of* 10^{2000} *configurations ?"* Indeed you need to find a "collective computational strategy" that produces some kind of "emergent" behaviour, something that brings a superlinear speed-up when you use a population of interacting agents, instead of a single one (this idea has been further explored in the following years, see [678] and references therein). For us, it was recombination and the interspersed phases of cooperation and competition which was the secret of our early success, and we planned to exploit them on some "cool" applications.

Before Mike left Caltech, when his contract expired at the end of November, I showed him a drawing and a new set of rules for a modification of the basic population structure (the ring topology). I was proposing to use a ternary tree instead. It took more than two years to finally start experimenting with that idea, when I returned to Argentina, but I recall how early we concluded that it was the way to go. The ternary tree structure was also crafted with the transputers in mind, it was adopted in many of our papers (for instance, see [631]), and it has proved to be a robust strategy (see for instance the experiments in [94]).

I spent most of the year 1989, while still at Caltech, engaged in full-time student duties while I was also experimenting with the basic MA code as implemented by Mike during our short collaboration (which lasted less than one month!). Mike had ported it to the Intel Hypercubes, SUN workstations and, of course, the Meiko, in a couple of days. It showed linear speed-ups and we were very happy in all respects. We put together our first technical report [673]. On his way back to Edinburgh, Mike shared Thanksgiving Dinner with Lawrence Davis in Cambridge, MA (later Davis was the President of Tica Associates) and told him about our results. Mike then soon embarked on PhD studies and defended a thesis that had produced some remarkable theoretical results that are still highly influential [675, 676].

17.4 Landscapes and the Correlation of Local Optima

"But why does this thing work so well?" I spent a lot of time at Caltech running simulations to try to figure out the reasons. Of course, some of the features were there by design, but a few others were unexpected. For instance, a 16-agents population was performing more-or-less equally to one with 128 agents. In fact, it was perhaps slightly better to use 16. Initially, I was not obtaining quantitative data, but to qualitatively understand how to make it faster and better, without losing the key ingredients of its success. Typically, on instances having up to 500 cities (including the famous `att532.tsp` instance) our runs were ending close to one percent of

the optimal tour length. The algorithm was remarkably insensitive to the annealing schedule that the agents were using and it was not very dependent on random mutation. This was something that I expected due to its design, but the level of its robustness was still a surprise to me. Clearly, the key was the synergy between the local search methods we were using (a subset of the 3-opt "moves" for the TSP) and the recombination method.

I was soon sharing the excitement of these discoveries with some people (both my fellow students and my lecturers). One of my lecturers, J. Stephen Judd, who really had a lot of patience with me, was sceptical at best. He was a good help all the time and together with Yaser Abu-Mostafa they were great mentors in Computer Science. Stephen told me he had tried GAs before and in his experience *"GAs do not work"*. He could not see much difference still between GAs and our approach, for lack of understanding of the area I guess. But he still had an open mind after seeing our results on the TSP; he also gave me a working manuscript illustrating his experiments. He tried to use GAs (no local search at all there, a real GA, *alla* Holland) for a variation of one of his loading problems in neural nets [437]. He gave me a manuscript of his experiments and he challenged me to "make it work". He respected L. Darrell Whitley's work and I think he had a curiosity to see if somebody else could make GAs work where he could not. Judd's incomplete but long manuscript was a very interesting read, and when I finished I went to talk with him. My "diagnosis" was that the problem was with the recombination operator he was using. He had tried a 1-point crossover for a binary encoding (or a uniform crossover, I do not recall, but it was a simple mechanism only). For the problem he was using, indeed the GA reduced to nothing more than a random search. He needed, in my view, something more sophisticated to deal with the *depth* of these networks. I had a similar experience while trying to help Shailesh Hegde, another of my classmates, improve his GAs for learning in neural networks [600]. In Judd's problem, a manuscript which I have unfortunately recently lost after so many years. I recall that the problem was inherent to the depth of the network, one of the parameters of its architecture.

Judd's attempt on GAs gave me the impression that the type of recombination should be highly problem-dependent and, most worrisome, instance-dependent. For these problems, it seemed that a recombination operator should utterly take into consideration that there is a strong dependence between attributes encoding for solutions. Given two solutions (encoded as a structure that possesses a linear ordering of attributes) the general heuristic rule in evolutionary algorithms is that a recombination operator must preserve the "common" parts. However, in this case a special consideration should be given to start from the end that contains the attributes of which the others depend. In Judd's loading problem, the values of attribute b would strongly depend on all the values given to attributes a if $a < b$. In my mind, I linked this with my previous readings about ultrametricity. For instance, if you have a set of words of arbitrary length (finite or infinite) over some alphabet Σ, a distance between two different words x, y can be defined as $d(x,y) = 2^{-n}$, where n is the first symbol at which the words differ. This distance measure is ultrametric. This said, a recombination operator that acts on these words and preserves the common parts until the first symbol on which they differ will, by construction, regardless of which algorithm is used to complete the process, always produce a new and different word which will be at most at the same distance to one of both parents. A given problem, I thought, may have different representations that have ultrametricity properties, and among those, one is perhaps very closely "correlated" with the cost function, so that closer local minima have very similar objective function values. Many years later this concept would be efficiently exploited by Peter Merz in his now classical *"Distance Preserving Crossover"* for the TSP.

I spotted a connection. If the solutions used as input for the recombination algorithm were the product of highly optimized configurations, they may have some sort of "quasi-ultrametric" structure. If a high correlation with the values of the objective functions exists this combination could be exploited. This was something at which perhaps a local search heuristic could be intrinsically limited because this structure involves some global property of the feasible solutions. If a recombination operator could somehow be designed to exploit this structure, then the stochastic nature of these population-based heuristics could help us explore this "reduced" configuration space. In essence, the individual optimizing processes after recombination act as a "repair" mechanism, a kind of "projection", that brings us back to an element of a subset of configurations that can be stochastically searched by recombination to reinitiate the search process. There were, however, some extra conditions needed. If we were seeking the minimal cost solutions for an optimization problem, we would like some kind of "smoothness" to be present. If two highly optimized feasible solutions were obtained by some individual search strategy, the closer they were (according to a given metric) the more similar their costs should be. Local minima should then be highly correlated for this strategy to work better than completely random search.

As early as January 1989, I started to put all these ideas onto a LaTeX file, working on it incrementally every day, while at the same time I was a full-time student in Physics and conducting research in other areas. I knew from the start that it would be "big", and probably I would never be able to publish it, so I called the file "bigone.tex" and started to work at including all the ideas and evidence that I had to the inner workings of our MAs. However, it still was not called "memetic algorithms". During that winter, and while discussing what I was doing with a fellow classmate, Scott John, in a Kung-Fu course I was taking, he recommended the book of Richard Dawkins, *The Selfish Gene*. After reading Dawkins' discussion on cultural evolution and memes, I thought that by calling this new type of method *'memetic'*, I might liberate researchers from the unnecessary corset imposed to their creativity by "emulating" biological evolution. I had no other intention, as it has been reported somewhere, to emulate "cultural evolution". If it seems I have accidentally done so in some manuscript, I would regret my improper use of words. I may have dropped my guard by saying "some aspect" of cultural evolution (like the tightly regulated evolution of a martial art), but how to emulate such an "inherently unregulated" thing like that escapes my imagination today. My only quest was really to motivate researchers to think outside of the box, not to bring a new, slightly wider corset on creativity, on which to restrict the development of new methods. I also chose to introduce the word *'Towards'* in the title of that report, as I knew that it would take a lot of time, decades perhaps, until the full development of some of these ideas would materialize. In some sense, there are many in that technical report that are still developing, brewing in different places around the world, so I still think we are moving towards memetic algorithms. A lot is yet to be done in memetic computing.

Shailesh Hegde convinced me to apply for a student grant that finally paid us both to attend the Third International Conference in Genetic Algorithms in Fairfax, Virginia in which he presented a later highly cited paper [600]. I was not presenting anything, not even an abstract there, but I guess I was accepted as we were coming from a respectable university and that was part of the outreach aspect of the conference. A very generous offer by the organizers. We arrived in early June and we stayed with some of Shailesh's friends in Washington D.C., watching together in horror the final outcome of the Tiananmen Square protests on TV. Then the conference started. To be honest, everything had a glimpse of irrelevance for me after all these events. However, in retrospect, having been at that conference was very useful and very relevant. I met Christopher L. Huntley and Donald E. Brown as well as Martina Gorges-Schleuter. I found that Mike Norman and I were not alone in our quest, and I decided I would report on the similarities I found between our approaches in the technical report I had been preparing since January.

At the conference, however, there was a session with an open debate on *"the future of GAs"* or something like that. I may be making this up, but at a certain point of the debate that was the main theme. At question time, somebody from the audience stood up and said that in his opinion, we should all move towards new types of hybrid algorithms, and he said: *"I have heard of very interesting work on hybrid algorithms being done at Caltech"*, etc. and I was kind of astonished. He was talking about our work! How did he know? Who was this person? Curiously, he was immediately, severely, and undeservedly in my opinion, "reprimanded" for such an undogmatic thought, by a highly regarded member of the GA community who now, 20 years later, proclaims to be keen on hybrids. My mother told me a lesson many years ago: *"never argue with a fool in public because half of the people listening may not be able to tell the difference"*. I am not saying that anybody there was a fool, they all may have reasons to argue against a change in the dogma, including political ones or a hidden agenda. But I also want to note that I also live by Richard Feynman's rule: *"The first principle is that you must not fool yourself - and you are the easiest person to fool"*. The fool in question might actually have been me in that discussion, I did not have much evidence after all. So, basically, after all this philosophical "méandre" that does not really lead anywhere, I would say, simply, that on the occasion, I just "chickened out" and kept my mouth closed. I did not defend the reprimanded early-adopter, early-defender of "hybrids". I should have done it, and over the years I regretted that. However, I did what my conscience and curiosity obliged me to do. After the debate finished, I approached the champion of the cause to introduce myself and to thank him for mentioning our work. To my surprise, the person in question was Lawrence Davis, the researcher that Mike had visited on his way back to Edinburgh seven months before. We chatted a bit, not much as the conference was finishing, and we both went back to our cities, to do what I think is far more constructive for a researcher. Davis went to Cambridge, Massachusets, and later presented his perspective on the importance of hybrids in *Handbook of Genetic Algorithms* (already cited more than 5,000 times) [188]. In my case I returned to Pasadena and completed the technical report I had been working on since January, now adapting it to also include the work of Huntley and Gorges-Scheleuter, giving them, as well as everybody supporting this view, deserved praise. I tried to find the good working features, the connections [615]. I paid particular attention to cite other researchers that pave the way before us like Kase and Nishiyama [444], Brady [81] and Kauffman and Levin [450], as well as those I met there like Gorges-Scheleuter, Huntley and many others. I felt, however, that I was very alone in my quest, and until June 2007, although I participated in dozens of program committees for meetings on every continent, I did not return to another Evolutionary Computation conference. I sometimes wonder if that experience could have had some influence, but I think ultimately the results, and not sterile political debates, should lead the way. I proactively championed the field, always requesting high standards for MAs whenever I could, either by giving formal and informal advice, either as reviewer, editor, or just by pointing at relevant literature from my web pages. Hybrids were attacked from all sides, and what is now considered the norm was considered the exception that had to prove its worth.

17.5 Hierarchical Objective Functions and Memetic Algorithms That Run on a "Segment"

When I returned to La Plata, things could have not been worse for science. I still had opportunities overseas, but by the Rotary Fellowship agreement I signed stated that I should stay in my own country for at least one year. In January 1990, I received a letter from Fox who sent me a preprint of his new book and an invitation to join him in Syracuse (*"when the dust settles"* said the letter) to finish my PhD under his supervision. The "dust" referred to the long process he was initiating, as he was to be moving on in one year.

When I left Caltech, Fox gave me a "blank cheque": *"If you can find a way to 'log in' from Argentina, you can use the computers at no cost."* That was really something, as in 1990 the group was establishing the Intel Touchstone Delta. But Argentina had no full Internet access. All universities were dialling in to a modem at the Ministry of Foreign Affairs, and from that single connection we, as thousands of others, had limited email contact with the world. The Internet map was very, very different at the time[1]. This means, you should forget about any niceties we enjoy daily. When there was a problem with your 'Inbox' you should travel to Buenos Aires to fix the problem at the Ministry! And sometimes, return the next day again, as the problem may have reoccurred (and iterate)...

The morale of the researchers could not have been lower. Inflation in 1989 was around 5,000 percent and 1,300 percent in 1990. The Minister of Finance, who

---

[1] http://www.worldmapper.org/posters/worldmapper_map335_ver5.pdf

"controlled" inflation by artificially creating a fixed dollar-argentinean peso exchange rate later gave a clear sign of the type of support he was giving to science. When the university academics united in protest due to the low salaries, the Minister on national TV famously asked scientists *"to go wash dishes"* (this means in Argentina that you should do the only useful thing you are capable of doing). If this support of science was coming from a former PhD in Economics from Harvard University, the readers can easily imagine which type of support we had from the rest of the government. He was also instrumental on the lack of Internet access(see: *The Internet in Argentina: Study and Analysis of Government Policy*)[2].

In the middle of these macro-chaotic working conditions, there was a lab at University of La Plata that had the opportunity to develop an incipient parallel computing program which I could lead. The idea was that I would direct three part-time students to develop computer programs on a T800 transputer system that was coming to the lab thanks to an established cooperation agreement that the head of the lab had with INRIA, Rocquencourt. We did receive the system, which we soon sarcastically nicknamed "the segment" (which was, after all, correct, it is just a hypercube topology of dimension one). It was also the ironic homage to the short-sightedness of our lab's head, who was answering my request for expansion of the system with his favourite one-liner: *"why do you need more processors if you have two? You are already working in parallel computing"*.

We did a lot of things to try to make the place habitable for a computational-intensive research group, like connecting old Textronix 4010 monitors (or other monitors made for oscilloscopes or some other uses) to help us use the existing Digital MicroVAX, etc. We also received three or four PCs, and that was basically the infrastructure we had to work for the next six years. I developed a research program in which we did both research in MAs as well as running on *"the segment"*, which was, after all, of little help. However, I systematically designed MAs that could produce linear speed-up running on the computers like the ones I had at Caltech, yet they can be simulated on PC systems and deliver good performance against other methods.

In late 1990, I gave my first talk at the EPFL, invited by Dominique de Werra, following a recommendation of Edoardo, who was there completing his PhD with him. I got only one person very interested in my presentation on MAs, Daniel Costa. This was, after all, the place in Europe where Tabu Search was reigning supreme. Before that, I visited Virasoro in Rome, who thanked me for sending him a year earlier my technical report (something I did it before leaving Caltech), and he said: *"somebody here liked your work a lot"*. I think he was referring to Filippo Menczer, but he was not there at the time. Those two colleagues provided the earliest citations of our work in MAs [146, 572].

In the period from 1990 to early 1993, I was mainly concerned with a few key problems on where to develop MAs, elucidating why they were working, or ascertaining how to perfect them. They were the TSP (still my "mentor") [625], the *Quadratic Assignment Problem* [109], and *The Binary Perceptron Learning*

---

[2] http://www.isoc.org/oti/articles/0599/chaumeil.html

Problem [616]. I had worked on these three problems with J. Fernando Fontanari while at Caltech. Together, we wrote a short, simple yet important paper in which we tested if a "deterministic update" rule was giving the same performance as the probabilistic update rule. We worked in a typical "Mythbuster-mode", pure scepticism at work. Our results indicated that the deterministic update was giving, typically, results that were of the same quality as the ones of the probabilistic update (when we were "thermalizing", e.g. when we used long runs at a constant temperature). We also thought that when you were not thermalizing, the indication was that this method was faster (actually, I was already experimenting with variants in deterministic update in our MA algorithm developed with Mike Norman at the time to speed-up the local search). We considered with Fernando that the work was perhaps publishable, providing a definite gravestone in simulated annealing. Indeed, one of the referees was extremely positive, saying that *"this is the paper I was waiting for!"*, or something of that note. We presented our results to John Hopfield in one of his group's internal seminar sessions and he was interested (which means that he asked a couple of questions at the end, something that was considered by some members of his group as a sign that he was really interested). Our paper was accepted very soon, but published more than a year later, after the journal confused it with a new submission after a change of title requested by a referee. Today, the original paper on simulated annealing in Science still reigns supreme in terms of citations (Google Scholar counts more than 20,510), while our "mythbuster" killer manuscript still runs far behind (only 47 so far!). However, within these, we have some interesting results, including a proof of optimality of our proposal among an infinite number of alternative strategies [284]. In addition, it is attributed to have an important role in a record-breaking method for disk-packing problems[3]. Johannes Schneider's method was rated by Time Magazine as *"one of the best inventions of 2009"* and uses our deterministic update proposal [641, 642].

Our paper in SA with Fontanari links with this story about MAs due to an interesting finding. When we plotted a curve for what is analogous to the *"specific heat"* as a function of the temperature during the annealing run, we noted some sudden changes when the temperature goes below some particular values. The quotation marks indicate that we were using the same functional form for the specific heat but the deterministic update. We also soon discovered that these values were not problem dependent, but *instance* dependent (i.e. directly associated with the specific distance matrix that we have as input). This behaviour indicated to us that, for the local search operator we were using, we were in the presence of some sort of "clustering" of solutions in the configuration space we were visiting. It was also indicating that when the typical fluctuations of the tour length at a given temperature were smaller, the closer we were getting to low length tours. There was, in some sense, a hierarchical cost structure [616] clearly revealed by the deterministic update rule. We concluded: *"Summarizing, our results indicate that the stochasticity of the updating rule in the simulated annealing algorithm does not play a major role in the search of near-optimal minima. It seems to us that the smoothening of*

---
[3] http://www.newscientist.com/article/dn16716

the cost function landscape at high temperature and the gradual definition of the minima during the cooling process are the fundamental ingredients for the success of simulated annealing." This was in line with my perceived qualitative runs with MAs. Once you were trapped in one area of configuration space, and for a particular threshold value, there were configurations that you were not able to reach.

The simplicity of our argumention uncovered consequences which are deep. Our findings seem to indicate a serious problem in the understanding of SA and the existence of a misinformation that perpetuates until today. In a minimization problem, our current MAs overcome the problems of SA by exploring low cost configurations very efficiently via, for instance, Tabu Search or a fast individual optimization algorithm, and by using recombination they can provide "long jumps" in configuration space. Tabu Search could be associated in Physics with some particular type of self avoiding random walk. The MAs would avoid getting trapped in less promising regions in configuration space by stochastic features present in the recombination operators. Challenged by one of the referees of [616] to give an example of those hierarchical cost functions, I chose to include a section on proteins and their energy landscape. Indeed, after several years and the success of methods like *Autodock*[4] that belongs to the MA paradigm (a "Lamarckian Genetic Algorithm"). I am happy I chose such an example. I also presented results on the Binary Perceptron Learning problem, a problem that Edoardo Amaldi had proven NP-complete[16], hoping to make a case of the importance of correlation of local optima and landscapes for MAs, like in my technical report of 1989 and on the use of different Tabu Search processes on each agent.

17.6 A Royal Visit to Argentina

In March 1993, when I was considering that I should leave Argentina to continue my work overseas as I could not see any future continuing there, I got an email from Mike Norman. He was finishing his PhD in Edinburgh and was considering his options that were plentiful, from writing a book on Parallel Fortran to come to Argentina for six months under the auspices of a Royal Society Fellowship. The latter was certainly the best option for me too, and we worked towards it. If he could get it, it would have been the first time since the Falkands/Malvinas war that a researcher would be sent from the UK to Argentina, on a re-established research cooperation between the Royal Society and CONICET (Argentina). After all, we had a "seemingly continuous" research cooperation that started in 1988 at Caltech. In 1992, we had published a joint conference paper at the Parallel Computing and Transputer Applications (PACTA'92) conference in Barcelona [622] that introduced MAs in Europe, and I had visited him twice in Edinburgh in November 1991 and September 1992. At PACTA '92 the Teraflop Grand Challenge was the major topic of discussions[5], but our paper with Mike has become the most cited of that conference and

---
[4] http://autodock.scripps.edu/
[5] http://www.chilton-computing.org.uk/inf/transputers/p011.htm

the one that introduced MAs in Europe. Mike obtained the desired fellowship and he travelled to Argentina with his partner, Ruth Thomas, in September 1993.

Having Mike in Argentina required a bit of preparation, mainly on the research front. Our collaboration had been sporadic but extremely productive. Without his talents, I might have not been able to do anything of value in MAs at Caltech. In November 1991, I visited him and I gave a seminar at the Department of Computer Science, The University of Edinburgh. He hosted me for two weeks and together we developed the *Strategic Edge Crossover* (SEX) [622] recombination that we incorporated to the old code. The extraordinarily leap in performance of MAs we observed by just changing the *Order Crossover* to the SEX showed us that investing in better algorithm designs was always better than fruitless tuning around parameters in heuristic optimization methods. The prospect of having Mike in Argentina was very exciting, and I knew that it would postpone, perhaps for a couple of years, my attempts to leave my country and the precarious research environment I was still working in.

During the six months before September, Mike and I worked towards making the visit very profitable. In late 1991, I had written a letter to David S. Johnson, of AT&T, asking for reprints of his works in SA, and any other preprint he could consider useful to send regarding my research interests. He was very kind in including with his response not only his (pre/re)-reprints, but also sending several papers by Jon Bentley (of *Programming Perls* fame), including his work on *kd-trees* for semidynamic point sets [59]. [Thanks David, once again, for that brilliant package of information you sent me!] What a joy it was to read all those papers! I understood then that there would not be any future progress in MAs for the TSP unless we started to efficiently use these data structures and clever local search algorithms. Supported by the extraordinary breakthroughs that Bentley was giving to the field we decided with Mike to incorporate these new data structures to the code. Via email we exchanged information, pieces of source code or pseudocode, and while Mike was preparing himself for his South American adventure we emailed on a daily basis discussing these future MAs.

We hosted Mike and Ruth for a week in a in Buenos Aires, in a small apartment that my family had in a posh neighbourhood called Recoleta. They were really enjoying the city and a lot of things amused them, like the red mailboxes that were almost identical to those of London, in a city with hints of Paris, Madrid, and Austro-Hungarian buildings. I recall they were also amused by some of the local pubs that had noticeboards in front of them saying things like: *"Tonight: Freddie Mercury"*. Mike couldn't stop laughing hard at those, *"Pablo, another one I am pretty sure is dead, should I enter and tell the owner?"*, he used to say. They brought a laptop and he developed two versions of the MA for the TSP, "cheddar" and "stilton" (the latter identical to the former but with fancy though very useful graphics). They incorporated all that we knew was best practice in MAs at the time, a ternary tree structure for the population composed of 13 agents, the *Strategic Edge Crossover* recombination algorithm, neighbourhood searches based on 2-opt and special versions of the 3-opt, all powered by a clever use of the *kd-tree* data structures of Bentley. In September 1993, I am pretty sure we had the world's fastest MA for the TSP, and probably one of the world's best heuristics, and it was running on a laptop! In 1992, a previous MA [625] that we developed with a one student from La Plata, Fernando Tinetti as part of his Degree Thesis, was already extremely competitive against state of the art branch-and-cut based methods. I tested this at EPFL during a visit and it was clear that, although our MA was not an exact method, they could constitute a great upper-bounding method that could work well together with an exact search procedure. The new code we developed with Mike while in Argentina was much faster, orders of magnitude faster. The use of Bentley's special data structures, plus all the other advances we introduced in the first three months in Argentina, allowed us to obtain optimal solutions for 100-cities problems in a few seconds on a laptop, while we had been unable to do that at Caltech four years before even with computers that had costed two million US dollars.

During those first days, we incorporated some of the TSP instances that I had been designing thanks to Lindenmayer systems (L-Systems). The purpose was to have a controlled set of testbed instances for the experimental evaluation of heuristics. These instances, later named *"fractal instances of the TSP"* [552, 553], would allow us to uncover key mechanisms of the MA, and its implementation in the code, that could be considered strengths and potential weaknesses [623], and complement the use of the TSPLIB instances. We used several of these fractal instances to test our MAs while developing the codes.

It turned out that this research on these instances, and this new original idea, took a lot of our time during those six months. We identified a very interesting property of the real number 0.714782700791294...[674], later recognized as one of the only 164 "fundamental mathematical constants". From more than 215,000,000 mathematical constants calculated with more than 2 billion digits, our constant (named "TSP constant"[6]) has been selected as "essential" for the elaborated mathematics behind Simon Plouffe's Inverter, establishing one important opening conjecture in Computer Science[7], namely that our constant may indeed be the elusive "TSP constant" introduced by Beardwood, Halton and Hammersley, first proposed in 1959 in a manuscript published in the Proc. of the Cambridge Philosophical Society [674]. During all those six months we also worked on the MA, improving it bit by bit, sometimes with unexpectedly bad results, as when we tried to replace the recombination algorithm by other more elaborate methods that did not perform so well. It was, however, becoming conceptually simpler and simpler, and very elegant, without any of those ad hoc parameters which unfortunately plague many metaheuristic procedures.

Towards the end of Mike's period in La Plata, it was clear that the tasks ahead were exceeding the capacity we currently had to handle them. There were too many subprojects to be completed, and the research group allocated for parallel computing, that I was directing, had vanished by lack of funds. I decided to place a small advertisement at a couple of the university's billboards asking for undergraduates who would like to volunteer some time to do some research in my area. With the

---

[6] http://pi.lacim.uqam.ca/eng/table_en.html

[7] http://mathworld.wolfram.com/TravelingSalesmanConstants.html

brief CVs and academic transcripts I received, it was difficult to choose one, but there was at least for me a clear winner. With a previous background in Engineeering I was not entirely sure, he had recently moved to informatics. Mike interceded and said, after I interviewed him: *"Get him! He is keen."* That student was Natalio Krasnogor, who worked with me first as a volunteer and then as a Degree Thesis student. Later, in turn, Natalio convinced me to bring David Pelta to our set of undergraduate volunteers, and although we worked on-and-off in MAs during that time, we built an incipient bioinformatics group. Natalio became very involved in research in MAs. He completed a PhD in MAs supervised by Jim Smith at University of West of England (Bristol) in 2002, and he is founding editor of *Memetic Computing*, having organized Workshops on Memetic Algorithms since 2000 and being a champion of the cause in Europe and the USA. Natalio is now Professor of Applied Interdisciplinary Computing at the University of Nottingham in the UK and is having a brilliant international career with great emphasis in Evolutionary Computation, MAs and their application. David is Associate Professor at University of Granada, Spain, and has also been very active in the whole field of Evolutionary Algorithms. This said, I guess I have little to add but thank all members of this chain of people who almost invariably have appeared in my life through being highly recommended by somebody else, Fox recommended Mike, who recommended Natalio, who recommended David. I guess I have been extremely lucky to work with them all. Follow the heart-felt good recommendations, may be the moral. In turn, I may have been also part of the chain as I recommended to both Natalio and David the preprints of Bill Hart, at Sandia Labs, who had been working on methods for toy problems in the area of protein folding [369, 370, 371] and while still in Argentina I directed Natalio on heuristics for the TSP and David in these problems on models of protein folding. I understand Natalio visited Hart in 1996 and introduced MAs there, which blended well with their interest in evolutionary algorithms [372, 376]. I also understand that *AutoDock*, since version 4.0 if not before that, uses the optimization routines of the ACRO optimization library developed by Hart.

Mike and Ruth returned to Scotland in March 1994, to watch the BBC on a TV instead of listening to it on a shortwave radio when the interference allowed it. We all had many unfinished manuscripts. Mike and I continued collaborating, time permitting, finishing several papers during the next years. Ruth wrote part of her first book of short stories while in La Plata, and completed it in Edinburgh [882]. Mike started a consultancy company and established Makespan Ltd. and later Scapa Technologies, two very highly successful commercial enterprises that kept him very busy for many years. Some of our work in MAs was surpassed in terms of performance very soon after, in 1996, by Peter Merz, preventing us from having a chance to capitalize on that small time-window of opportunity in 1993-94 to publish our own results. But, on the other hand, it was really good to see Peter Merz coming out, as a rising star bringing innovative ideas and great results into the field.

17.7 To Brazil, without the Beaches

We finally obtained full Internet access at the universities by mid 1994. I recalled logging in with Natalio at a server at the ETHZ. We were trying to "get in" using "ping" everyday (the whole of Argentina had IP numbers and an internal intranet). One day in June finally it was working. We are now on the World Wide Web! What a thrill it was when it was finally working. We were reprimanded by the local authorities: *"How do you dare to use the Internet if it has not been officially inaugurated?"* we were told. Yes... Argentina "inaugurated" the Internet. With access to information, and the ability to provide information uploading .ps files with our work, we could get out of the ostracism imposed by the lack of physical presence on the international circles. How could we use it?

In 1994, after Mike returned to Edinburgh, he told me about two papers that Nicholas Radcliffe and Patrick Surry had published. I knew that there was some trouble ahead. Although they do not explicitly contradict our definition of MA, their formalization included the restrictive idea that recombination has as input parent solutions which are locally optimal with respect to some neighbourhood function. We had never used such a restrictive definition, as we had viewed this constraint in the algorithm design as rather unproductive since 1988. The papers, however, were giving a very clear and strong message. They showed the benefits of MAs, and were supporting the use of *Forma Analysis*, to reflect on the design of evolutionary operators [751, 752]. Mike introduced MAs to Radcliffe, a colleague working at the EPCC in 1989. The three of us discussed MAs during my visit to Edinburgh in 1991 and I had adopted Forma Analysis for semantical discussion of properties of the operators as a useful description. A student we co-directed with Mike in 1992 on a summer project I prepared, Reimar Hofmann, applied this formalism and developed it as a predictive tool for the a priori evaluation of operators. He did a fantastic job, showed clearly the benefits of our MA approach and then wrote an interesting thesis with the results of his further studies [387, 388].

In 1995 I met Regina Berretta, my wife, who at the time was a second year PhD candidate at the prestigious Universidade Estadual de Campinas (UNICAMP), Brazil. We met on an Operations Research Summer School. She was working on metaheuristics for production planning problems. I also started to have a part-time job as Visiting Professor at Tandil, around 400 kms from La Plata. A year later, I moved to UNICAMP as Visiting Professor and we married in Brazil in 1996 and lived in Campinas, which contrary to what my readers may daydream, was not close to the beach. The contract I had was for six months (and I was on leave from my post in Argentina), and at the end of this period there was a big question mark in our lives. How will I get a salary? I wrote a project to obtain a grant that would pay my PhD salary to study the possible links, or show the existence of them, between the theory of approximability in computer science and the informal notions of "hardness" in Evolutionary Computing [367]. With that project I obtained a PhD scholarship from FAPESP, Brazil, and I became a colleague of some of the same graduate students that the previous semester had been my students (often attending the same courses). I finally followed the advice of Joshua "Shuki" Bruck

17 Memetic Algorithms: The Untold Story

who four years earlier seriously summoned me to get a PhD if I wanted to have a life in academia. In 1997 I met two other graduate students (Luciana Buriol and Alexandre Mendes), and we all had Paulo Morelato França as supervisor; we started a fruitful research collaboration that remains active up to the present. Our work tried to support the use of MAs in Operations Research and Production Planning [279, 280, 281, 282, 302, 573, 574, 575, 626, 627, 628].

At UNICAMP I also discovered, now as a student, how big the field of MAs was becoming. I was taking one course that had the opportunity for us to present some research on a topic related to the Theory of Computation. A few months before, I had met Pavel Pudlak, who gave a seminar at the Department of Computing, and after we had a post-seminar discussion. He gave me a copy of something he was writing on a novel idea *"Genetic Turing Machines"* [743, 744]. During one class, I proposed to the lecturer that I would study this subject who immediately counter-suggested: *"I propose you something better; why don't you study Memetic Turing Machines? Have you ever heard about memetic algorithms?"* I thought my lecturer was teasing me; but that was not the case. I had started the Memetic Algorithms' Home Page in February 1996, so two years later a lot of people had already discovered the subject (and each other). I knew that nobody had proposed such a thing as a Memetic Turing Machines. However the whole experience brought back home an interesting feeling. I had the impression that now the genius was definitely out of the bottle, clearly MAs did not belong to a few of us any more.

It was now clear, by mid-1998, that a large number of people, including ourselves, would start to experiment with MAs in their role as heuristic problem solvers in a large number of different settings. There would be a lot of experimental results, more practical application sought in novel problem areas, and perhaps an increasing void in theory. However, it also alarmed me that there would be a widening gap on a direction that I wanted MAs to have right from the beginning: the hybridization with exact methods from mathematical programming, logic and constraint programming and artificial intelligence techniques. The final three sentences of [615] were:

> "Memetic algorithms are not a new heuristic that can be chosen to be applied in an optimization problem. They are not motivated to replace present heuristics. Instead they are a framework to exploit all previous knowledge about the problem, combining methods to improve their performance."

so that, of course, includes the use of exact algorithms. I decided to look on the web for some people who may be heading in that direction and I found a researcher in Málaga, Spain, (Carlos Cotta) who had been discussing these issues in his PhD thesis and his published manuscripts [147, 160, 162]. To my surprise, he was also fluent in Radcliffe and Surry's *Forma Analysis* [166]. These factors allowed us to have a useful dialogue from the start and we developed a collaboration via email. It took two years until I was able to finally visit him in Málaga in 2000, after our interaction increased later in 1999.

Some of these concerns on this widening gap are evident in the chapters that I contributed, either as author or as editor, to *"New Ideas in Optimization"*, written during the second part of 1998 [136, 617]. Peter Merz, who kindly accepted my request, contributed a highly cited chapter on the relationship of *"fitness distance correlation"* and MA design [585]. I thought that he was the key person to write that chapter at the time. Peter not only mastered how to create a useful recombination operator that worked well with the Lin-Kernighan heuristic for the TSP, he was also providing impressive results in other problems as well. His chapter would then link a long list of discussions and results linked to my own work, as well as predecessors like Kaufmann, Kirkpatrick, and the work done by Hofmann, Radcliffe and Surry, and Terry Jones (who I met after I gave a talk at the Santa Fe Institute in April of 1994) [435]. I also invited one of my students, Diana Holstein [390], to write a chapter since we were working on an area that I wanted to expand, that of the adaptation of individual search processes by each agent. It does seem, however, that this line of research has been developing more in non-linear than in combinatorial optimization.

With Regina, we chose to present work on the *Number Partitioning Problem* [60, 62]. Although a fully-polynomial time approximation scheme exists for this problem, it did appear to be hard for a variety of metaheuristic methods. The reasons behind this failure, possibly related to a poor selection of neighbourhood function and representation, was explored. It has been included as a challenge for the MA field practitioners, to help overcome the limitations and to promote new creative developments. The use of Tabu Search, to overcome the limitations of Simulated Annealing, indicated that this is a possibility, but the problem has become a standard challenge to other algorithmic approaches. We considered it paradigmatic of the type of problems on which theoretical and practical advances can lead to improvements in our understanding of how to create better evolutionary algorithms of the memetic type.

In *"New Ideas in Optimization"* I also started some other more ambitious theoretical endeavours that, probably due to lack of my own talent, but certainly due to lack of time and funding, I was not able to develop as I think they deserve to be. I tried to build two research directions: one was related to a computational complexity class denominated *Polynomial Merger algorithms* [617]. The other one was related to a way of designing recombination operators based on a worst-case analysis, borrowing elements from the analysis of off-line algorithms [136]. The idea was to use *Competitive Analysis* and *Comparative Analysis* to give tight theoretical bounds for the problem of the design of recombination operators with proven worst-case performance. More than a decade after publication of this book, these two ideas have not yet been developed by the community, and I feel that theoreticians will find a fertile ground to be further developed in the context of the *theory of parameterized complexity*. Next I will link this story with the next chapter in our lives, when we took our research "down under".

17.8 Fixed-Parameter Tractability, and the Complexity of Recombination

By the end of 1999, I was sure that there were severe obstacles to reconcile what is considered 'hard' for the theory of approximability in computational complexity (which is a well-established formal concept) with what is considered 'hard' for people working in Evolutionary Computation. In some sense, the answer to one of the questions I posed myself as a thesis project in Brazil was *'No'*; these notions of hardness were not compatible at all. The problem we were studying with Regina, called NUMBER PARTITIONING, was a case in point [60]. The problem is in class FPTAS, i.e. there exists a fully-polynomial time approximation scheme for it. This means it is considered on the "lowest level" for a decision problem that is the NP-complete class, yet it seems "hard" for Evolutionary Computation and other metaheuristics. It is, however, clear that the "hardness" for EC is usually biased by the limitations of the skills of the practitioner instead of a clear mathematical tight bounding classification. On the other hand, the notion of having a *fptas* or a *polynomial-time approximation scheme*, does not necessarily guarantee that it may be a useful algorithm for some practical application. There was no bridge between theory and practice in the horizon.

I was searching for an alternative theoretical perspective that could put some kind of new perspective on these matters when I found on the web an article by R. Downey, M. Fellows and U. Stege called *"Computational Tractability: the view from Mars"* [225]. To my surprise, the perspective from Mars was also my own! My wife is the witness of how much I enjoyed reading each line of that paper. Finally, I found a few authors that, regardless what "the community" would do with them after that manuscript was published, were spelling out, loud and clear, the deficiencies of research in Theoretical Computer Science at the time. They did it with a passionate stand, for a change of perspective on the status quo of Theoretical Computer Science, stagnated in a single minded exercise of proving NP-completeness and inapproximability results and then *"walk away"*. I hope my readers will not misunderstand me on this point. Research in NP-completeness of a new problem is useful, absolutely essential I would say, but it should not produce a single minded breed of theoreticians, incapable of recognizing the amount of positive results of empirically well-performing heuristics and exact algorithms (or their hybridizations). Unfortunately, many of our computer science theoreticians have been indoctrinated, in a fundamentalist way, that this research is always the only relevant thing, while practical aspects of computing are of secondary importance.

The "View from Mars" paper included something that also hit me as of very important consequence. The authors were highlighting the role of *treewidth*. They said: *"Important distribution parameters may also arise in ways that are not at all obvious. Thorup, for example, has shown that the flow graphs of structured programs for the major computer languages have treewidth $k \leq 7$ [884]."* After all, these *"important distributional parameters"* which *"are not at all obvious"*, are the ones that, when understood, can also lead to important insights about how to develop efficient evolutionary algorithms and metaheuristics. *Would they have a way* *to identify them in a more systematic way ?* Could the theory of *"parameterized complexity"* [100, 220, 222, 223, 226] help in providing tools for the "efficient" implementation of key algorithms in memetic algorithms, like optimal recombination algorithms under some constraints and variable-depth local neighbourhood searches?

I knew that there might be an important connection. I can remark here again that the success of the highly performing CONCORDE code for the TSP (a joint project of David Applegate, Robert E. Bixby, Vaek Chvátal, and William J. Cook), is also based on the notion of treewidth. In 2003, Cook and Seymour presented the case in their *"Tour Merging via Branch-decomposition"* paper [143], although reference to their recombination algorithm existed on other manuscripts well before that date. They showed that by first creating a sparse weighted graph (with the set of cities as vertices and the union of the edges of ten high-quality tours on a TSP instance as edge set), it is sometimes possible to improve on the best of these ten tours by finding another one that belongs to this graph and that has the minimum total tour length. This was very important as although this new problem is still NP-hard it can be solved, via dynamic programming algorithms, in time proportional to the size of graph and a constant of proportionality that grows with the branchwidth of the graph. The concepts of *treewidth* and *branchwidth* are related, and they are intrinsically linked today to many important results in the theory of parameterized complexity.

For me all these concepts were like pieces of a puzzle which I started to collect more than a decade before. In 1991, when I first met Nick Radcliffe in Edinburgh, during a meeting I had with him and Mike Norman, I suggested that, perhaps, the problems we were facing in designing "optimal" recombination operators could be linked to computational complexity. *"Perhaps, the problem of finding a recombination operator that is optimal, under certain conditions, is NP-hard"* I said. We all agreed that it was a good target for a research agenda, to systematically identify these new NP-hard problems arising from our metier; but that was an ill-posed question: what did *"optimal"* really mean? I had no clear mathematical model at the time to frame the discussion.

This said, after I finished reading "the view from Mars" paper, I was motivated to send an email to Fellows and Downey, to try to entice their interest in the development of a theory for recombination algorithms based on parameterized complexity. Judd taught me well about how to entice the interest of theoretical computer scientists, with a clean cut problem. I sent an email to them that was very brief and said something like this:

> "Dear Profs. Fellows and Downey, I have the following problem: I have as input an undirected weighted graph $G(V,E,W)$, and I also have as input two Hamiltonian cycles $C1$ and $C2$ of G, such that $C1 \neq C2$, and without losing generality $Length(C1) \leqslant Length(C2)$. The question is then, "Does there exist C, which is another Hamiltonian cycle of G, such that $C \neq C1, C2$; and $Length(C) < Length(C2)$? Can parameterized complexity serve to address this type of problem?"

In essence, this is what my proposal of a class for polynomial merger algorithms was all about. Can we find out under which conditions the problem of creating

17 Memetic Algorithms: The Untold Story 301

another feasible solution *better than the worst of the solutions given*, is solvable in polynomial-time? If there is a small parameter that restricts the input, under which conditions/properties of the input can we find a fixed-parameter tractable algorithm?

Their answer did not take much time. Downey immediately thought that there may be a neat formalization of this generic set of problems in terms of the k-STEP HALTING PROBLEM FOR NONDETERMINISTIC TURING MACHINES which is the following: *"given a nondeterministic Turing machine M (with unrestricted nondeterminism and alphabet size) and a positive integer parameter k, is it possible for M to halt in at most k steps, starting with an empty tape?"*. I do not recall exactly his email, but it made some sense at the time and involved two machines with different halting time steps k and k' and the question involved the existence of a third machine. Fellows was eager to frame the source of these problems; he was interested on the logistics side: *"Where did this problem come from?"* I told him that I can produce, eventually, hundreds of those, as they were coming naturally from the polynomial merger algorithms class, a proposal for a new computational complexity class for which I had no complete problem. After several emails during the next two months he said, *"Come here ! Too much email, I'll buy you a ticket and we can discuss in Victoria."* Those five or six days that I spent in British Columbia were very useful. I had a great first-hand tutorial on parameterized complexity and we had a chance to explore the possible synergies between fixed-parameter tractability and MAs.

One key event that helped to understand the potential of the combination techniques occurred when Fellows proposed to discuss the MAXIMUM LEAF SPANNING TREE problem. In this problem, we are given an undirected connected graph and the task is to find a subgraph of the graph that is spanning tree and, from those, find one that has a maximum number of leaves possible. Mike asked me to illustrate how to create a MA for this problem (the game was basically, I give you any NP-optimization problem, can you always create an MA for it, or something of that sort). With two different greedy heuristics, I quickly generated two feasible solutions for a given instance of the problem, two trees which turned out to have eight and nine leaves respectively. I then proceeded like in other combinatorial optimization problems, in which recombination is achieved by looking only at the information produced by the solutions (analogous to the merging of TSP tours of Cook and Seymour or our *Strategic Edge Recombination* for the TSP). Running the same greedy heuristics now on the graph formed by the union of the edges of the two trees, led to two solutions with eleven and fourteen leaves, dramatically improving the previous results. None of these things would surprise somebody working on MAs today. However, the interesting bit for me came right after. During the afternoon (as we had the discussion during the morning), Mike was intrigued by the procedure and tried to see if the recombination could have been done better, not by the greedy procedures I used but by another kind of algorithm, hopefully an exact. He noticed that the sparseness of this graph now made evident that some *reduction rules* could have been used. Mike, and his wife and collaborator Fran Rosamond, who was working with us during the whole day, quickly started to annotate all the reduction rules that were coming naturally by inspection of the graph. Some of these rules were easy to mathematically prove to be correct/safe, others would require more effort, but what struck us all was how quickly the theses of many theorems for the correctness of these reductions rules were "begging to be proved". The sparseness of the graph, and perhaps I should better say its treewidth, were the key ingredients behind this process.

Later the same year, in June 2000, I finally was able to visit Carlos Cotta in Spain for a couple of weeks. That was great, the visit promised to be very productive; we finally had the opportunity to talk face-to-face. But, *which face?* I discovered in the airport that I had been working with him for two years, yet I had never seen a picture of him. On his home page, he had a picture of David Duchovny, of X-Files fame at the time instead of his own. Universities were more liberal in their web policies then. Carlos would get from time to time the occasional email: *"You won't believe this Professor, but you look very much alike a famous actor that works in a science fiction TV program..."*. When all arriving people at the airport finally met their relatives and friends and left, and only the two of us remained; we shook hands went to an Irish pub for an Irish beer (so Andalusian a beginning).

During the two weeks, Carlos and I discussed several methodological improvements to MAs, mainly involving the incorporation of exact algorithms, and I told him about my experience in Victoria just a few months before. We discussed several ideas about how to improve MAs and the possible links with the theory of parameterized complexity.

A key issue that arose, which we recognized early in our first meeting, was that it could be possible that with parameterized complexity we could finally answer several questions regarding the complexity of recombination. This was a long quest for me, and perhaps we could establish some sort of lower bound on the complexity of recombination under certain circumstances. If we took a problem of interest, for instance the TSP, we noticed that many recombination operators for this problem have been established on the basis of attributes/features which were required to be either present or absent in the newly generated solutions (usually via polynomial-time algorithms). This seemed to be a generic template, a design pattern of sorts. For instance, the *Distance Preserving Crossover* of Merz explicitly included all edges that are present in both parents and avoided those edges that were present in one but not the other parent. This means that, although randomized, it still had the same characteristic design pattern, that of systematically allowing attributes to be present or absent. Merz' DPX did not require the retrieval of the "optimal tour" among all those that had this characteristic. Assuming that this could be done in polynomial-time for two tours, the natural question was *to identify the computational complexity of having, let's say, k tours, and finding the* optimal tour *under some giving constraints.*

What are these general constraints? We noticed that this issue is very problem specific, so apart from generic membership constraints, it is difficult to have a core problem that encompasses a template for recombination; the generic pattern for recombination. We thus resort to what is core for the polynomial merger class. In the basic definition, we are given p feasible solutions and we want to create a new one which is, different from the others, and is better than the worst. This means that we can assign different labels to the samples and, without losing generality, assume that one of the p solutions is the worst one (or the one with information to be avoided) and the others with information to be preserved. If that is the case, the task of identifying which features are to be preserved or avoided, can be viewed as a special case of the k-Feature Set as defined by Davies and Russell, who proved it NP-complete [187]. This clearly linked to my discussion with Radcliffe and Norman in 1991; there are multi-parent recombination problems which are NP-hard, but it could be that some of them are *fixed-parameter tractable*. After initially trying to prove this, what we then ended up proving is that it is unlikely that the problem is in class FPT (the class of problems for which a fixed-parameter tractable algorithm exists) if the number of attributes is the parameter. Instead, we proved that k-Feature Set is $W[2]$-complete, meaning that only by discovering some kind of special structure that restricts the instances to be of a particular subclass we can we hope to find a subproblem that is in FPT (unless the $FPT \neq W[2]$ conjecture is proven false). A year later we completed a manuscript entitled *"On the parameterized complexity of multiparent recombination"*, which, due to the referees request, has finally been published as *"The k-Feature Set problem is $W[2]$-complete"* [156], without the important emphasis that the result has on the complexity of recombination. Later in 2005, we published another manuscript at an international conference where we had the opportunity to include this important information as a guide for the development of a theoretical approach to prove tight lower-bounds on the complexity of recombination in a general scenario [159].

After my trip to Europe and our discussions with Carlos, I returned to Brazil, with the conviction that there was a lot of potential in linking memetic algorithms with parameterized complexity research. In particular, I considered that fixed-parameter tractable algorithms had a role in the development of exact solutions for subproblems that originate in questions related to recombination. To some extent, fpt algorithms also could play a role in local search. I considered that they were key for problems that arise in merging feasible solutions. These ideas, which I still believe have a lot of potential, were later included in the thesis I defended at UNICAMP, a year later. Today, several results involving parameterized complexity and local search exist (see for instance the work of Dániel Marx [497, 559, 560, 561, 562, 563], Daniel Lokshtanov [253, 275], Stephan Szeider [865, 866] and references therein). While some researchers who are active in parameterized complexity and also in memetic algorithms are joining forces, like Gregory Gutin, Daniel Karapetyan and Natalio Krasnogor for instance [343, 344, 345, 346, 442] the systematic design of recombination algorithms using fpt algorithmics it is still in its infancy.

17.9 Newcastle, Australia, and Biomedical Research Closer to the Beach

In late 1999, I read in FAPESP's newsletter that a new lab was being set up at the Ludwig Institute in Sao Paulo which, at an estimated cost of 1 million US dollars, would allow cancer researchers to investigate *"thousands of genes at the same time"*. I am talking about the first generation of microarray technologies that were finally arriving to Brazil. During the next two years, while working on the *k-Feature Set* as a model of recombination problems for Evolutionary Computation, I also looked at the practical applications that it can have for the analysis of gene expression in cancer and other human diseases [168].

With this double interest which presents a duality of objectives but still relates to some basic core problems that are common to MAs and biomedical research, we moved from Campinas (UNICAMP), Brazil to The University of Newcastle, Australia, in September 2002. We are now relatively near the beautiful vineyards of the Hunter Valley and very close indeed to pristine beaches. My wife then joined as a Faculty member in March 2003, and on the basis of our joint work we consolidated the Newcastle Bioinformatics Initiative (NBI). We moved due to the insistence of Mike Fellows, who thought that we could develop this joint program of research in parameterized complexity, memetic algorithms and bioinformatics. The truth is that the latter started almost 48 hours after arrival. Two days after arrival, we were still with no credit card, no tax number, no bank account, no house, no car, nothing really, our Pro Vice-Chancellor was already asking me to write a project to establish a group in bioinformatics. Thanks to a grant I put together during my first month in my new job, I obtained funds to establish the NBI, a three year project that would start in 2003, and aimed at bringing together the some research "silos" in biotechnology and medical research on a common theme. We joined the Hunter Medical Research Institute (HMRI). One of the terms of reference of the NBI was to work so that, after the three year period, it should have enough research outcomes to justify the creation of a Centre dedicated to the theme. In June 2006, I led the proposal of a new project in collaboration with other HMRI and university researchers. Our bid was successful and in January 2007 I established the *Centre for Bioinformatics, Biomarker Discovery and Information-based Medicine*. Today we are also one of the seven research programs of the HMRI, the third medical research institute of New South Wales.

As expected, much of the development of memetic algorithms and computational complexity studies were linked to the research in bioinformatics and biomarker discovery. When I received the referees reports of the first national competitive grant that we submitted from Newcastle, one of the reviewers wrote something like this: *"The Traveling Salesman Problem is not one of Australia's National Priorities"*. My "true love" is a dusty Cinderella here. Concerned with continue the development of MAs, but with the added constraint of having to support my research group only from external competitive grant, our research has become increasingly more concentrated on applications of MAs. A number of different ideas on experimental testing of improvements on our basic MAs are still pending to be fully developed. In spite of these new circumstances, the new ideas we develop in MAs are nevertheless incorporated into our more recent publications [167, 404, 405, 406, 545, 576, 629, 631].

As a consequence, during the recent years we have proposed a number of MAs with novel ideas for very challenging problems. For instance, in another collaboration with Carlos Cotta, we proposed an interesting approach for the problem of finding a Minimum Weight Ultrametric Tree from distance matrices [157]. We selected this problem area as we felt at the time that there were not many MAs developed for problems which involve problem representations in which feasible solutions are trees. Given an instance of the problem, by first using an MA to solve some variants of the Minimum Weight Hamiltonian Path Problem on the input distance matrix, we first obtain a probably optimal solution of this NP-hard problem. By restricting the solution space of the original problem to only those trees that have leaves that can be ordered we drastically prune the space of possible solutions. In that way we guide the search performed by an actual clustering algorithm. With a high-quality feasible solution that acts as an excellent upper bound, we can return to the original unconstrained problem to prove optimality or to improve on the solution already obtained. This approach has been shown to be highly effective in practice, allowing for the first time the scaling by one order of magnitude the size of instances solved for this problem. We conjecture that this is an important new design tool for MAs. Ideally, the population of agents can be using different "guiding hints" (i.e. different alternative high-quality orderings of the leaves) and help to guide the exact methods in different ways. An alternative approach, in which a co-evolution of the ordering of the leaves and the evolution of the topologies of the trees is performed in a different time scale seems also promising and worth exploring. We have applied this methodology to the analysis of mitochondrial DNA as well as for gene expression obtained from microarray studies of cancer samples.

We have continued with the development of MAs for other combinatorial problems, like the Asymmetric TSP [94], and problems involving ordering in bioinformatics [397, 577, 630, 778]. In most cases we have continued developing proven ideas involving population structure (a ternary tree, for example), the use of diversification and intensification local search procedures based on Tabu Search, and the hybridization with exact algorithms and other heuristics.

17.10 Future Opportunities (if We Constrain the Beast)

Writing these notes has not been what I expected. I thought it would have been something of an easy ride, but it was not. It is hard to condense so many ideas and so many years in a few pages, so unfortunately I am leaving a lot outside these lines. I am giving my apologies to a large number of collaborators and projects which I have been unable to include in this discussion in which I tried to have *"bounded treewidth"*. But at the end of this chapter, as I said in the introductory comments, again I do not feel that we have reached a conclusion, a "grand finale". I have the feeling that the best in MAs is still to come. And that the obligations and the challenges of people that practice MAs are immense [158] but that the future is indeed promising for those that will master these techniques and take the ideas to their full potential. Collectively, over two decades, we may have proven a first case for MAs on many problems, and in particular on the TSP, but we still have myriad other problems waiting for important advances. We still have a gap on the design of efficient and smarter recombination algorithms.

Although these gaps exists, it is true that we are also just discovering how to exploit structural parameters in our problems of interest. We can take for instance one of the concepts I discussed before, *treewidth*. We are just at the beginning of understanding how to explore it well, yet this beginning has lasted for more than two decades [71, 71, 512, 934]. I am struck at how many times I have encountered this concept within my research in MAs, and in parameterized complexity, and, even more surprisingly, in my formal education in computer science at Caltech. Yet I am also surprised how little we have explored it properly; my mistake. I do not hope this to be repeated so this section aims to tell the story, with the hope that new researchers can develop it further.

I recalled that, back in 1989, I had a conversation one day with J. Stephen Judd in his office. I was teasing him: *"I don't think you are very popular among machine learning researchers. You go there, and 'Bang!' you prove their decision problems are NP-compete and then walk away. That's not going to create a lot of friends in applied conferences"*. Of course, Stephen was not "just" doing that. It was true that he did prove "negative" results of that sort, but he was also proving "positive" results, algorithms that could run in polynomial-time for particular network structures. In his thesis, he studied families of *"shallow"* architectures that are defined as having bounded depth and unbounded width, and he defined the "support cone interaction graph" (SCI) which allows one to distinguish the tractable from the intractable subcases. He tried to stress the importance of this concept, and indeed he achieved that, and at least in my case I have learnt the lesson well. During his lectures, he gave us a homework on which we had to program an algorithm for machine learning one task, and we experienced first-hand the difficulty inherent to crossing this boundary (between a problem that is NP-complete and a special case for which a tractable loading/learning algorithm was easy available). In his thesis, he showed that when the treewidth is $O(log n)$, the learning problem can be solved in polynomial time; and when the treewidth is $n^{\Omega(1)}$ the problem becomes NP-complete even if the SCI graph is a simple 2-dimensional planar grid. However, it was not until I found it again being central in parameterized complexity and also in the tour merging approaches of the CONCORDE team for the TSP, that the full-impact of its utility became obvious to me.

I will say something that will probably shock both theoreticians and practitioners. In essence, at the core, both MAs and *fixed-parameter tractablity* researchers are striving to find some form of intrinsic structural characteristic of the class of instances. This is a common theme. Rolf Neidermeier and other researchers are proposing the "deconstruction" of NP-hardness proofs as a way to identify the parameters that inherently are present in a problem and that can "push the problem" to be intractable [479] and recognizes the need of merging these paradigms when he proposes in his book *"that topics such as local search or evolutionary and memetic algorithms will become new subjects of fixed-parameter studies"* [666]. These two algorithmic streams do not differ in that one is deterministic and the other is all randomized; in complete MAs (CONCORDE being a case in point), we stop when we have actually proved that we have obtained the optimal solution; an implicit exhaustive enumeration scheme is necessarily being employed to guarantee optimality. In *fpt* algorithms, generally we have a data reduction step (which in Operations Research has been traditionally called *'preprocessing'*) based on reduction rules [72, 252]. As such, that step must be always attempted in MAs, provided that the problem is in the FPT class. We take that lesson from parameterized complexity, but, in turn, we should increase the awareness among the theoretical community that the problems of finding optimal recombination schemes are a great niche for combined theoretical and practical computing research. It is curious that the fpt-community is neglecting this area, as the preprocessing mechanisms are far more useful for these subproblems that appear in MA implementations than in dealing with the original problem. This is an area, on which, after all these years, we still have a lot to do and where I would be glad to see more researchers both theoreticians and practitioners shaking hands and productively collaborating.

Reading again the conclusions of Judd's thesis, I found a paragraph which is remarkable and perhaps suits well the discussion we have here. He wrote:

> "Whatever the case, our underlying assumption is that complexity analysis (and specifically the P vs. NP distinction) provides a means to narrow down the things that biological machines do and how they do it. Our strategy is to take the general NP-complete problem and add architectural constraints, and search for polynomial-time loading problems. We feel very safe in assuming that the brain cannot be solving any NP-hard problem, and we feel secure in assuming further that evolution would have found efficient ways to utilize the available hardware. Ergo brain mechanisms are likely to be described by decision problems found 'just below' the level of NP-completeness. Hence the general outline and thrust of our research program."

Judd's quest was to find *"a general methodology of how connectionist networks should be constructed"* for machine learning applications. He is indeed a pioneer in this process of "deconstructing" NP-hardness proofs. We do have a similar quest: *how can we establish a general and systematic methodology that guides how MA algorithms should be designed*. In particular, given an NP-hard optimization problem which is not in class FPT, it may be possible that there exists identifiable FPT problems that naturally appear in the design of optimal recombination operators and local search techniques. Thus *fpt* algorithms could be used, in concert with randomized methods, and under some circumstances, for more systematic exploration when needed.

The lessons learned in the past two decades give now a general idea of how to progress. Here is where the past can also enlighten the future. When confronted with an NP-hard problem, we can look at both efficient procedures, coming from approximation algorithms and *fpt* algorithmics as tools that can provide efficient solutions "just below" the full complexity of the problem [73]. These special cases can provide us with powerful individual search and recombination algorithms. The "tour merging" procedures for the TSP as described by Cook and Seymour are a case in point. While in general the complexity of recombination is $W[2]$-hard (recall that we have proven that k-Feature Set is $W[2]$-complete), given an NP-hard problem, this does not mean that, by looking at treewidth/hypertree width, local treewidth [206, 207, 272, 351] pathwidth [74, 274], branchwidth [216, 273, 384], rankwidth [89, 398], cliquewidth [88, 265], q-branched tree treewidth [274], tree-length/branch-length [898], boolean-width [3, 58], NLC-width [341, 644], bounded degeneracy [567] etc. we can not provide efficient algorithms that provide optimal recombination methods for many problems.

There will be some challenges ahead in this route, of course. One of them is that, and one of likely criticisms, is that in general, given an instance of an NP-hard problem, we are not given as input these values of parameters. Even the task of computing them, like *cliquewidth* or *NLC-width*, has already been proven to be NP-hard in the general case [254, 342]. This means that special attention should be given to the way of encoding feasible solutions of the original NP-hard problem such that it can give rise to other NP-hard problems of bounded, and hopefully small, parameters, so that optimal recombination strategies can be found. I expect that we can soon have researchers exploring this space.

More than two decades from my own beginnings in this area of research, starting from the unconventional perspective of spin-glasses and ultrametricty, it is interesting to note that problems in that area are now solved with algorithms that can be categorized as "memetic" [415]. In some sense, we have come full circle, as Parisi states are now searched with this technique in spin glasses. I do, however, note another thing: that it took us years of mathematics and computer science to recognize that, in many problems of interest, for many of our implicit enumerative schemes based on some forms of tree-based search, it was some parameter related with some "width" notion that was " the source" of increased computational complexity. Take the case of CONCORDE, for instance, but this is true for other problems as well. This said, although sometimes I feel we have learned a lot in the past two decades, I also feel we have learnt in a hard way a lesson that was pretty obvious from the start.

I am also convinced that MAs have not lost a single bit of their innovation. I said in 1989 referring to a MA: *"For it the network really is the computational device"* in a paragraph where I predicted MAs could be perfect for what we now call "grids" or "clouds". Their time in history is right now. The revolutionary social role of MAs stays the same. They have always aimed at bringing the best of each field of expertise and act as a framework for an algorithmic engineering process when we need to address a problem. As a framework for collaboration, MAs role is intact. It is not like other methodologies, based on a theme and neglecting the others. It is a philosophical stand asking for a different collaborative perspective on algorithm design, and a such they remain novel and "aggressively useful".

The number of heuristic, metaheuristic and parallel techniques that are converging to MAs increases day by day. And I think that, while research on things like deterministic preprocessing kernelization will continue for decades, and that fpt-algorithmics will continue to grow, they will not replace the need of randomization to provide upper and lower bounding schemes to the implicit enumerative schemes. Clearly, the design of these schemes should be revisited, and co-evolution of deterministic and randomized methods [181], together co-evolution of instances [5] to "train" deterministic methods will be used. The right balance of randomization and determinism, and the synergistic collaboration of algorithmic methods and theoretical results, will keep on fueling the development of MAs and bring them to higher achievements.

I can't wait to see the developments of the next two decades!

Acknowledgements. I would like to thank all people named in this chapter, as well as many other colleagues and collaborators for fruitful discussions over the years which I could unfortunately include here due to space constraints. In particular, I would like to thank Mike Norman and Regina Berretta for their help in developing MAs and for critical reviews of parts of this chapter. Sincere thanks also go to Elena Prieto and Daniel Johnstone for proof-reading and comments on an early draft. Many thanks also to Ferrante and Carlos, with my apologies for not being the best co-editor that their efforts deserved in three very complicated years for me. In addition, I would like to thank them for giving me the freedom to present these notes in a highly non-standard way.

References

1. Aarts, E.: Boltzmann machines for travelling salesman problems. European Journal of Operational Research (1989)
2. Ackley, D., Hinton, G., Sejnowski, T.: A learning algorithm for boltzmann machines. Cognitive Science (1985)
3. Adler, I., Bui-Xuan, B.-M., Rabinovich, Y., Renault, G., Telle, J.A., Vatshelle, M.: On the Boolean-Width of a Graph: Structure and Applications. In: Thilikos, D.M. (ed.) WG 2010. LNCS, vol. 6410, pp. 159–170. Springer, Heidelberg (2010)
4. Adra, S., Hamody, A., Griffin, I., Fleming, P.: A hybrid multi-objective evolutionary algorithm using an inverse neural network for aircraft control system design. In: [116], pp. 1–8 (2005)
5. Ahammed, F., Moscato, P.: Evolving L-systems as an intelligent design approach to find classes of difficult-to-solve traveling salesman problem instances. In: Di Chio, C. (ed.) EvoApplications 2011, Part I. LNCS, vol. 6624, pp. 1–11. Springer, Heidelberg (2011)
6. Aickelin, U., Burke, E.K., Li, J.: An estimation of distribution algorithm with intelligent local search for rule-based nurse rostering. Journal of the Operational Research Society 58, 1574–1585 (2007)
7. Alabau, M., Idoumghar, L., Schott, R.: New hybrid genetic algorithms for the frequency assignment problem. In: International Conference on Tools with Artificial Intelligence, pp. 136–142. IEEE Press, Los Alamitos (2001)
8. Alabau, M., Idoumghar, L., Schott, R.: New hybrid genetic algorithms for the frequency assignment problem. IEEE Transactions on Broadcasting 48(1), 27–34 (2002)
9. Ali, W., Topchy, A.P.: Memetic optimization of video chain designs. In: [201], pp. 869–882 (2004)
10. Alidaee, B., Kochenberger, B.G., Ahmadian, A.: 0–1 Quadratic Programming Approach for the Optimal Solution of Two Scheduling Problems. International Journal of Systems Science 25, 401–408 (1994)
11. Alkhamis, T., Hasan, M., Ahmed, M.: A Simulated Annealing for the Unconstrained Quadratic Pseudo-Boolean Function. European Journal of Operational Research 108, 641–652 (1998)
12. Allen, B.L., Steel, M.: Subtree transfer operations and their induced metrics on evolutionary trees. Annals of Combinatorics 5(1), 1–15 (2001)
13. Alon, U.: An Introduction to Systems Biology – Design Principles of Biological Circuits. Mathematical and Computational Biology Series, vol. 10. Chapman & Hall/Crc (2006)

14. Altenberg, L.: Fitness Distance Correlation Analysis: An Instructive Counterexample. In: [33], pp. 57–64 (1997)
15. Althaus, E., Klau, G.W., Kohlbacher, O., Lenhof, H.-P., Reinert, K.: Integer linear programming in computational biology. In: Albers, S., Alt, H., Näher, S. (eds.) Efficient Algorithms. LNCS, vol. 5760, pp. 199–218. Springer, Heidelberg (2009)
16. Amaldi, E.: On the complexity of training perceptrons. In: Kohonen, T., Mäkisara, K., Simula, O., Kangas, J. (eds.) Artificial Neural Networks, pp. 55–60. Elsevier science publishers B.V, Amsterdam (1991)
17. Amaldi, E., Nicolis, S.: Stability-capacity diagram of a neural network with ising bonds. Journal de Physique (1989)
18. Amini, M.M., Alidaee, B., Kochenberger, G.A.: A Scatter Search Approach to Unconstrained Quadratic Binary Programs. In: [145], pp. 317–329 (1999)
19. Andreatta, A., Ribeiro, C.: Heuristics for the phylogeny problem. Journal of Heuristics 8, 429–447 (2002)
20. Angel, E., Zissimopoulos, V.: Autocorrelation Coefficient for the Graph Bipartitioning Problem. Theoretical Computer Science 191, 229–243 (1998)
21. Angel, E., Bampis, E., Gourves, L.: A dynasearch neighborhood for the bicriteria traveling salesman problem. In: Gandibleux, X., et al. (eds.) Metaheuristics for Multiobjective Optimisation. Lecture Notes in Economics and Mathematical Systems, vol. 535, pp. 153–176. Springer, Heidelberg (2004)
22. Angeline, P.: Adaptive and self-adaptive evolutionary computations. In: Computational Intelligence, pp. 152–161. IEEE Press, Los Alamitos (1995)
23. Antoun, G., El-Nozahi, M., Fikry, W., Abbas, H.: A hybrid genetic algorithm for MOSFET parameter extraction. In: IEEE Canadian Conference on Electrical and Computer Engineering, vol. 2, pp. 1111–1114. IEEE Press, Los Alamitos (2003)
24. Applegate, D., Bixby, R., Chvátal, V., Cook, B.: Finding Cuts in the TSP (A preliminary report). Technical Report 95-05, DIMACS (1995)
25. Applegate, D., Bixby, R., Chvátal, V., Cook, W.: Finding Tours in the TSP. Tech. Rep. Report Number 99885, Research Institute for Discrete Mathematics, University of Bonn, Germany (1999)
26. Applegate, D., Cook, W., Rohe, A.: Chained Lin-Kernighan for Large Traveling Salesman Problems. INFORMS Journal on Computing 15(1), 82–92 (2003)
27. Arora, S.: Polynomial Time Approximation Schemes for Euclidean Traveling Salesman and Other Geometric Problems. Journal of the ACM 45(5), 753–782 (1998)
28. Auger, A., Hansen, N.: A restart CMA evolution strategy with increasing population size. In: [116], pp. 769–1776 (2005)
29. Baba, N.: A hybrid algorithm for finding a global minimum. International Journal of Control 37(5), 930–942 (1983)
30. Babu, B., Chakole, P., Mubeen, J.H.S.: Multiobjective differential evolution (MODE) for optimization of adiabatic styrene reactor. Chemical Engineering Science 60(17), 4822–4837 (2005)
31. Bäck, T.: Self adaptation in genetic algorithms. In: Varela, F., Bourgine, P. (eds.) Toward a Practice of Autonomous Systems: Proceedings of the 1st European Conference on Artificial Life, pp. 263–271. MIT Press, Cambridge (1992)
32. Bäck, T.: Evolutionary Algorithms in Theory and Practice. Oxford University Press, New York (1996)
33. Bäck, T. (ed.): Seventh International Conference on Genetic Algorithms. Morgan Kaufmann, San Mateo (1997)

34. Bäck, T., Hoffmeister, F.: Adaptive search by evolutionary algorithms. In: Ebeling, W., Peschel, M., Weidlich, W. (eds.) Models of Self-organization in Complex Systems. Mathematical Research, vol. 64, pp. 17–21. Akademie-Verlag (1991)
35. Bäck, T., Fogel, D.B., Michalewicz, Z.: Handbook of evolutionary computation. Springer, Heidelberg (1989)
36. Bäck, T., Fogel, D., Michalewicz, Z.: Evolutionary Computation 1: Basic Algorithms and Operators. Institute of Physics Publishing (2000)
37. Bäck, T., Fogel, D., Michalewicz, Z.: Evolutionary Computation 2: Advanced Algorithms and Operators. Institute of Physics Publishing (2000)
38. Baicher, G., Turton, B.: Comparative study for optimisation of causal IIR perfect reconstruction filter banks. In: [112], pp. 974–977 (2000)
39. Balas, E., Simonetti, N.: Linear Time Dynamic-Programming Algorithms for New Classes of Restricted TSPs: A Computational Study. INFORMS Journal on Computing 13(1), 56–75 (2000)
40. Bambha, N.K., Bhattacharyya, S.S., Teich, J., Zitzler, E.: Systematic integration of parameterized local search into evolutionary algorithms. IEEE Transactions on Evolutionary Computation 8(2), 137–155 (2004)
41. Banzhaf, W., Nordin, P., Keller, R., Francone, F.: Genetic Programming: An Introduction. Morgan Kaufmann, San Francisco (1998)
42. Banzhaf, W., et al. (eds.): Genetic and Evolutionary Computation Conference – GECCO 1999. Morgan Kaufmann, Orlando (1999)
43. Bärecke, T., Detyniecki, M.: Memetic algorithms for inexact graph matching. In: [118], pp. 4238–4245 (2007)
44. Barkat Ullah, A.S.S.M., Sarker, R., Cornforth, D., Lokan, C.: AMA: A new approach for solving constrained real-valued optimization problems. Soft Computing 13(8-9), 741–762 (2009)
45. Barkat Ullah, A.S.S.M., Sarker, R., Lokan, C.: An agent-based memetic algorithm (AMA) for nonlinear optimization with equality constraints. In: [120], pp. 70–77 (2009)
46. Barr, A., Feigenbaum, E.: Handbook of Artificial Intelligence. Morgan Kaufmann, New York (1981)
47. Bartz-Beielstein, T.: Experimental Research in Evolutionary Computation—The New Experimentalism. Natural Computing Series. Springer, Heidelberg (2006)
48. Bartz-Beielstein, T., Markon, S.: Tuning search algorithms for real-world applications: A regression tree based approach. Tech. Rep. CI-172/04, Collaborative Research Centre 531, University of Dortmund, Germany (2004)
49. Bartz-Beielstein, T., Lasarczyk, C., Preuß, M.: Sequential parameter optimization. In: [116], pp. 773–780 (2005)
50. Basseur, M.: Design of cooperative algorithms for multi-objective optimization: application to the flow-shop scheduling problem. 4OR: A Quarterly Journal of Operations Research 4(3), 255–258 (2006)
51. Batenburg, K.: An evolutionary algorithm for discrete tomography. Discrete Applied Mathematics 151(1–3), 36–54 (2005)
52. Battiti, R., Amaldi, E., Koch, C.: Computing optical flow across multiple scales: an adaptive coarse-to-fine strategy. International Journal of Computer Vision (1991)
53. Bazzoli, A., Tettamanzi, A.G.B.: A memetic algorithm for protein structure prediction in a 3D-lattice HP model. In: Raidl, G.R., Cagnoni, S., Branke, J., Corne, D.W., Drechsler, R., Jin, Y., Johnson, C.G., Machado, P., Marchiori, E., Rothlauf, F., Smith, G.D., Squillero, G. (eds.) EvoWorkshops 2004. LNCS, vol. 3005, pp. 1–10. Springer, Heidelberg (2004)

54. Beasley, J.E.: OR-Library: Distributing Test Problems by Electronic Mail. Journal of the Operational Research Society 41(11), 1069–1072 (1990)
55. Beasley, J.E.: Heuristic Algorithms for the Unconstrained Binary Quadratic Programming Problem. Tech. rep., Management School, Imperial College, London, UK (1998)
56. Belew, R.K., Booker, L.B. (eds.): Fourth International Conference on Genetic Algorithms. Morgan Kaufmann, San Diego (1991)
57. Bellman, R.: Dynamic Programming. Princeton University Press, Princeton (1957)
58. Belmonte, R., Vatshelle, M.: On graph classes with logarithmic boolean-width. CoRR abs/1009.0216 (2010)
59. Bentley, J.: Experiments on traveling salesman heuristics. In: Proceedings of the 1st Annual ACM-SIAM Symposium on Discrete Algorithms, pp. 91–99 (1990)
60. Berretta, R., Moscato, P.: The number partitioning problem: An open challenge for evolutionary computation? In: [145], pp. 261–278 (1999)
61. Berretta, R., Rodrigues, L.F.: A memetic algorithm for a multistage capacitated lot-sizing problem. International Journal of Production Economics 87(1), 67–81 (2004)
62. Berretta, R., Cotta, C., Moscato, P.: Enhancing the performance of memetic algorithms by using a matching-based recombination algorithm: Results on the number partitioning problem. In: Resende, M., Pinho de Sousa, J. (eds.) Metaheuristics: Computer-Decision Making, pp. 65–90. Kluwer Academic Publishers, Boston (2003)
63. Berretta, R., Costa, W., Moscato, P.: Combinatorial optimization models for finding genetic signatures from gene expression datasets. In: Keith, J.M. (ed.) Bioinformatics, Volume II: Structure, Function and Applications, Methods in Molecura Biology, ch. 19, pp. 363–378. Humana Press (2008)
64. Beume, N., Naujoks, B., Emmerich, M.: SMS-EMOA: Multiobjective selection based on dominated hypervolume. European Journal of Operational Research 181(3), 1653–1669 (2007)
65. Beyer, H.G.: The Theory of Evolution Strategies. Springer, Heidelberg (2001)
66. Binder, K., Young, A.: Spin glasses: Experimental facts, theoretical concepts, and open questions. Reviews of Modern Physics (1986)
67. Birattari, M., Yuan, Z., Balaprakash, P., Stützle, T.: F-race and iterated f-race: An overview. In: Bartz-Beielstein, T. (ed.) Empirical Methods for the Analysis of Optimization Algorithms, Natural Computing, pp. 311–336. Springer, Heidelberg (2010)
68. Bishop, C.M.: Neural Networks for Pattern Recognition. Oxford University Press, Oxford (1995)
69. Blesa, M.J., Blum, C., Cotta, C., Fernández, A.J., Gallardo, J.E., Roli, A., Sampels, M. (eds.): HM 2008. LNCS, vol. 5296. Springer, Heidelberg (2008)
70. Blum, C., Cotta, C., Fernández, A.J., Gallardo, J.E.: A probabilistic beam search approach to the shortest common supersequence problem. In: Cotta, C., van Hemert, J. (eds.) EvoCOP 2007. LNCS, vol. 4446, pp. 36–47. Springer, Heidelberg (2007)
71. Bodlaender, H.L.: Some classes of graphs with bounded treewidth. Bulletin of the EATCS 36, 116–125 (1988)
72. Bodlaender, H.L.: On reduction algorithms for graphs with small treewidth. In: van Leeuwen, J. (ed.) WG 1993. LNCS, vol. 790, pp. 45–56. Springer, Heidelberg (1994)
73. Bodlaender, H.L.: Improved self-reduction algorithms for graphs with bounded treewidth. Discrete Applied Mathematics 54(2-3), 101–115 (1994)
74. Bodlaender, H.L., Kloks, T.: Efficient and constructive algorithms for the pathwidth and treewidth of graphs. Journal of Algorithms 21(2), 358–402 (1996)
75. Boese, K.: Cost versus Distance in the Traveling Salesman Problem. Tech. Rep. TR-950018, UCLA Computer Science Department, Los Angeles, CA (1995)

References

76. Boese, K., Kahng, A.B., Muddu, S.: A new adaptive multi-start technique for combinatorial global optimization. Operations Research Letters 16(2), 101–113 (1994)
77. Boettcher, S., Percus, A.G.: Extremal optimization: Methods derived from co-evolution. In: [42], pp. 825–832 (1999)
78. Booker, A.J., Dennis, J.E., Frank, P.D., Serafini, D.B., Torczon, V., Trosset, M.W.: A rigorous framework for optimization of expensive functions by surrogates. Structural Optimization 17(1), 1–13 (1999)
79. Boudia, M., Prins, C.: A memetic algorithm with dynamic population management for an integrated production-distribution problem. European Journal of Operational Research 195(3), 703–715 (2009)
80. Boudia, M., Prins, C., Reghioui, M.: An effective memetic algorithm with population management for the split delivery vehicle routing problem. In: Bartz-Beielstein, T., Blesa Aguilera, M.J., Blum, C., Naujoks, B., Roli, A., Rudolph, G., Sampels, M. (eds.) HCI/ICCV 2007. LNCS, vol. 4771, pp. 16–30. Springer, Heidelberg (2007)
81. Brady, R.M.: Optimization Strategies Gleaned from Biological Evolution. Nature 317, 804–806 (1985)
82. Branke, J.: Memory enhanced evolutionary algorithms. In: [111], pp. 1875–1882 (1999)
83. Branke, J., Middendorf, M., Schneider, F.: Improved heuristics and a genetic algorithm for finding short supersequences. OR-Spektrum 20, 39–45 (1998)
84. Branke, J., Deb, K., Miettinen, K., Słowiński, R. (eds.): Multiobjective Optimization, Interactive and Evolutionary Approaches. LNCS, vol. 5252. Springer, Heidelberg (2008)
85. Buhmann, M.D.: Radial Basis Functions Theory and Implementations. Cambridge Monographs on Applied and Computational Mathematics, vol. 12. Cambridge University Press, Cambridge (2003)
86. Bui, T.G., Moon, B.R.: A New Genetic Approach for the Traveling Salesman Problem. In: Proceedings of the First IEEE Conference on Evolutionary Computation, pp. 7–12. IEEE Press, Los Alamitos (1994)
87. Bui, T.N., Moon, B.R.: GRCA: a hybrid genetic algorithm for circuit ratio-cut partitioning. IEEE Transactions on Computer-Aided Design of Integrated Circuits and Systems 17(3), 193–204 (1998)
88. Bui-Xuan, B.M., Telle, J.A., Vatshelle, M.: Feedback vertex set on graphs of low cliquewidth. In: Fiala, J., Kratochvíl, J., Miller, M. (eds.) IWOCA 2009. LNCS, vol. 5874, pp. 113–124. Springer, Heidelberg (2009)
89. Bui-Xuan, B.M., Telle, J.A., Vatshelle, M.: H-join decomposable graphs and algorithms with runtime single exponential in rankwidth. Discrete Applied Mathematics 158(7), 809–819 (2010)
90. Bull, L.: Artificial symbiology. PhD thesis, University of the West of England (1995)
91. Bull, L.: Evolutionary computing in multi agent environments: Partners. In: [33], pp. 370–377 (1997)
92. Bull, L., Fogarty, T.: Horizontal gene transfer in endosymbiosis. In: Langton, C., Shimohara, K. (eds.) Proceedings of the 5th International Workshop on Artificial Life: Synthesis and Simulation of Living Systems (ALIFE 1996), pp. 77–84. MIT Press, Cambridge (1997)
93. Bull, L., Holland, O., Blackmore, S.: On meme-gene coevolution. Artificial Life 6(3), 227–235 (2000)
94. Buriol, L., França, P., Moscato, P.: A new memetic algorithm for the asymmetric traveling salesman problem. Journal of Heuristics 10(5), 483–506 (2004)
95. Burke, E., Smith, A.: A multi-stage approach for the thermal generator maintenance scheduling problem. In: [111], pp. 1085–1092 (1999)

96. Burke, E., Smith, A.: Hybrid evolutionary techniques for the maintenance scheduling problem. IEEE Transactions on Power Systems 15(1), 122–128 (2000)
97. Burke, E., Kendall, G., Soubeiga, E.: A tabu search hyperheuristic for timetabling and rostering. Journal of Heuristics 9(6), 451–470 (2003)
98. Burke, E.K., De Causmaecker, P., De Maere, G., Mulder, J., Paelinck, M., Berghe, G.V.: A multi-objective approach for robust airline scheduling. Computers and Operations Research 37, 822–832 (2010)
99. Cadieux, S., Tanizaki, N., Okamura, T.: Time efficient and robust 3-D brain image centering and realignment using hybrid genetic algorithm. In: 36th SICE Annual Conference, pp. 1279–1284. IEEE Press, Los Alamitos (1997)
100. Cai, L., Chen, J.: On fixed-parameter tractability and approximability of NP-hard optimization problems. In: 2nd Israel Symposium on Theory of Computing and Systems, pp. 118–126. IEEE Comp. Soc. Press, Natanya (1993)
101. Cao, S., Zheng, J., Huang, W., Yang, G., Sun, Y., Wang, B.: Identification of strain hysteresis model for giant magnetostrictive actuators using a hybrid genetic algorithm. In: International Conference on Electrical Machines and Systems, vol. 3, pp. 2009–2012 (2005)
102. Cao, S., Wang, B., Zheng, J., Huang, W., Sun, Y., Yang, Q.: Modeling dynamic hysteresis for giant magnetostrictive actuator using hybrid genetic algorithm. IEEE Transactions on Magnetics 42(4), 911–914 (2006)
103. Caponio, A., Neri, F.: Integrating cross-dominance adaptation in multi-objective memetic algorithms. In: Goh, C., Ong, Y., Tan, K. (eds.) Multi-Objective Memetic Algorithms. Studies in Computational Intelligence, vol. 171, pp. 325–351. Springer, Heidelberg (2009)
104. Caponio, A., Cascella, G.L., Neri, F., Salvatore, N., Sumner, M.: A fast adaptive memetic algorithm for on-line and off-line control design of PMSM drives. IEEE Transactions on System Man and Cybernetics-part B, special issue on Memetic Algorithms 37(1), 28–41 (2007)
105. Caponio, A., Neri, F., Cascella, G.L., Salvatore, N.: Application of memetic differential evolution frameworks to PMSM drive design. In: [119], pp. 2113–2120 (2008)
106. Caponio, A., Neri, F., Tirronen, V.: Super-fit control adaptation in memetic differential evolution frameworks. Soft Computing 13(8-9), 811–831 (2009)
107. Carr, R., Hart, W., Krasnogor, N., Hirst, J., Burke, E.: Alignment of protein structures with a memetic evolutionary algorithm. In: [504], pp. 1027–1034 (2002)
108. Carrano, E.G., Souza, B.B., Neto, O.M., Takahashi, R.H.C.: An immune inspired memetic algorithm for power distribution system design under load evolution uncertainties. In: [119], pp. 3252–3258 (2008)
109. Carrizo, J., Tinetti, F., Moscato, P.: A computational ecology for the quadratic assignment problem. In: Proceedings of the 21st Meeting on Informatics and Operations Research, SADIO, Buenos Aires (1992)
110. Cattolico, M. (ed.): Genetic and Evolutionary Computation Conference – GECCO 2006. ACM Press, Seattle (2006)
111. CEC, IEEE Congress on Evolutionary Computation 1999, IEEE Press, Washington DC (1999)
112. CEC, IEEE Congress on Evolutionary Computation 2000, IEEE Press, San Diego (2000)
113. CEC, IEEE Congress on Evolutionary Computation 2002, IEEE Press, Honolulu (2002)
114. CEC, IEEE Congress on Evolutionary Computation 2003, IEEE Press, Canberra (2003)
115. CEC, IEEE Congress on Evolutionary Computation 2004, IEEE Press, Portland (2004)

116. CEC, IEEE Congress on Evolutionary Computation 2005, IEEE Press, Edinburgh (2005)
117. CEC, IEEE Congress on Evolutionary Computation 2006, IEEE Press, Vancouver (2006)
118. CEC, IEEE Congress on Evolutionary Computation 2007, IEEE Press, Singapore (2007)
119. CEC, IEEE Congress on Evolutionary Computation 2008, IEEE Press, Hong Kong (2008)
120. CEC, IEEE Congress on Evolutionary Computation 2009, IEEE Press, Trondheim (2009)
121. CEC, IEEE Congress on Evolutionary Computation 2010, IEEE Press, Barcelona (2010)
122. Cerny, V.: A thermodynamical aprroach to the traveling salesman problem. Journal of Optimization, theory and Application 45(1), 41–51 (1985)
123. Chelouah, R., Siarry, P.: Genetic and Nelder–Mead algorithms hybridized for a more accurate global optimization of continuous multiminima functions. European Journal of Operational Research 148(2), 335–348 (2003)
124. Chen, J.H., Chen, J.H.: Multi-objective memetic approach for flexible process sequencing problems. In: [784], pp. 2123–2128 (2008)
125. Cheng, H.C., Chiang, T.C., Fu, L.C.: Multiobjective job shop scheduling using memetic algorithm and shifting bottleneck procedure. In: IEEE Symposium on Intelligence in Scheduling, pp. 15–21. IEEE Press, Los Alamitos (2009)
126. Chootinan, P., Chen, A.: Constraint handling in genetic algorithms using a gradient-based repair method. Computers & Operations Research 33(8), 2263–2281 (2006)
127. Chu, P.C., Beasley, J.E.: A genetic algorithm for the multidimensional knapsack problem. Journal of Heuristics 4, 63–86 (1998)
128. Clark, D., Westhead, D.: Evolutionary algorithms in computer-aided molecular design. Journal of Computer-aided Molecular Design 10(4), 337–358 (1996)
129. Clerc, M., Kennedy, J.: The particle swarm-explosion, stability and convergence in a multidimensional complex space. IEEE Transactions on Evolutionary Computation 6(1), 58–73 (2002)
130. Cobb, H., Grefenstette, J.: Genetic algorithms for tracking changing environments. In: Forrest, S. (ed.) ICGA 1993, pp. 529–530. Morgan Kaufmann, San Mateo (1993)
131. Coello Coello, C., Van Veldhuizen, D.A., Lamont, G.: Evolutionary Algorithms for Solving Multi-Objective Problems. Kluwer Academic Publishers, Dordrecht (2002)
132. Coello Coello, C.A.: A comprehensive survey of evolutionary-based multiobjective optimization techniques. Knowledge and Information Systems 1, 269–308 (1998)
133. Coello Coello, C.A.: Constraint-handling using an evolutionary multiobjective optimization technique. Civil engineering and environmental systems 17(4), 319–346 (2000)
134. Coello Coello, C.A., Pulido, G.T., Lechuga, M.: Handling multiple objectives with particle swarm optimization. IEEE Transactions on Evolutionary Computation 8(3), 256–279 (2004)
135. Coello Coello, C.A., Hernández Aguirre, A., Zitzler, E. (eds.): EMO 2005. LNCS, vol. 3410. Springer, Heidelberg (2005)
136. Coll, P., Durán, G., Moscato, P.: On worst-case and comparative analysis as design principles for efficient recombination operators: A graph coloring case study. In: [145], pp. 279–294 (1999)
137. Collet, P., Fonlupt, C., Hao, J.-K., Lutton, E., Schoenauer, M. (eds.): EA 2001. LNCS, vol. 2310. Springer, Heidelberg (2002)

138. Congram, R.: Polynomially searchable exponential neighbourhoods for sequencing problems in combinatorial optimisation. PhD thesis. University of Southampton, Faculty of Mathematical Studies (2000)
139. Congram, R., Potts, C., van de Velde, S.: An iterated dynasearch algorithm for the single-machine total weighted tardiness scheduling problem. INFORMS Journal on Computing 14(1), 52–67 (2002)
140. Conn, A.R., Scheinberg, K., Toint, P.L.: On the convergence of derivative-free methods for unconstrained optimization. In: Iserles, A., Buhmann, M.D. (eds.) Approximation Theory and Optimization: Tributes to M.J.D. Powell, pp. 83–108. Cambridge University Press, Cambridge (1997)
141. Conn, A.R., Scheinberg, K., Toint, P.L.: Recent progress in unconstrained nonlinear optimization without derivatives. Mathematical Programming 79, 397–414 (1997)
142. Conn, A.R., Gould, N., Toint, P.L.: Trust-Region Methods. SIAM, Philadelphia (2000)
143. Cook, W., Seymour, P.D.: Tour merging via branch-decomposition. INFORMS Journal on Computing 15(3), 233–248 (2003)
144. Cordón, O., Damas, S., Santamaría, J., Martí, R.: Scatter search for the 3D point matching problem in image registration. INFORMS Journal on Computing 20(1), 55–68 (2008)
145. Corne, D., Dorigo, M., Glover, F. (eds.): New Ideas in Optimization. McGraw-Hill, Maidenhead (1999)
146. Costa, D.: An evolutionary tabu search algorithm and the NHL scheduling problem. INFOR 33(3), 161–178 (1995)
147. Cotta, C.: A study of hybridisation techniques and their application to the design of evolutionary algorithms. AI Communications 11(3-4), 223–224 (1998)
148. Cotta, C.: Protein structure prediction using evolutionary algorithms hybridized with backtracking. In: Mira, J., Álvarez, J.R. (eds.) IWANN 2003. LNCS, vol. 2687, pp. 321–328. Springer, Heidelberg (2003)
149. Cotta, C.: A comparison of evolutionary approaches to the shortest common supersequence problem. In: Cabestany, J., Prieto, A.G., Sandoval, F. (eds.) IWANN 2005. LNCS, vol. 3512, pp. 50–58. Springer, Heidelberg (2005)
150. Cotta, C.: Hybrid evolutionary algorithms for protein structure prediction in the HPNX model. In: Computational Intelligence, Theory and Applications. Advances in Soft Computing, vol. 2, pp. 525–534. Springer, Heidelberg (2005)
151. Cotta, C.: Memetic algorithms with partial lamarckism for the shortest common supersequence problem. In: [602], pp. 84–91 (2005)
152. Cotta, C.: On the Application of Evolutionary Algorithms to the Consensus Tree Problem. In: Raidl, G.R., Gottlieb, J. (eds.) EvoCOP 2005. LNCS, vol. 3448, pp. 58–67. Springer, Heidelberg (2005)
153. Cotta, C.: Scatter search with path relinking for phylogenetic inference. European Journal of Operational Research 169(2), 520–532 (2005)
154. Cotta, C., Fernández, A.: Memetic algorithms in planning, scheduling, and timetabling. In: Dahal, K., Tan, K., Cowling, P. (eds.) Evolutionary Scheduling. Studies in Computational Intelligence, vol. 49, pp. 1–30. Springer, Heidelberg (2007)
155. Cotta, C., Moscato, P.: Inferring phylogenetic trees using evolutionary algorithms. In: [580], pp. 720–729 (2002)
156. Cotta, C., Moscato, P.: The k-feature set problem is W[2]-complete. Journal of Computer and Systems Science 67(4), 686–690 (2003)
157. Cotta, C., Moscato, P.: A memetic-aided approach to hierarchical clustering from distance matrices: Application to phylogeny and gene expression clustering. Biosystems 72(1-2), 75–97 (2003)

References

158. Cotta, C., Moscato, P.: Evolutionary computation: Challenges and duties. In: Menon, A. (ed.) Frontiers of Evolutionary Computation, pp. 53–72. Kluwer Academic Press, Boston (2004)
159. Cotta, C., Moscato, P.: The parameterized complexity of multiparent recombination. In: Proceedings of MIC 2005 - The 6th Metaheuristics International Conference, Vienna, Austria, pp. 237–242 (2005)
160. Cotta, C., Troya, J.: A hybrid genetic algorithm for the 0-1 multiple knapsack problem. In: Smith, G., Steele, N., Albrecht, R. (eds.) Artificial Neural Nets and Genetic Algorithms 3, pp. 251–255. Springer, Wien (1998)
161. Cotta, C., Troya, J.: Optimal discrete recombination: Hybridising evolution strategies with the A* algorithm. In: Mira, J. (ed.) IWANN 1999. LNCS, vol. 1607, pp. 58–67. Springer, Heidelberg (1999)
162. Cotta, C., Troya, J.: On the influence of the representation granularity in heuristic forma recombination. In: Carroll, J., Damiani, E., Haddad, H., Oppenheim, D. (eds.) ACM Symposium on Applied Computing 2000, pp. 433–439. ACM Press, New York (2000)
163. Cotta, C., Troya, J.M.: Using a Hybrid Evolutionary-A* Approach for Learning Reactive Behaviours. In: Oates, M.J., Lanzi, P.L., Li, Y., Cagnoni, S., Corne, D.W., Fogarty, T.C., Poli, R., Smith, G.D. (eds.) EvoIASP 2000, EvoWorkshops 2000, EvoFlight 2000, EvoSCONDI 2000, EvoSTIM 2000, EvoTEL 2000, and EvoROB/EvoRobot 2000. LNCS, vol. 1803, pp. 347–356. Springer, Heidelberg (2000)
164. Cotta, C., Troya, J.: Embedding branch and bound within evolutionary algorithms. Applied Intelligence 18(2), 137–153 (2003)
165. Cotta, C., Aldana, J., Nebro, A., Troya, J.: Hybridizing genetic algorithms with branch and bound techniques for the resolution of the TSP. In: Pearson, D., Steele, N., Albrecht, R. (eds.) Artificial Neural Nets and Genetic Algorithms 2, pp. 277–280. Springer, Wien (1995)
166. Cotta, C., Alba, E., Troya, J.: Utilising dynastically optimal forma recombination in hybrid genetic algorithms. In: [240], pp. 305–314 (1998)
167. Cotta, C., Mendes, A., García, F., França, P., Moscato, P.: Applying memetic algorithms to the analysis of microarray data. In: Raidl, G.R., Cagnoni, S., Cardalda, J.J.R., Corne, D.W., Gottlieb, J., Guillot, A., Hart, E., Johnson, C.G., Marchiori, E., Meyer, J.-A., Middendorf, M. (eds.) EvoIASP 2003, EvoWorkshops 2003, EvoSTIM 2003, EvoROB/EvoRobot 2003, EvoCOP 2003, EvoBIO 2003, and EvoMUSART 2003. LNCS, vol. 2611, pp. 22–32. Springer, Heidelberg (2003)
168. Cotta, C., Sloper, C., Moscato, P.: Evolutionary search of thresholds for robust feature set selection: Application to the analysis of microarray data. In: Raidl, G.R., Cagnoni, S., Branke, J., Corne, D.W., Drechsler, R., Jin, Y., Johnson, C.G., Machado, P., Marchiori, E., Rothlauf, F., Smith, G.D., Squillero, G. (eds.) EvoWorkshops 2004. LNCS, vol. 3005, pp. 21–30. Springer, Heidelberg (2004)
169. Cowling, P.I., Kendall, G., Soubeiga, E.: A hyperheuristic approach to scheduling a sales summit. In: Burke, E., Erben, W. (eds.) PATAT 2000. LNCS, vol. 2079, pp. 176–190. Springer, Heidelberg (2001)
170. Cox, M., Bowden, N., Moscato, P., Berretta, R., Scott, R.I., Lechner-Scott, J.S.: Memetic algorithms as a new method to interpret gene expression profiles in multiple sclerosis. In: Abstracts of the 23rd Congress of the European Committee for Treatment and Research in Multiple Sclerosis and the 12th Annual Conference of Rehabilitation in Multiple Sclerosis, Prague, Czech Republic, vol. 13(suppl. 2), p. S205 (2007)
171. Créput, J.C., Koukam, A.: The memetic self-organizing map approach to the vehicle routing problem. Journal of Soft Computing 12, 1125–1141 (2008)

172. Cressie, N.A.C.: Statistics for Spatial Data. Wiley, Chichester (1993)
173. Crutchley, D., Zwolinski, M.: Using evolutionary and hybrid algorithms for dc operating point analysis of nonlinear circuits. In: [113], pp. 753–758 (2002)
174. da Silva Maximiano, M., Vega-Rodriguez, M.A., Gomez-Pulido, J., Sanchez-Perez, J.: A hybrid differential evolution algorithm to solve a real-world frequency assignment problem. In: International Multiconference on Computer Science and Information Technology, pp. 201–205 (2008)
175. Dahal, K., Burt, G., McDonald, J., Galloway, S.: GA/SA-based hybrid techniques for the scheduling of generator maintenance in power systems. In: [112], pp. 567–574 (2000)
176. Dai, J., Wu, G., Wu, Y., Zhu, G.: Helicopter trim research based on hybrid genetic algorithm. In: World Congress on Intelligent Control and Automation, pp. 2007–2011 (2008)
177. Dantas, M., da C Brito, L., de Carvalho, P.: Biobjective hybrid evolutionary algorithm applied to resonator filters of arbitrary topology. In: IEEE International Conference on Electronics, Circuits and Systems, pp. 296–299. IEEE Press, Los Alamitos (2006)
178. Dantas, M.J.P., da C. Brito, L., de Carvalho, P.H.P.: Multi-objective memetic algorithm applied to the automated synthesis of analog circuits. In: Sichman, J.S., Coelho, H., Rezende, S.O. (eds.) IBERAMIA 2006 and SBIA 2006. LNCS (LNAI), vol. 4140, pp. 258–267. Springer, Heidelberg (2006)
179. Dantzig, G.B., Fulkerson, D.R., Johnson, S.M.: Solution of a Large-Scale Traveling Salesman Problem. Operations Research 2, 393–410 (1954)
180. Das, S., Konar, A., Chakraborty, U.K.: Annealed differential evolution. In: [118], pp. 1926–1933 (2007)
181. David-Tabibi, O., Koppel, M., Netanyahu, N.S.: Genetic algorithms for automatic search tuning. ICGA Journal 33(2), 67–79 (2010)
182. Davidon, W.C.: Variable metric method for minimization. Tech. Rep. ANL-5990, Argonne National Laboratory (1959)
183. Davidor, Y.: Epistasis Variance: Suitability of a Representation to Genetic Algorithms. Complex Systems 4(4), 369–383 (1990)
184. Davidor, Y., Ben-Kiki, O.: The interplay among the genetic algorithm operators: Information theory tools used in a holistic way. In: [551], pp. 75–84 (1992)
185. Davidor, Y., Männer, R., Schwefel, H.-P. (eds.): PPSN 1994. LNCS, vol. 866. Springer, Heidelberg (1994)
186. Davies, D.L., Bouldin, D.W.: A cluster separation measure. IEEE Transactions on In Pattern Analysis and Machine Intelligence PAMI-1, 224–227 (1979)
187. Davies, S., Russell, S.: NP-completeness of searches for smallest possible feature sets. In: Greiner, R., Subramanian, D. (eds.) AAAI Symposium on Intelligent Relevance, pp. 41–43. AAAI Press, New Orleans (1994)
188. Davis, L.: Handbook of Genetic Algorithms. Van Nostrand Reinhold Computer Library, New York (1991)
189. Day, W.: The complexity of computing metric distances between partitions. Mathematical Social Sciences 1(1), 269–287 (1981)
190. Day, W.: Computationally difficult problems in phylogeny systematics. Journal of Theoretic Biology 103, 429–438 (1983)
191. Day, W.: Computational complexity of inferring phylogenies from dissimilarity matrices. Bulletin of Mathematical Biology 49(4), 461–467 (1987)
192. De Falco, I.: An introduction to evolutionary algorithms and their application to the aerofoil design problem–Part II: The Results. In: van den Braembussche, R., Manna, M. (eds.) Inverse Design and Optimisation Methods, Von Karman Institute for Fluid Dynamics (1997)

References

193. De Jong, K.: Evolutionary Computation: A Unified Approach. MIT Press, Cambridge (2006)
194. de Queiroz, L., Lyra, C.: A genetic approach for loss reduction in power distribution systems under variable demands. In: [117], pp. 2691–2698 (2006)
195. Deb, K.: An efficient constraint handling method for genetic algorithms. Computer Methods in Applied Mechanics and Engineering 186, 311–338 (2000)
196. Deb, K.: Multi-objective Optimization using Evolutionary Algorithms, pp. 147–149. John Wiley and Sons LTD, Chichester (2001)
197. Deb, K., Goel, T.: Controlled elitist non-dominated sorting genetic algorithms for better convergence. In: [971], pp. 67–81 (2001)
198. Deb, K., Goel, T.: A hybrid multi-objective evolutionary approach to engineering shape design. In: [971], pp. 385–399 (2001)
199. Deb, K., Agrawal, S., Pratab, A., Meyarivan, T.: A fast elitist non-dominated sorting genetic algorithm for multi-objective optimization: NSGA-II. In: [796], pp. 849–858 (2000)
200. Deb, K., Pratap, A., Agarwal, S., Meyarivan, T.: A fast and elitist multiobjective genetic algorithm: NSGA-II. IEEE Transactions on Evolutionary Computation 6(2), 182–197 (2002)
201. Deb, K., et al. (eds.): GECCO 2004. LNCS, vol. 3102. Springer, Heidelberg (2004)
202. Dechter, R.: Bucket elimination: A unifying framework for reasoning. Artificial Intelligence 113(1-2), 41–85 (1999)
203. DeGroot, M.H.: Optimal Statistical Decisions. McGraw-Hill Book Company, New York (1970)
204. DeJong, K.A.: An analysis of the behavoir of a class of genetic adaptive systems. PhD thesis, University of Michigan, Ann Arborn, MI, USA (1975)
205. Delvecchio, G., Lofrumento, C., Neri, F., Sylos Labini, M.: A fast evolutionary-deterministic algorithm to study multimodal current fields under safety level constraints. COMPEL: International Journal for Computation and Mathematics in Electrical and Electronic Engineering 25(3), 599–608 (2006)
206. Demaine, E.D., Hajiaghayi, M.T.: Fast algorithms for hard graph problems: Bidimensionality, minors, and local treewidth. In: Pach, J. (ed.) GD 2004. LNCS, vol. 3383, pp. 517–533. Springer, Heidelberg (2005)
207. Demaine, E.D., Fomin, F.V., Hajiaghayi, M.T., Thilikos, D.M.: Bidimensional parameters and local treewidth. SIAM J. Discrete Math. 18(3), 501–511 (2004)
208. Dengiz, B., Altiparmak, F., Belgin, O.: A hybrid ant colony optimization approach for the design of reliable networks. In: [118], pp. 1118–1125 (2007)
209. Di Gaspero, L., Schaerf, A.: Neighborhood portfolio approach for local search applied to timetabling problems. Journal of Mathematical Modeling and Algorithms 5(1), 65–89 (2006)
210. Di Gesù, V., Lo Bosco, G., Millonzi, F., Valenti, C.: A memetic algorithm for binary image reconstruction. In: Brimkov, V.E., Barneva, R.P., Hauptman, H.A. (eds.) IWCIA 2008. LNCS, vol. 4958, pp. 384–395. Springer, Heidelberg (2008)
211. Di Gesù, V., Lo Bosco, G., Millonzi, F., Valenti, C.: Discrete tomography reconstruction through a new memetic algorithm. In: Giacobini, M., Brabazon, A., Cagnoni, S., Di Caro, G.A., Drechsler, R., Ekárt, A., Esparcia-Alcázar, A.I., Farooq, M., Fink, A., McCormack, J., O'Neill, M., Romero, J., Rothlauf, F., Squillero, G., Uyar, A.Ş., Yang, S. (eds.) EvoWorkshops 2008. LNCS, vol. 4974, pp. 347–352. Springer, Heidelberg (2008)
212. Dias, J., Captivo, M., Clímaco, J.: A memetic algorithm for multi-objective dynamic location problems. Journal of Global Optimization 42(2), 221–253 (2008)

213. Dill, K.: Dominant forces in protein folding. Biochemistry 29, 7133–7155 (1990)
214. Doerner, K., Gutjahr, W., Hartl, R., Strauss, C., Stummer, C.: Pareto ant colony optimization: A metaheuristic approach to multiobjective portfolio selection. Annals of Operations Research 131(1-4), 79–99 (2004)
215. Dorigo, M., Stützle, T.: Ant Colony Optimization. MIT Press, Cambridge (2004)
216. Dorn, F., Telle, J.A.: Semi-nice tree-decompositions: The best of branchwidth, treewidth and pathwidth with one algorithm. Discrete Applied Mathematics 157(12), 2737–2746 (2009)
217. Dorne, R., Hao, J.K.: A new genetic local search algorithm for graph coloring. In: [240], pp. 745–754 (1998)
218. Dorronsoro, B., Alba, E., Luque, G., Bouvry, P.: A self-adaptive cellular memetic algorithm for the dna fragment assembly problem. In: [119], pp. 2656–2663 (2008)
219. dos Santos Coelho, L., Rodrigues Coelho, A., Krohling, R.: Parameters tuning of multivariable controllers based on memetic algorithm: fundamentals and application. In: IEEE International Symposium on Intelligent Control, pp. 752–757. IEEE Press, Los Alamitos (2002)
220. Downey, R., Fellows, M.: Fixed parameter tractability and completeness III: Some structural aspects of the W-hierarchy. In: Ambos-Spies, K., Homer, S., Schöning, U. (eds.) Complexity Theory: Current Research, pp. 166–191. Cambridge University Press, Cambridge (1993)
221. Downey, R., Fellows, M.: Fixed parameter tractability and completeness I: Basic theory. SIAM Journal of Computing 24, 873–921 (1995)
222. Downey, R., Fellows, M.: Fixed-parameter tractability and completeness II: On completeness for $W[1]$. Theoretical Computer Science 141(1-2), 109–131 (1995)
223. Downey, R., Fellows, M.: Parameterized computational feasibility. In: Clote, P., Remmel, J. (eds.) Feasible Mathematics II, pp. 219–244. Birkhäuser, Basel (1995)
224. Downey, R., McCartin, C.: Some new directions and questions in parameterized complexity. In: Calude, C.S., Calude, E., Dinneen, M.J. (eds.) DLT 2004. LNCS, vol. 3340, pp. 12–26. Springer, Heidelberg (2004)
225. Downey, R., Fellows, M., Stege, U.: Computational Tractability: The View From Mars. Bulletin of the European Association for Theoretical Computer Science 69, 73–97 (1999)
226. Downey, R., Fellows, M., Stege, U.: Parameterized Complexity: A framework for systematically confronting computational intractability. In: Contemporary Trends in Discrete Mathematics: From DIMACS to DIMATIA to the future. AMS-DIMACS Proceedings Series, pp. 49–99. AMS, Providence (1999)
227. Droste, S., Jansen, T., Wegener, I.: On the analysis of the (1+1) evolutionary algorithm. Theoretical Computer Science 276, 51–81 (2002)
228. Droste, S., Jansen, T., Wegener, I.: Optimization with randomized search heuristics—the (A)NFL theorem, realistic scenarios, and difficult functions. Theoretical Computer Science 287(1), 131–144 (2002)
229. Duvigneau, R., Praveen, C.: Meta-modeling for robust design and multi-level optimization. In: Proceedings of the 42nd AAAF Congress on Applied Aerodynamics, AAAF, Sophia-Antipolis, France (2007)
230. Dzubera, J., Whitley, D.: Advanced Correlation Analysis of Operators for the Traveling Salesman Problem. In: [185], pp. 68–77 (1994)
231. Edwards, S., Anderson, P.: Theory of spin glasses. Journal of Physics F: Metal Physics (1975)

References

232. Egea, J.A., Balsa-Canto, E., Garćia, M.S.G., Ranga, J.R.: Dynamic optimization of nonlinear processes with an enhanced scatter search method. Journal of Industrial Chemical Engineering Research 48, 4388–4401 (2009)
233. Ehrgott, M.: Multicriteria Optimization. Springer, Heidelberg (2005)
234. Eiben, A.: Multiparent recombination. In: Bäck, T., Fogel, D., Michalewicz, Z. (eds.) Evolutionary Computation 1: Basic Algorithms and Operators, pp. 289–307. Institute of Physics Publishing (2000)
235. Eiben, A., Michalewicz, Z.: Evolutionary Computation. IOS Press, Amsterdam (1998)
236. Eiben, A., Aarts, E., van Hee, K.: Global convergence of genetic algorithms: A markov chain analysis. In: [802], pp. 4–12 (1991)
237. Eiben, A., Hinterding, R., Michalewicz, Z.: Parameter control in evolutionary algorithms. IEEE Transactions on Evolutionary Computation 3(2), 124–141 (1999)
238. Eiben, A., Michalewicz, Z., Schoenauer, M., Smith, J.: Parameter control in evolutionary algorithms. In: Lobo, F., Lima, C., Michalewicz, Z. (eds.) Parameter Setting in Evolutionary Algorithms, pp. 19–46. Springer, Heidelberg (2007)
239. Eiben, A.E., Smith, J.E.: Introduction to Evolutionary Computing. Springer, Berlin (2003)
240. Eiben, A.E., Bäck, T., Schoenauer, M., Schwefel, H.-P. (eds.): PPSN 1998. LNCS, vol. 1498. Springer, Heidelberg (1998)
241. Eichfelder, G.: Adaptive Scalarization Methods in Multiobjective Optimization. Springer, Heidelberg (2008)
242. Eisen, M., Spellman, P., Brown, P., Botstein, D.: Cluster analysis and display of genome-wide expression patterns. National Academy of Sciences 95(25), 14,863–14,868 (1998)
243. Elmohamed, M.A.S., Coddington, P.D., Fox, G.C.: A comparison of annealing techniques for academic course scheduling. In: Burke, E.K., Carter, M. (eds.) PATAT 1997. LNCS, vol. 1408, pp. 92–112. Springer, Heidelberg (1998)
244. Emmerich, M., Beume, N., Naujoks, B.: An EMO algorithm using the hypervolume measure as selection criterion. In: [135], pp. 62–76 (2005)
245. Englert, M., Röglin, H., Vöcking, B.: Worst case and probabilistic analysis of the 2-Opt algorithm for the TSP. In: Bansal, N., Pruhs, K., Stein, C. (eds.) 18th ACM-SIAM Symposium on Discrete Algorithms (SODA 2007), pp. 1295–1304. SIAM, Philadelphia (2007)
246. Eshelman, L.: The CHC Adaptive Search Algorithm: How to Have Safe Search When Engaging in Nontraditional Genetic Recombination. In: Rawlings, G.J.E. (ed.) Foundations of Genetic Algorithms, pp. 265–283. Morgan Kaufmann, San Francisco (1991)
247. Eshelman, L.J., Schaffer, J.D.: Preventing premature convergence in genetic algorithms by preventing incest. In: [56], pp. 115–122 (1991)
248. Falkenauer, E.: Genetic algorithms and grouping problems. John Wiley & Sons, Inc., New York (1998)
249. Fallahi, A.E., Prins, C., Calvo, R.W.: A memetic algorithm and a tabu search for the multi-compartment vehicle routing problem. Computers & Operations Research 35(5), 1725–1741 (2008)
250. Fan, S.K.S., Liang, Y.C., Zahara, E.: A genetic algorithm and a particle swarm optimizer hybridized with Nelder–Mead simplex search. Computers & Industrial Engineering 50(4), 401–425 (2006)
251. Fawaz, Z., Xu, Y., Behdinan, K.: Hybrid evolutionary algorithm and application to structural optimization. Structural and Multidisciplinary Optimization 30(3), 219–226 (2005)

252. Fellows, M.R.: Recent developments in the theory of pre-processing. In: Atallah, M., Li, X.-Y., Zhu, B. (eds.) FAW-AAIM 2011. LNCS, vol. 6681, pp. 4–5. Springer, Heidelberg (2011)
253. Fellows, M.R., Rosamond, F.A., Fomin, F.V., Lokshtanov, D., Saurabh, S., Villanger, Y.: Local search: Is brute-force avoidable? In: Boutilier, C. (ed.) IJCAI, pp. 486–491 (2009)
254. Fellows, M.R., Rosamond, F.A., Rotics, U., Szeider, S.: Clique-width is NP-complete. SIAM J. Discrete Math. 23(2), 909–939 (2009)
255. Felsenstein, J.: Evolutionary trees from dna sequences: a maximum likelihood approach. Journal of Molecular Evolution 17, 368–376 (1981)
256. Felsenstein, J.: Inferring phylogenies. Sinauer Associates, Inc., Publishers, Sunderland (2003)
257. Ferentinos, K., Tsiligiridis, T.: A memetic algorithm for dynamic design of wireless sensor networks. In: [118], pp. 2774–2781 (2007)
258. Fernandez, E., Grana, M., Cabello, J.: An instantaneous memetic algorithm for illumination correction. In: [115], pp. 1105–1110 (2004)
259. Festa, P.: On some optimization problems in molecular biology. Mathematical Biosciences 207(2), 219–234 (2007)
260. Fialho, Á., Da Costa, L., Schoenauer, M., Sebag, M.: Dynamic multi-armed bandits and extreme value-based rewards for adaptive operator selection in evolutionary algorithms. In: Stützle, T. (ed.) LION 3. LNCS, vol. 5851, pp. 176–190. Springer, Heidelberg (2009)
261. Fidanova, S., Alba, E., Molina, G.: Memetic simulated annealing for the GPS surveying problem. In: Margenov, S., Vulkov, L.G., Waśniewski, J. (eds.) NAA 2008. LNCS, vol. 5434, pp. 281–288. Springer, Heidelberg (2009)
262. Fischer, T., Bauer, K., Merz, P.: A distributed memetic algorithm for the routing and wavelength assignment problem. In: [781], pp. 879–888 (2008)
263. Fischetti, M., Lodi, A.: Local branching. Mathematical Programmming 98, 23–47 (2003)
264. Fitch, W.M.: Towards defining course of evolution: minimum change for a specified tree topology. Systematic Zoology 20, 406–416 (1971)
265. Flarup, U., Lyaudet, L.: On the expressive power of permanents and perfect matchings of matrices of bounded pathwidth/cliquewidth. Theory Comput. Syst. 46(4), 761–791 (2010)
266. Fletcher, R., Powell, M.J.D.: A rapidly convergent descent method for minimization. The Computer Journal 6(2), 163–168 (1963)
267. Fleurent, C., Ferland, J.A.: Object-oriented implementation of heuristic search methods for graph coloring. In: Johnson, D.S., Trick, M.A. (eds.) Cliques, Coloring, and Satisfiability: Second DIMACS Implementation Challenge. DIMACS Series in Discrete Mathematics and Theoretical Computer Science, vol. 26, pp. 619–652. American Mathematical Society, Providence (1996)
268. Fogel, D.: Evolving artificial intelligence. PhD thesis, University of California (1992)
269. Fogel, D.: Evolutionary Computation. IEEE Press, Los Alamitos (1995)
270. Fogel, D.: Evolutionary Computation: the Fossil Record. IEEE Press, Los Alamitos (1998)
271. Fogel, L., Owens, A., Walsh, M.: Artificial Intelligence through Simulated Evolution. Wiley, Chichester (1966)

272. Fomin, F.V., Thilikos, D.M.: Dominating sets and local treewidth. In: Di Battista, G., Zwick, U. (eds.) ESA 2003. LNCS, vol. 2832, pp. 221–229. Springer, Heidelberg (2003)
273. Fomin, F.V., Thilikos, D.M.: Branchwidth of graphs. In: Kao, M.Y. (ed.) Encyclopedia of Algorithms. Springer, Heidelberg (2008)
274. Fomin, F.V., Fraigniaud, P., Nisse, N.: Nondeterministic graph searching: From pathwidth to treewidth. Algorithmica 53(3), 358–373 (2009)
275. Fomin, F.V., Lokshtanov, D., Raman, V., Saurabh, S.: Fast local search algorithm for weighted feedback arc set in tournaments. In: Fox, M., Poole, D. (eds.) Proceedings of the Twenty-Fourth AAAI Conference on Artificial Intelligence, AAAI 2010, Atlanta, Georgia, USA, July 11-15, AAAI Press, Menlo Park (2010)
276. Fonseca, C., Fleming, P.: An overview of evolutionary algorithms in multiobjective optimisation. Evolutionary Computation 3(1), 1–16 (1995)
277. Foulds, L., Graham, R.: The steiner problem in phylogeny is NP-complete. Advances in Applied Mathematics 3(1), 43–49 (1982)
278. Fourer, R., Gay, D.M., Kernighan, B.W.: AMPL: A Modeling Language for Mathematical Programming, 2nd edn. Duxbury Press Brooks Cole Publishing Co. (2003)
279. Franca, P., Gupta, J., Mendes, A., Moscato, P., Veltink, K.: Evolutionary algorithms for flowshop scheduling with family setups. Computers and Industrial Engineering 48(3), 491–506 (2005)
280. França, P., Mendes, A., Moscato, P.: Algoritmos Meméticos e o Sequenciamento em Máquina Simples com Setup Times e Restrições de Datas de Entrega. In: XXX SOBRAPO - Simpósio Brasileiro de Pesquisa Operacional, Curitiba, PR, Brasil, de Novembro 25-27, Sociedade Brasileira de Pesquisa Operacional, pp. 315–316 (1998), extended abstract
281. França, P., Moscato, P., Müller, F., Mendes, A., Buriol, L.: O Projeto MemePool: Um Framework para Otimização Combinatória. In: XXX SOBRAPO- Simpósio Brasileiro de Pesquisa Operacional, Curitiba, PR, Brasil, de Novembro 25-27, Sociedade Brasileira de Pesquisa Operacional, pp. 20–21 (1998), extended abstract
282. França, P.M., Mendes, A., Moscato, P.: A memetic algorithm for the total tardiness single machine scheduling problem. European Journal of Operational Research 132(1), 224–242 (2001)
283. Frank, P.D.: Global modeling for optimization. SIAM SIAG/OPT Views-and-News 7, 9–12 (1995)
284. Franz, A., Hoffmann, K.H., Salamon, P.: A best possible strategy for finding ground states. Physical Review Letters 86, 5219–5222 (2001)
285. Freisleben, B., Merz, P.: A genetic local search algorithm for solving symmetric and asymmetric traveling salesman problems. In: Proceedings of the 1996 IEEE International Conference on Evolutionary Computation, Nagoya, Japan, pp. 616–621. IEEE Press, Los Alamitos (1996)
286. Freisleben, B., Merz, P.: New genetic local search operators for the traveling salesman problem. In: [909], pp. 890–899 (1996)
287. French, A., Robinson, A., Wilson, J.: Using a hybrid genetic-algorithm/branch and bound approach to solve feasibility and optimization integer programming problems. Journal of Heuristics 7(6), 551–564 (2001)
288. Fukunaga, A.: Automated discovery of local search heuristics for satisfiability testing. Evolutionary Computation 16(1), 31–61 (2008)
289. Galinier, P., Hao, J.K.: Tabu search for maximal constraint satisfaction problems. In: Smolka, G. (ed.) CP 1997. LNCS, vol. 1330, pp. 196–208. Springer, Heidelberg (1997)

290. Galinier, P., Hao, J.K.: Hybrid evolutionary algorithms for graph coloring. Journal of Combinatorial Optimization 3(4), 379–397 (1999)
291. Galinier, P., Hao, J.K.: A general approach for constraint solving by local search. Journal of Mathematical Modelling and Algorithms 3(1), 73–88 (2004)
292. Galinier, P., Hertz, A., Zufferey, N.: An adaptive memory algorithm for the k-coloring problem. Discrete Applied Mathematics 156(2), 267–279 (2008)
293. Gallardo, J.E., Cotta, C., Fernández, A.J.: A hybrid model of evolutionary algorithms and branch-and-bound for combinatorial optimization problems. In: [116], pp. 2248–2254 (2005)
294. Gallardo, J.E., Cotta, C., Fernández, A.J.: Solving the multidimensional knapsack problem using an evolutionary algorithm hybridized with branch and bound. In: [602], pp. 21–30 (2005)
295. Gallardo, J.E., Cotta, C., Fernández, A.J.: A memetic algorithm with bucket elimination for the still life problem. In: Gottlieb, J., Raidl, G.R. (eds.) EvoCOP 2006. LNCS, vol. 3906, pp. 73–85. Springer, Heidelberg (2006)
296. Gallardo, J.E., Cotta, C., Fernández, A.J.: A multi-level memetic/exact hybrid algorithm for the still life problem. In: [783], pp. 212–221 (2006)
297. Gallardo, J.E., Cotta, C., Fernández, A.J.: On the hybridization of memetic algorithms with branch-and-bound techniques. IEEE Transactions on Systems, Man and Cybernetics, part B 37(1), 77–83 (2007)
298. Gallardo, J.E., Cotta, C., Fernández, A.J.: Reconstructing phylogenies with memetic algorithms and branch-and-bound. In: Bandyopadhyay, S., Maulik, U., Wang, J. (eds.) Analysis of Biological Data: A Soft Computing Approach, pp. 59–84. World Scientific, Singapore (2007)
299. Gallardo, J.E., Cotta, C., Fernández, A.J.: Solving weighted constraint satisfaction problems with memetic/exact hybrid algorithms. Journal of Artificial Intelligence Research 35, 533–555 (2009)
300. Gallo, G., Hammer, P.L., Simeone, B.: Quadratic Knapsack Problems. Mathematical Programming 12, 132–149 (1980)
301. Gandibleux, X., Morita, H., Katoh, N.: The supported solutions used as a genetic information in a population heuristic. In: [971], pp. 429–442 (2001)
302. Garcia, V., França, P.M., Mendes, A., Moscato, P.: A parallel memetic algorithm applied to the total tardiness machine scheduling problem. In: 20th International Parallel and Distributed Processing Symposium. IEEE Press, Los Alamitos (2006)
303. Garey, M.R., Johnson, D.S.: Computers and Intractability, A Guide to the Theory of NP-Completeness. W. H. Freeman and Company, New York (1979)
304. Garret, D., Dasgupta, D.: An empirical comparison of memetic algorithm strategies on the multiobjective quadratic assignment problem. In: IEEE Symposium on Computational intelligence in multi-criteria decision-making, pp. 80–87. IEEE Press, Los Alamitos (2009)
305. Gaspar-Cunha, A., Vieira, A.: A multi-objective evolutionary algorithm using neural networks to approximate fitness evaluations. International Journal of Computers, Systems and Signals 6(1), 18–36 (2005)
306. Geman, S., Geman, D.: Stochastic relaxation, gibbs distributions, and the bayesian restoration of images. IEEE Trans. Pattern Anal. Mach. Intell. 6, 721–741 (1984)
307. Giannakoglou, K.C.: Design of optimal aerodynamic shapes using stochastic optimization methods and computational intelligence. International Review Journal Progress in Aerospace Sciences 38(1), 43–76 (2002)

308. Glover, F.: Heuristics for integer programming using surrogate constraints. Decision Sciences 8(1), 156–166 (1977)
309. Glover, F.: Tabu search – part I. ORSA Journal of Computing 1(3), 190–206 (1989)
310. Glover, F.: Tabu search – part II. ORSA Journal of Computing 2(1), 4–31 (1989)
311. Glover, F.: Tabu search for nonlinear and parametric optimization (with links to genetic algorithms). Discrete Applied Mathematics 49(1–3), 231–255 (1994)
312. Glover, F.: Tabu search and adaptive memory programming advances, applications and challenges. In: Barr, R., Helgason, R., Kennington, J. (eds.) Interfaces in Computer Science and Operations Research, vol. 7, pp. 1–75. Springer, Heidelberg (1997)
313. Glover, F.: A template for scatter search and path relinking. In: Hao, J.-K., Lutton, E., Ronald, E., Schoenauer, M., Snyers, D. (eds.) AE 1997. LNCS, vol. 1363, pp. 13–54. Springer, Heidelberg (1998)
314. Glover, F.: Scatter search and path relinking. In: [145], pp. 291–316 (1999)
315. Glover, F., Kochenberger, G. (eds.): Handbook of Metaheuristics. Kluwer Academic Publishers, Boston (2003)
316. Glover, F., Kochenberger, G.A.: Critical event tabu search for multidimensional knapsack problems. In: Osman, I., Kelly, J. (eds.) Metaheuristics: The Theory and Applications, pp. 407–425. Kluwer Academic Publishers, Dordrecht (1996)
317. Glover, F., Laguna, M.: Tabu Search. Kluwer Academic Publishers, Dordrecht (1997)
318. Glover, F., Kochenberger, G., Alidaee, B., Amini, M.: Tabu Search with Critical Event Memory: An Enhanced Application for Binary Quadratic Programs. In: Voss, S., Martello, S., Osman, I., Roucairol, C. (eds.) Meta-Heuristics - Advances and Trends in Local Search Paradigms for Optimization, pp. 83–109. Kluwer Academic Publishers, Dordrecht (1998)
319. Glover, F., Kochenberger, G.A., Alidaee, B.: Adaptive Memory Tabu Search for Binary Quadratic Programs. Management Science 44(3), 336–345 (1998)
320. Glover, F., Laguna, M., Martí, R.: Fundamentals of scatter search and path relinking. Control and Cybernetics 39(3), 653–684 (2000)
321. Glover, F., Laguna, M., Martí, R.: Scatter search. In: Ghosh, A., Tsutsui, S. (eds.) Advances in Evolutionary Computation: Theory and Applications. Natural Computing Series, pp. 519–537. Springer, Heidelberg (2003)
322. Goëffon, A., Richer, J.M., Hao, J.K.: A distance-based information preservation tree crossover for the maximum parsimony problem. In: [783], pp. 761–770 (2006)
323. Goëffon, A., Richer, J.M., Hao, J.K.: Progressive tree neighborhood applied to the maximum parsimony problem. IEEE/ACM Transactions on Computational Biology and Bioinformatics 5(1), 136–145 (2008)
324. Goh, C.K., Tan, K.C.: Evolving the tradeoffs between pareto-optimality and robustness in multi-objective evolutionary algorithms. In: Yang, S., Ong, Y.S., Jin, Y. (eds.) Evolutionary Computation in Dynamic and Uncertain Environments. SCI, pp. 457–478. Springer, Heidelberg (2007)
325. Goldberg, D.E.: Genetic algorithms in search, optimization and machine learning. Addison-Wesley, Reading (1989)
326. Goldberg, D.E.: The Design of Innovation: Lessons from and for Competent Genetic Algorithms. Kluwer Academic Publishers, Norwell (2002)
327. Goldberg, D.E., Richardson, J.: Genetic algorithms with sharing for multimodal function optimization. In: [333], pp. 41–49 (1987)
328. Goldberg, D.E., Segrest, P.: Finite markov chain analysis of genetic algorithms. In: [333], pp. 1–8 (1987)

329. Goldberg, D.E., Voessner, S.: Optimizing global-local search hybrids. In: [42], pp. 220–228 (1999)
330. Gómez Ravetti, M., Rosso, O.A., Berretta, R., Moscato, P.: Uncovering molecular biomarkers that correlate cognitive decline with the changes of hippocampus' gene expression profiles in alzheimer's disease. PLoS ONE 5(4), e10, 153 (2010)
331. Gorges-Schleuter, M.: ASPARAGOS: An Asynchronous Parallel Genetic Optimization Strategy. In: [794], pp. 422–427 (1989)
332. Greenberg, H.J., Hart, W.E., Lancia, G.: Opportunities for combinatorial optimization in computational biology. INFORMS Journal on Computing 16(3), 211–231 (2004)
333. Grefenstette, J. (ed.): Second International Conference on Genetic Algorithms. Morgan Kaufmann, San Mateo (1987)
334. Grefenstette, J.J.: Incorporating Problem Specific Knowledge into Genetic Algorithms. In: Davis, L. (ed.) Genetic Algorithms and Simulated Annealing. Research Notes in Artificial Intelligence, pp. 42–60. Morgan Kaufmann Publishers, San Francisco (1987)
335. Greffenstette, J.: Optimisation of control parameters for genetic algorithms. IEEE Transactions on Systems, Man and Cybernetics 16, 122–128 (1986)
336. Grimbleby, J.B.: Hybrid genetic algorithms for analogue network synthesis. In: [111], pp. 1781–1787 (1999)
337. Gross, D., Harris, C.M.: Fundamentals of Queueing Theory. Wiley, NY (1985)
338. Guéret, C., Prins, C., Sevaux, M.: Applications of optimisation with Xpress-MP. Dash Optimization (2002)
339. Guerra-Salcedo, C., Chen, S., Whitley, D., Smith, S.: Fast and accurate feature selection using hybrid genetic strategies. In: [111], pp. 177–184 (1999)
340. Guimaraes, F., Lowther, D., Ramirez, J.: Analysis of the computational cost of approximation-based hybrid evolutionary algorithms in electromagnetic design. IEEE Transactions on Magnetics 44(6), 1130–1133 (2008)
341. Gurski, F.: Characterizations for restricted graphs of NLC-width 2. Theor. Comput. Sci. 372(1), 108–114 (2007)
342. Gurski, F., Wanke, E.: Minimizing NLC-width is NP-complete. In: Kratsch, D. (ed.) WG 2005. LNCS, vol. 3787, pp. 69–80. Springer, Heidelberg (2005)
343. Gutin, G., Karapetyan, D.: A memetic algorithm for the multidimensional assignment problem. In: Stützle, T., Birattari, M., Hoos, H.H. (eds.) SLS 2009. LNCS, vol. 5752, pp. 125–129. Springer, Heidelberg (2009)
344. Gutin, G., Karapetyan, D.: A selection of useful theoretical tools for the design and analysis of optimization heuristics. Memetic Computing 1(1), 25–34 (2009)
345. Gutin, G., Karapetyan, D.: A memetic algorithm for the generalized traveling salesman problem. Natural Computing 9(1), 47–60 (2010)
346. Gutin, G., Karapetyan, D., Krasnogor, N.: Memetic algorithm for the generalized asymmetric traveling salesman problem. In: Krasnogor, N., Nicosia, G., Pavone, M., Pelta, D.A. (eds.) NICSO. SCI, vol. 210, pp. 199–210. Springer, Heidelberg (2007)
347. Haas, O., Burnham, K., Mills, J., Reeves, C., Fisher, M.: Hybrid genetic algorithms applied to beam orientation in radiotherapy. In: Proccedings of 4th European Conference on Intelligent Techniques and Soft Computing, Verlag Mainz, vol. 3, pp. 2050–2055 (1996)
348. Haas, O., Burnham, K., Mills, J.: Optimization of beam orientation in radiotherapy using planar geometry. Physics in Medicine and Biology 43(8), 2179–2193 (1998)

349. Haidari, S., Farsangi, M., Nezamabadi-pour, H., Lee, K.Y.: Design of supplementary controller for HVDC using memetic algorithm with population management. In: IEEE Power and Energy Society General Meeting - Conversion and Delivery of Electrical Energy in the 21st Century, pp. 1–6. IEEE Press, Los Alamitos (2008)
350. Hajela, P., Lin, C.Y.: Genetic search strategies in multicriterion optimal design. Structural Optimization 4, 99–107 (1992)
351. Hajiaghayi, M., Nishimura, N.: Subgraph isomorphism, log-bounded fragmentation, and graphs of (locally) bounded treewidth. J. Comput. Syst. Sci. 73(5), 755–768 (2007)
352. Hamiez, J.P., Hao, J.K.: Scatter search for graph coloring. In: [137], pp. 168–179 (2002)
353. Hamiez, J.-P., Robet, J., Hao, J.-K.: A tabu search algorithm with direct representation for strip packing. In: Cotta, C., Cowling, P. (eds.) EvoCOP 2009. LNCS, vol. 5482, pp. 61–72. Springer, Heidelberg (2009)
354. Hammel, U., Bäck, T.: Evolution strategies on noisy functions, how to improve convergence properties. In: [185], pp. 159–168 (1994)
355. Han, C.W., Park, J.I.: SA-selection-based genetic algorithm for the design of fuzzy controller. International Journal of Control, Automation, and Systems 3(2), 236–243 (2005)
356. Handoko, S., Kwoh, C., Ong, Y., Lim, M.: A study on constrained ma using ga and sqp: Analytical vs. finite-difference gradients. In: [119], pp. 4031–4038 (2008)
357. Hansen, M.: Metaheuristics for multiple objective combinatorial optimization. PhD thesis, Technical University of Denmark (1998)
358. Hansen, M.: Tabu search for multiobjective combinatorial optimization: TAMOCO. Control and Cybernetics 29(3), 799–818 (2000)
359. Hansen, M.: Use of substitute scalarizing functions to guide a local search based heuristic: the case of moTSP. Journal of Heuristics 6(3), 419–430 (2000)
360. Hansen, N.: The CMA evolution strategy (2011), http://www.lri.fr/~hansen/cmaesintro.html (accessed on April 18, 2011)
361. Hansen, N., Ostermeier, A.: Adapting arbitrary normal mutation distributions in evolution strategies: The covariance matrix adaptation. In: IEEE Conference on Evolutionary Computation, pp. 312–317. IEEE Press, Piscataway (1996)
362. Hansen, N., Ostermeier, A.: Completely derandomized self-adaptation in evolution strategies. Evolutionary Computation 9(2), 159–195 (2001)
363. Hansen, P., Mladenović, N.: An introduction to variable neighborhood search. In: Voss, S., et al. (eds.) Metaheuristics, Advances and Trends in Local Search Paradigms for Optimization, pp. 433–458. Kluwer Academic Publishers, Dordrecht (1999)
364. Hansen, P., Mladenović, N.: Variable neighborhood search. In: Glover, F., Kochenberger, G. (eds.) Handbook of Metaheuristics, pp. 145–184. Kluwer Academic Publishers, Boston (2002)
365. Harada, K., Ikeda, K., Kobayashi, S.: Hybridization of genetic algorithm and local search in multiobjective function optimization: recommendation of GA then LS. In: [110], pp. 667–674 (2006)
366. Hart, W.: Adaptive global optimization with local search. PhD thesis, University of California, San Diego, CA (1994)
367. Hart, W., Belew, R.: Optimizing an arbitrary function is hard for the genetic algorithm. In: [56], pp. 190–195 (1991)

368. Hart, W., Belew, R.K.: Optimization with genetic algorithm hybrids that use local search. In: Belew, R.K., Mitchell, M. (eds.) Adaptive Individuals in Evolving Populations: Models and Algorithms, Santa Fe Institute Studies in the Sciences of Complexity, ch. 27, pp. 483–496. Addison-Wesley, Reading (1995)
369. Hart, W., Istrail, S.: Fast protein folding in the hydrophobic-hydrophilic model within three-eighths of optimal. Journal of Computational Biology 3(1), 53–96 (1996)
370. Hart, W., Istrail, S.: Lattice and off-lattice side chain models of protein folding: Linear time structure prediction better than 86% of optimal. Journal of Computational Biology 4(3), 241–259 (1997)
371. Hart, W., Istrail, S.: Robust proofs of NP-hardness for protein folding: General lattices and energy potentials. Journal of Computational Biology 4(1), 1–20 (1997)
372. Hart, W., Istrail, S.: Locally-adaptive and memetic evolutionary pattern search algorithms. Evolutionary Computaton 11(1), 29–52 (2003)
373. Hart, W., Rosin, C., Belew, R., Morris, G.: Improved evolutionary hybrids for flexible ligand docking in autodock. In: Floudas, C.A., Pardalos, P.M. (eds.) Optimization in Computational Chemistry and Molecular Biology, Nonconvex Optimization and Its Applications, vol. 40, pp. 209–230. Springer, Heidelberg (2000)
374. Hart, W., Krasnogor, N., Smith, J.E.: Special issue on memetic algorithms. Evolutionary Computation 12(3) (2004)
375. Hart, W., Krasnogor, N., Smith, J.: Recent advances in memetic algorithms. STUDFUZZ, vol. 166. Springer, Heidelberg (2005)
376. Hart, W., Krasnogor, N., Smith, J.E.: Memetic evolutionary algorithms. In: [375], pp. 3–27 (2005)
377. Hasan, K., Sarker, R., Essam, D.: Evolutionary scheduling with rescheduling option for sudden machine breakdowns. In: [121], pp. 1913–1920 (2010)
378. Hasan, S., Sarker, R., Essam, D., Cornforth, D.: Memetic algorithms for solving jobshop scheduling problems. Memetic Computing 1(1), 69–83 (2009)
379. Hazrati, N., Rashidi-Nejad, M., Gharaveisi, A.A.: Pricing and allocation of spinning reserve and energy in restructured power systems via memetic algorithm. In: Conference on Power Engineering, 2007 Large Engineering Systems, pp. 234–238 (2007)
380. Helsgaun, K.: An Effective Implementation of the Lin-Kernighan Traveling Salesman Heuristic. European Journal of Operational Research 126(1), 106–130 (2000)
381. Herdy, M.: Reproductive isolation as strategy parameter in hierarchically organized evolution strategies. In: [551], pp. 2–9 (1992)
382. Herrera, F., Lozano, M., Sánchez, A.M.: A taxonomy for the crossover operator for real-coded genetic algorithms: An experimental study. International Journal of Intelligent Systems 18, 309–338 (2003)
383. Herrera, F., Lozano, M., Molina, D.: Continuous scatter search: An analysis of the integration of some combination methods and improvement strategies. European Journal of Operational Research 169(2), 450–476 (2006)
384. Hicks, I.V.: Branchwidth and branch decompositions. In: Floudas, C.A., Pardalos, P.M. (eds.) Encyclopedia of Optimization, pp. 332–339. Springer, Heidelberg (2009)
385. Hidalgo, J., Lanchares, J., Ibarra, A., Hermida, R.: A hybrid evolutionary algorithm for multi-FPGA systems design. In: Euromicro Symposium on Digital System Design, pp. 60–67 (2002)
386. Hinterding, R., Michalewicz, Z., Eiben, A.: Adaptation in evolutionary computation: A survey. In: 1997 IEEE Conference on Evolutionary Computation. IEEE Press, Los Alamitos (1997)

387. Hofmann, R.: Parallel evolutionary trajectories. Research Report SS92-11, Edinburgh Parallel Computing Centre (1992)
388. Hofmann, R.: Examinations on the algebra of genetic algorithms. Master's thesis, Technische Universität München, Institut fü Informatik (1993)
389. Holland, J.H.: Adaptation and artificial systems. University of Michigan Press (1975)
390. Holstein, D., Moscato, P.: Memetic algorithms using guided local search: A case study. In: [145], pp. 235–244 (1999)
391. Hooke, R., Jeeves, T.A.: Direct search solution of numerical and statistical problems. Journal of the ACM 8, 212–229 (1961)
392. Hoos, H.H., Stützle, T.: SATLIB: An online resource for research on SAT. In: Gent, I., Maaren, H., Walsh, T. (eds.) SAT 2000, pp. 283–292. IOS Press, Amsterdam (2000), http://www.satlib.org
393. Hoos, H.H., Stützle, T.: Stochastic Local Search: Foundations and Applications. Morgan Kaufmann Publishers, San Francisco (2004)
394. Horjik, W., Manderick, B.: The Usefulness of Recombination. In: Proceedings of the European Conference on Artificial Life. Springer, Heidelberg (1995)
395. Horn, J., Goldberg, D.E., Deb, K.: Long path problems. In: [185], pp. 149–158 (1994)
396. Houck, C., Joines, J., Kay, M., Wilson, J.: Empirical investigation of the benefits of partial lamarckianism. Evolutionary Computation 5(1), 31–60 (1997)
397. Hourani, M., Berretta, R., Mendes, A., Moscato, P.: Genetic signatures for a rodent model of parkinson's disease using combinatorial optimization methods. In: Keith, J.M. (ed.) Bioinformatics: Structure, Function and Applications, Methods in Molecura Biology, vol. II, pp. 379–392. Humana Press (2008)
398. Hung, L.-J., Kloks, T.: Classifying rankwidth **k**-DH-graphs. In: Ablayev, F., Mayr, E.W. (eds.) CSR 2010. LNCS, vol. 6072, pp. 195–203. Springer, Heidelberg (2010)
399. Husbands, P., Mill, F.: Simulated coevolution as the mechanism for emergent planning and scheduling. In: [56], pp. 264–270 (1991)
400. Hwang, C.L., Paidy, S., Yoon, K., Masud, A.: Mathematical programming with multiple objectives: A tutorial. Computers and Operations Research 7, 5–31 (1980)
401. Igel, C., Toussaint, M.: A no-free-lunch theorem for non-uniform distributions of target functions. Journal of Mathematical Modelling and Algorithms 3(4), 313–322 (2004)
402. Il-Seok, O., Jin-Seon, L., Byung-Ro, M.: Hybrid genetic algorithms for feature selection. IEEE Transactions on Pattern Analysis and Machine Intelligence 26, 1424–1437 (2004)
403. Im, C.H., Jung, H.K., Kim, Y.J.: Hybrid genetic algorithm for electromagnetic topology optimization. IEEE Transactions on Magnetics 39(5), 2163–2169 (2003)
404. Inostroza-Ponta, M., Berretta, R., Mendes, A., Moscato, P.: An automatic graph layout procedure to visualize correlated data. In: Bramer, M. (ed.) IFIP AI. IFIP, vol. 217, pp. 179–188. Springer, Heidelberg (2006)
405. Inostroza-Ponta, M., Mendes, A., Berretta, R., Moscato, P.: An integrated QAP-based approach to visualize patterns of gene expression similarity. In: Randall, M., Abbass, H.A., Wiles, J. (eds.) ACAL 2007. LNCS (LNAI), vol. 4828, pp. 156–167. Springer, Heidelberg (2007)
406. Inostroza-Ponta, M., Berretta, R., Moscato, P.: Qapgrid: A two level qap-based approach for large-scale data analysis and visualization. PLoS ONE 6(1), 6(1), e14, 468 (2011)
407. Isaacs, A., Ray, T., Smith, W.: Memetic algorithm for dynamic bi-objective optimization problems. In: [120], pp. 1707–1713 (2009)

408. Ishibuchi, H., Murata, T.: Multi-objective genetic local search algorithm. In: Fukuda, T., Furuhashi, T. (eds.) 1996 International Conference on Evolutionary Computation, pp. 119–124. IEEE Press, Nagoya (1996)
409. Ishibuchi, H., Murata, T.: Multi-objective genetic local search algorithm and its application to flowshop scheduling. IEEE Transactions on Systems, Man and Cybernetics - Part C: Applications and Reviews 28(3), 392–403 (1998)
410. Ishibuchi, H., Narukawa, K.: Some issues on the implementation of local search in evolutionary multiobjective optimization. In: [201], pp. 1246–1258 (2004)
411. Ishibuchi, H., Yoshida, T., Murata, T.: Balance between genetic search and local search in memetic algorithms for multiobjective permutation flowshop scheduling. IEEE Transactions on Evolutionary Computation 7(2), 204–223 (2003)
412. Ishibuchi, H., Narukawa, K., Tsukamoto, N., Nojima, Y.: An empirical study on similarity-based mating for evolutionary multiobjective combinatorial optimization. European Journal of Operational Research 188(1), 57–75 (2008)
413. Ishibuchi, H., Hitotsuyanagi, Y., Tsukamoto, N., Nojima, Y.: Use of biased neighborhood structures in multiobjective memetic algorithms. Soft Computing 13(8–9), 795–810 (2009)
414. Ivănescu, P.L.: Some Network Flow Problems Solved with Pseudo-Boolean Programming. Operations Research 13, 388–399 (1965)
415. Iyama, Y., Matsubara, F.: Ground-state properties of a heisenberg spin glass model with a hybrid genetic algorithm. Journal of the Physical Society of Japan 78(1), 014,703 (2009)
416. Jansen, T., Wegener, I.: On the choice of the mutation probability for the (1+1) EA. In: [796], pp. 89–98 (2000)
417. Jansen, T., De Jong, K.A., Wegener, I.: On the choice of the offspring population size in evolutionary algorithms. Evolutionary Computation 13, 413–440 (2005)
418. Jaszkiewicz, A.: Multiple objective metaheuristic algorithms for combinatorial optimization. PhD thesis, Poznan University of Technology, habilitation thesis (2001)
419. Jaszkiewicz, A.: Genetic local search for multi-objective combinatorial optimization. European Journal of Operational Research 137, 50–71 (2002)
420. Jaszkiewicz, A.: On the performance of multiple objective genetic local search on the 0/1 knapsack problem. a comparative experiment. IEEE Transactions on Evolutionary Computation 6(4), 402–412 (2002)
421. Jaszkiewicz, A.: A comparative study of multiple-objective metaheuristics on the biobjective set covering problem and the pareto memetic algorithm. Annals of Operations Research 131(1-4), 135–158 (2004)
422. Jaszkiewicz, A., Branke, J.: Interactive multiobjective evolutionary algorithms. In: [84], pp. 179–193 (2008)
423. Jaszkiewicz, A., Zielniewicz, P.: Efficient adaptation of the pareto memetic algorithm to the multiple objective travelling salesperson problem. In: 7th International conference on MultiObjective Programming and Goal Programming, MOPFP 2006 (2006)
424. Jin, Y.: A comprehensive survey of fitness approximation in evolutionary computation. Soft Computing - A Fusion of Foundations, Methodologies and Applications 9(1), 3–12 (2005)
425. Jin, Y., Branke, J.: Evolutionary optimization in uncertain environments. IEEE Transactions on Evolutionary Computation 9(3), 305–317 (2005)
426. Jin, Y., Olhofer, M., Sendhoff, B.: A framework for evolutionary optimization with approximate fitness functions. IEEE Transactions on evolutionary computation 6(5), 481–494 (2002)

References

427. Johnson, D.S.: Local Optimization and the Traveling Salesman Problem. In: Paterson, M. (ed.) ICALP 1990. LNCS, vol. 443, pp. 446–461. Springer, Heidelberg (1990)
428. Johnson, D.S., McGeoch, L.A.: Experimental Analysis of Heuristics for the STSP. In: Gutin, G., Punnen, A. (eds.) The Traveling Salesman Problem and its Variations, Kluwer Academic Publishers, Dordrecht (2002)
429. Johnson, D.S., Papadimitriou, C.H., Yannakakis, M.: How easy is local search? Journal of Computer and System Sciences 37(1), 79–100 (1988)
430. Johnson, D.S., Aragon, C.R., McGeoch, L.A., Schevon, C.: Optimization by Simulated Annealing: An Experimental Evaluation; Part I, Graph Partitioning. Operations Research 37, 865–892 (1989)
431. Johnson, D.S., McGeoch, L.A., Schevon, C.: Optimization by simulated annealing: An experimental evaluation; part II, graph coloring and number partitioning. Operations Research 39(3), 378–406 (1991)
432. Jones, T.: Crossover, macromutation, and population-based search. In: Eshelman, L. (ed.) Sixth International Conference on Genetic Algorithms, pp. 73–80. Morgan Kaufmann, San Francisco (1995)
433. Jones, T.: Evolutionary algorithms, fitness landscapes and search. PhD thesis, University of New Mexico, USA (1995)
434. Jones, T.: One operator, one landscape. Tech. Rep. #95-02-025, Santa Fe Institute (1996)
435. Jones, T., Forrest, S.: Fitness Distance Correlation as a Measure of Problem Difficulty for Genetic Algorithms. In: Eshelman, L.J. (ed.) Proceedings of the 6th International Conference on Genetic Algorithms, pp. 184–192. Morgan Kaufmann, San Francisco (1995)
436. Juang, C.F.: A hybrid of genetic algorithm and particle swarm optimization for recurrent network design. IEEE Transactions on Systems, Man, and Cybernetics–Part B: Cybernetics 34(2), 997–1006 (2004)
437. Judd, J.S.: Neural network design and the complexity of learning. In: Neural Network Modeling and Connectionism. MIT Press, Cambridge (1990)
438. Kallel, L., Naudts, B., Reeves, C.R.: Properties of fitness functions and search landscapes. In: Kallel, L., Naudts, B., Rogers, A. (eds.) Theoretical Aspects of Evolutionary Computing. Natural Computing Series, pp. 175–206. Springer, Heidelberg (2001)
439. Karakasis, M.K., Giannakoglou, K.C.: On the use of surrogate evaluation models in multi-objective evolutionary algorithms. In: Proceedings of the European Conference on Computational Methods in Applied Sciences and Engineering, ECCOMAS 2004 (2004)
440. Karakasis, M.K., Koubogiannis, D., Giannakoglou, K.C.: Hierarchical distributed evolutionary algorithms in shape optimization. International Journal of Numerical Methods in Fluids 53(3), 455–469 (2007)
441. Karaoğlu, B., Topçuoğlu, H., Gürgen, F.: Evolutionary algorithms for location area management. In: Rothlauf, F., Branke, J., Cagnoni, S., Corne, D.W., Drechsler, R., Jin, Y., Machado, P., Marchiori, E., Romero, J., Smith, G.D., Squillero, G. (eds.) EvoWorkshops 2005. LNCS, vol. 3449, pp. 175–184. Springer, Heidelberg (2005)
442. Karapetyan, D., Gutin, G.: A new approach to population sizing for memetic algorithms: A case study for the multidimensional assignment problem. CoRR abs/1003.4314 (2010)
443. Kargupta, H., Ghosh, S.: Towards machine learning through genetic code-like transformations. Tech. rep., Computer Science and Electrical Engineering Department, University of Maryland Baltimore County (2001)

444. Kase, S., Nishiyama, N.: An Industrial Engineering Game Model for Factory Layout. The Journal of Industrial Engineering XV(3), 148–150 (1964)
445. Kask, K., Detcher, R.: A general scheme for automatic generation of search heuristics from specification dependencies. Artificial Intelligence 129, 91–131 (2001)
446. Katayama, K., Narihisa, H.: Performance of Simulated Annealing-based Heuristic for the Unconstrained Binary Quadratic Programming Problem. Tech. rep., Okayama University of Science, Dept. of Information and Computer Engineering, Okayama, Japan (1999)
447. Katayama, K., Narihisa, H.: Solving Large Binary Quadratic Programming Problems by Effective Genetic Local Search Algorithm. In: [932], pp. 643–650 (2000)
448. Katayama, K., Narihisa, H.: A Variant k-opt Local Search Heuristic for Binary Quadratic Programming. Trans. IEICE (A) J84-A(3), 430–435 (2001)
449. Katsumata, Y., Terano, T.: Bayesian optimization algorithm for multi-objective solutions: application to electric equipment configuration problems in a power plant. In: [114], pp. 1101–1107 (2003)
450. Kauffman, S., Levin, S.: Towards a general theory of adaptive walks on rugged landscapes. Journal of Theoretical Biology 128, 11–45 (1987)
451. Kauffman, S.A.: The Origins of Order: Self-Organization and Selection in Evolution. Oxford University Press, Oxford (1993)
452. Kauffman, S.A., Levin, S.: Towards a General Theory of Adaptive Walks on Rugged Landscapes. Journal of Theoretical Biology 128, 11–45 (1987)
453. Keller, R., Banzhaf, W.: Genetic programming using genotype-phenotype mapping from linear genomes into linear phenotypes. In: Koza, J., Goldberg, D., Fogel, D., Riolo, R. (eds.) 1st Annual Conference on Genetic Programming, pp. 116–122. MIT Press, Cambridge (1996)
454. Keller, R., Banzhaf, W.: The evolution of genetic code in genetic programming. In: [42], pp. 1077–1082 (1999)
455. Kelner, V., Capitanescu, F., Léonard, O., Wehenkel, L.: A hybrid optimization technique coupling an evolutionary and a local search algorithm. Journal of Computational and Applied Mathematics 215(2), 448–456 (2008)
456. Kendall, G., Cowling, P., Soubeiga, E.: Choice function and random hyperheuristics. In: Fourth Asia-Pacific Conference on Simulated Evolution and Learning, Singapore, pp. 667–671 (2002)
457. Kennedy, J., Eberhart, R.: Swarm Intelligence. Morgan Kaufmann Publishers, San Francisco (2001)
458. Kennedy, J., Eberhart, R.C.: Particle swarm optimization. In: IEEE International Conference on Neural Networks, pp. 1942–1948. IEEE Press, Los Alamitos (1995)
459. Kerdchuen, T., Ongsakul, W.: Optimal measurement placement for power system state estimation using hybrid genetic algorithm and simulated annealing. In: International Conference on Power System Technology, pp. 1–5 (2006)
460. Kernighan, B.W., Lin, S.: An efficient heuristic procedure for partitioning graphs. The Bell System Technical Journal 49(2), 291–307 (1970)
461. KhudaBukhsh, A., Xu, L., Hoos, H., Leyton-Brown, K.: SATenstein: Automatically building local search SAT solvers from components. In: 21st International Joint Conference on Artificial Intelligence, pp. 517–524. AAAI Press, Menlo Park (2009)
462. Kim, D.H., Abraham, A.: A hybrid genetic algorithm and bacterial foraging approach for global optimization and robust tuning of PID controller with disturbance rejection. In: Grosan, C., Abraham, A., Ishibuchi, H. (eds.) Hybrid Evolutionary Algorithms, pp. 171–199. Springer, Heidelberg (2007)

463. Kim, E.Y., Yang, B.S., Tan, A.C.C.: A hybrid evolutionary algorithm and its application to parameter identification of rolling elements bearings. In: Kosinski W (ed.) Advances in Evolutionary Algorithms, IN-TECH (2008)
464. Kim, S.S., Smith, A.E., Lee, J.H.: A memetic algorithm for channel assignment in wireless fdma systems. Computers and Operations Research 34(6), 1842–1856 (2007)
465. Kimura, S., Hatakeyama, M., Konagaya, A.: Inference of s-system models of genetic networks from noisy time-series data. Chem.-Bio. Informatics Journal 4(1), 1–14 (2004)
466. Kimura, S., et al.: Inference of S-system models of genetic networks using a cooperative coevolutionary algorithm. Bioinformatics 21(7), 1154–1163 (2005)
467. Kirkpatrick, S., Sherrington, D.: Infinite-ranged models of spin-glasses. Physical Review B (1978)
468. Kirkpatrick, S., Gelatt, C.D.J., Vecchi, M.P.: Optimization by simulated annealing. Science 220(4598), 671–680 (1983)
469. Klau, G., Ljubić, I., Moser, A., Mutzel, P., Neuner, P., Pferschy, U., Raidl, G., Weiskircher, R.: Combining a memetic algorithm with integer programming to solve the prize-collecting Steiner tree problem. In: [201], pp. 1304–1315 (2004)
470. Kleeman, M.P., Lamont, G.B., Cooney, A., Nelson, T.R.: A multi-tiered memetic multiobjective evolutionary algorithm for the design of quantum cascade lasers. In: Obayashi, S., Deb, K., Poloni, C., Hiroyasu, T., Murata, T. (eds.) EMO 2007. LNCS, vol. 4403, pp. 186–200. Springer, Heidelberg (2007)
471. Knowles, J., Corne, D.: A Comparison of Diverse Aproaches to Memetic Multiobjective Combinatorial Optimization. In: Wu, A.S. (ed.) Proceedings of the 2000 Genetic and Evolutionary Computation Conference Workshop Program, pp. 103–108 (2000)
472. Knowles, J., Corne, D.: M-PAES: A memetic algorithm for multiobjective optimization. In: [112], pp. 325–332 (2000)
473. Knowles, J., Corne, D.W.: Approximating the nondominated front using the pareto archived evolution strategy. Evolutionary Computation 8(2), 149–172 (2000)
474. Knowles, J.D., Oates, M., Corne, D.: Advanced multi-objective evolutionary algorithms applied to two problems in telecommunications. British Telephone Technology Journal 18(18), 51–65 (2000)
475. Koch, P., Kramer, O., Rudolph, G., Beume, N.: On the hybridization of SMS-EMOA and local search for continuous multiobjective optimization. In: Rothlauf, F. (ed.) GECCO 2009, pp. 603–610. ACM Press, Montreal (2009)
476. Koehler, J.R., Owen, A.B.: Computer experiments. In: Ghosh, S., Rao, C.R., Krishnaiah, P.R. (eds.) Handbook of Statistics, pp. 261–308. Elsevier, Amsterdam (1996)
477. Kohonen, T.: Self-Organizing Maps. Springer Series in Information Sciences, vol. 30. Springer, Heidelberg (1995)
478. Kolen, A., Pesch, E.: Genetic Local Search in Combinatorial Optimization. Discrete Applied Mathematics and Combinatorial Operations Research and Computer Science 48, 273–284 (1994)
479. Komusiewicz, C., Niedermeier, R., Uhlmann, J.: Deconstructing intractability - a multivariate complexity analysis of interval constrained coloring. J. Discrete Algorithms 9(1), 137–151 (2011)
480. Konak, A., Smith, A.: A hybrid genetic algorithm approach for backbone design of communication networks. In: [111], pp. 1817–1823 (1999)
481. Kononova, A.V., Hughes, K.J., Pourkashanian, M., Ingham, D.B.: Fitness diversity based adaptive memetic algorithm for solving inverse problems of chemical kinetics. In: [118], pp. 2366–2373 (2007)

482. Koo, W.T., Goh, C.K., Tan, K.C.: A predictive gradient strategy for multiobjective evolutionary algorithms in a fast changing environment. Journal of Soft Computing 2, 87–110 (2010)
483. Koza, J.: Genetic Programming. MIT Press, Cambridge (1992)
484. Koziel, S., Bandler, J.W., Madsen, K.: A space mapping framework for engineering optimization: theory and implementation. IEEE Transactions on Microwave Theory 54, 3721–3730 (2006)
485. Krarup, J., Pruzan, P.M.: Computer-Aided Layout Design. Mathematical Programming Study 9, 75–94 (1978)
486. Krasnogor, N.: Coevolution of genes and memes in memetic algorithms. In: Wu, A. (ed.) Proceedings of the 1999 Genetic and Evolutionary Computation Conference Workshop Program (1999)
487. Krasnogor, N.: Studies in the theory and design space of memetic algorithms. PhD thesis, University of the West of England (2002)
488. Krasnogor, N.: Self-generating metaheuristics in bioinformatics: The protein structure comparison case. Genetic Programming and Evolvable Machines 5(2), 181–201 (2004)
489. Krasnogor, N.: Toward robust memetic algorithms. In: [375], pp. 185–207 (2005)
490. Krasnogor, N., Gustafson, S.: A study on the use of "self-generation" in memetic algorithms. Natural Computing 3(1), 53–76 (2004)
491. Krasnogor, N., Smith, J.: A memetic algorithm with self-adaptive local search: TSP as a case study. In: [932], pp. 987–994 (2000)
492. Krasnogor, N., Smith, J.: Emergence of profitable search strategies based on a simple inheritance mechanism. In: [839], pp. 432–439 (2001)
493. Krasnogor, N., Smith, J.: A tutorial for competent memetic algorithms: model, taxonomy, and design issues. IEEE Transactions on Evolutionary Computation 9, 474–488 (2005)
494. Krasnogor, N., Smith, J.: Memetic algorithms: The polynomial local search complexity theory perspective. Journal of Mathematical Modelling and Algorithms 7(1), 3–24 (2008)
495. Krasnogor, N., Hart, W., Smith, J., Pelta, D.: Protein structure prediction with evolutionary algorithms. In: [42], pp. 1569–1601 (1999)
496. Krasnogor, N., Blackburne, B., Burke, E., Hirst, J.: Multimeme algorithms for proteine structure prediction. In: [580], pp. 769–778 (2002)
497. Krokhin, A.A., Marx, D.: On the hardness of losing weight. In: Aceto, L., Damgård, I., Goldberg, L.A., Halldórsson, M.M., Ingólfsdóttir, A., Walukiewicz, I. (eds.) ICALP 2008, Part I. LNCS, vol. 5125, pp. 662–673. Springer, Heidelberg (2008)
498. Kumar, D., Kumar, S., Rai, C.S.: Memetic algorithms for feature selection in face recognition. In: International Conference on Hybrid Intelligent Systems, pp. 931–934. IEEE Computer Society, Los Alamitos (2008)
499. Kumarappan, N., Mohan, M.: Hybrid genetic algorithm based fuel restricted real power optimization for utility system. In: [114], pp. 1294–1301 (2003)
500. Laguna, M., Martí, R.: Scatter Search. Methodology and Implementations in C. Kluwer Academic Publishers, Boston (2003)
501. Lancia, G.: Mathematical programming in computational biology: an annotated bibliography. Algorithms 1(2), 100–129 (2008)
502. Lancia, G., Carr, R., Walenz, B., Istrail, S.: 101 optimal pdb structure alignments: a branch-and-cut algorithm for the maximum contact map overlap problem. In: Fifth Annual International Conference on Computational Molecular Biology, RECOMB, pp. 193–202. ACM Press, New York (2001)

References

503. Land, M.W.S.: Evolutionary algorithms with local search for combinatorial optimization. PhD thesis, University of California, San Diego, CA (1998)
504. Langdon, W.B., et al. (eds.): Genetic and Evolutionary Computation Conference – GECCO 2002. Morgan Kaufmann, New York (2002)
505. Lapedes, A., Farber, R.: A self-optimizing, nonsymmetrical neural net for content addressable memory and pattern recognition. Physica D: Nonlinear Phenomena 22(1-3), 247–259 (1986)
506. Laughunn, D.J.: Quadratic Binary Programming. Operations Research 14, 454–461 (1970)
507. Lawler, E., Wood, D.: Branch and bounds methods: A survey. Operations Research 4(4), 669–719 (1966)
508. Lawler, E., Lenstra, J., Kan, A.R., Shmoys, D.: The Travelling Salesman Problem: A Guided Tour of Combinatorial Optimization. Wiley Interscience, Chichester (1985)
509. Lawrence, S.: Resource constrained project scheduling: an experimental investigation of heuristic scheduling techniques (supplement). Tech. rep., Graduate School of Industrial Administration, Carnegie-Mellon University, Pittsburgh, Pennsylvania (1984)
510. Le, M.N., Ong, Y.S., Jin, Y., Sendhoff, B.: Lamarckian memetic algorithms: local optimum and connectivity structure analysis. Memetic Computing 1(3), 175–190 (2009)
511. Lehre, P.K., Yao, X.: On the impact of the mutation-selection balance on the runtime of evolutionary algorithms. In: Jansen, T., Garibay, I., Wiegand, R.P., Wu, A.S. (eds.) Tenth ACM SIGEVO Workshop on Foundations of Genetic Algorithms (FOGA 2009), pp. 47–58. ACM, New York (2009)
512. Lepistö, T., Salomaa, A. (eds.): ICALP 1988. LNCS, vol. 317. Springer, Heidelberg (1988)
513. Leskinen, J., Neri, F., Neittaanmäki, P.: Memetic variation local search vs. life-time learning in electrical impedance tomography. In: Giacobini, M., Brabazon, A., Cagnoni, S., Di Caro, G.A., Ekárt, A., Esparcia-Alcázar, A.I., Farooq, M., Fink, A., Machado, P. (eds.) EvoWorkshops 2009. LNCS, vol. 5484, pp. 615–624. Springer, Heidelberg (2009)
514. Levine, J., Ducatelle, F.: Ant colony optimisation and local search for bin packing and cutting stock problems. Journal of the Operational Research Society 55(7), 705–716 (2004)
515. Li, B., Ong, Y.S., Le, M.N., Goh, C.K.: Memetic gradient search. In: [119], pp. 2894–2901 (2008)
516. Li, H., Zhang, Q.: Multiobjective optimization problems with complicated pareto sets, MOEA/D and NSGA-II. IEEE Transactions on Evolutionary Computation 13(2), 284–302 (2009)
517. Li, X., Li, B.: Synthesis of the shaped-beam array antennas using hybrid genetic algorithm. In: International Symposium on Antennas, Propagation and EM Theory, pp. 155–157 (2008)
518. Li, X., Liang, X.M.: A hybrid adaptive evolutionary algorithm for constrained optimization. In: Third International Conference on Intelligent Information Hiding and Multimedia Signal Processing, vol. 2, pp. 338–341 (2007)
519. Liang, Y., Leung, K.S., Mok, T.S.K.: Evolutionary drug scheduling model for cancer chemotherapy. In: [201], pp. 1126–1137 (2004)
520. Liang, Y., Leung, K.S., Mok, T.S.K.: Evolutionary drug scheduling models with different toxicity metabolism in cancer chemotherapy. Applied Soft Computing 8(1), 140–149 (2008)

521. Lim, D., Ong, Y.S., Lim, M.H., Jin, Y.: Single/Multi-objective inverse robust evolutionary design methodology in the presence of uncertainty. In: Yang, S., Ong, Y.S., Jin, Y. (eds.) Evolutionary Computation in Dynamic and Uncertain Environments. Studies in Computational Intelligence, vol. 51, pp. 437–456. Springer, Heidelberg (2007)
522. Lim, D., Jin, Y., Ong, Y.S., Sendhoff, B.: Generalizing surrogate-assisted evolutionary computation. IEEE Transactions on Evolutionary Computation 14(3), 329–355 (2010)
523. Lin, S.: Computer solutions of the traveling salesman problem. The Bell System Technical Journal 44(10), 2245–2269 (1965)
524. Lin, S., Kernighan, B.: An Effective Heuristic Algorithm for the Traveling Salesman Problem. Operations Research 21, 498–516 (1973)
525. Lin, S.C., Goodman, E.D., Punch, W.F.: A genetic algorithm approach to dynamic job shop scheduling problems. In: [33], pp. 481–488 (1997)
526. Linhart, H., Zucchini, W.: Model Selection. Wiley Series in Probability and Mathematical Statistics. Wiley-Interscience Publication, Hoboken (1986)
527. Liu, B., Ma, H., Zhang, X., Zhou, Y.: A memetic co-evolutionary differential evolution algorithm for constrained optimization. In: [118], pp. 2996–3002 (2007)
528. Liu, B.F., Chen, J.H., Hwang, S.F., Ho, S.Y.: MeSwarm: Memetic particle swarm optimization. In: Beyer, H.G., O'Reilly, U.M. (eds.) GECCO 2005, pp. 267–268. ACM Press, Washington DC (2005)
529. Liu, X., Wu, Y., Duan, J.: Optimal sizing of a series hybrid electric vehicle using a hybrid genetic algorithm. In: IEEE International Conference on Automation and Logistics, pp. 1125–1129. IEEE Press, Los Alamitos (2007)
530. Lo, C.C., Chang, W.H.: A multiobjective hybrid genetic algorithm for the capacitated multipoint network design problem. In: IEEE International Conference on Communications, vol. 3, pp. 1573–1576. IEEE Press, Los Alamitos (1999)
531. Lo, C.C., Chang, W.H.: A multiobjective hybrid genetic algorithm for the capacitated multipoint network design problem. IEEE Transactions on Systems, Man, and Cybernetics, Part B 30(3), 461–470 (2000)
532. Lodi, A., Allemand, K., Liebling, T.M.: An Evolutionary Heuristic for Quadratic 0–1 Programming. European Journal of Operational Research 119, 662–670 (1999)
533. Lourenço, H.R., Martin, O., Stützle, T.: Iterated Local Search. In: [315], pp. 321–353 (2003)
534. Lozano, J.A., Larrañaga, P., Inza, I., Bengoetxea, E.: Towards a New Evolutionary Computation. In: Advances on Estimation of Distribution Algorithms. Studies in Fuzziness and Soft Computing, vol. 192, Springer, Heidelberg (2006)
535. Lozano, M., García-Martínez, C.: Hybrid metaheuristics with evolutionary algorithms specializing in intensification and diversification: Overview and progress report. Computers & Operations Research 37(3), 481–497 (2010)
536. Lozano, M., Herrera, F., Krasnogor, N., Molina, D.: Real-coded memetic algorithms with crossover hill-climbing. Evolutionary Computation 12(3), 273–302 (2004)
537. Lü, Z., Hao, J.K.: A memetic algorithm for graph coloring. European Journal of Operational Research 200(1), 235–244 (2010)
538. Lü, Z., Glover, F., Hao, J.K.: A hybrid metaheuristic approach to solving the ubqp problem. European Journal of Operational Research 207(3), 1254–1262 (2010)
539. Lü, Z., Hao, J.K., Glover, F.: Neighborhood analysis: a case study on curriculum-based course timetabling. Journal of Heuristics 17(2), 97–118 (2010)
540. Luersena, M.A., Le Riche, R.: Globalized Nelder–Mead method for engineering optimization. Computers & Structures 82(23-26), 2251–2260 (2004)

541. Luke, S., Spector, L.: Evolving teamwork and coordination with genetic programming. In: Koza, J., Goldberg, D., Fogel, D., Riolo, R. (eds.) 1st Annual Conference on Genetic Programming, pp. 141–149. MIT Press, Cambridge (1996)
542. Lust, T., Jaszkiewicz, A.: Speed-up techniques for solving large-scale biobjective TSP. Computers and Operations Research 37, 521–533 (2010)
543. MacQueen, J.B.: Some methods for classification and analysis of multivariate observations. In: Proceedings of 5*th* Berkeley Symposium on Mathematical Statistics and Probability, pp. 281–297 (1967)
544. Madych, W.R.: Miscellaneous error bounds for multiquadric and related interpolators. Computers and Mathematics with Applications 24(12), 121–138 (1992)
545. Mahata, P., Costa, W., Cotta, C., Moscato, P.: Hierarchical clustering, languages and cancer. In: [779], pp. 67–78 (2006)
546. Mahfoud, S.W.: Crowding and preselection revisited. In: [551], pp. 27–36 (1992)
547. Mahfoud, S.W.: Niching methods. In: Bäck, T., Fogel, D.B., Michalewicz, Z. (eds.) Handbook of Evolutionary Computation, pp. C6.1:1–C6.1:4. Institute of Physics Publishing and Oxford University Press, Bristol (1997)
548. Mak, K.T., Morton, A.J.: Distances between Traveling Salesman Tours. Discrete Applied Mathematics and Combinatorial Operations Research and Computer Science 58, 281–291 (1995)
549. Malaguti, E., Monaci, M., Toth, P.: A metaheuristic approach for the vertex coloring problem. INFORMS Journal on Computing 20(2), 302–316 (2008)
550. Mallipeddi, R., Suganthan, P.N.: Problem definitions and evaluation criteria for the CEC, competition and special session on single objective constrained real-parameter optimization. Tech. rep., Nangyang Technological University, Singapore (2009)
551. Männer, R., Manderick, B. (eds.): Parallel Problem Solving from Nature II. Elsevier, Brussels (1992)
552. Mariano, A., Norman, M., Moscato, P.: Arbitrarily large planar ETSP instances with known optimal tours. Pesquisa Operacional 15(1,2), 89–96 (1995)
553. Mariano, A., Norman, M., Moscato, P.: Using L-systems to generate arbitrarily large instances of the euclidean traveling salesman problem with known optimal tours. In: XXVII Simposio Brasileiro de Pesquisa Operacional, Sociedade Brasileira de Pesquisa Operacional, Rio de Janeiro (1995)
554. Marida, K., Marshall, R.: Maximum likelihood estimation of models for residual covariance in spatial regression. Biometrika 71(1), 135–146 (1984)
555. Marinakis, Y., Marinaki, M.: A hybrid genetic - particle swarm optimization algorithm for the vehicle routing problem. Expert Systems with Applications 37(2), 1446–1455 (2010)
556. Marinakis, Y., Migdalas, A., Pardalos, P.M.: A Hybrid Genetic–GRASP Algorithm Using Lagrangean Relaxation for the Traveling Salesman Problem. Journal of Combinatorial Optimization 10(4), 311–326 (2005)
557. Martin, J.D., Simpson, T.W.: Use of Kriging models to approximate deterministic computer models. AIAA Journal 43(4), 853–863 (2005)
558. Martins, F.V.C., Carrano, E.G., Wanner, E.F., Takahashi, R.H.C., Mateus, G.R.: A dynamic multiobjective hybrid approach for designing wireless sensor networks. In: [120], pp. 1145–1152 (2009)
559. Marx, D.: Local search. Parameterized Complexity News 3, 7–8 (2008)
560. Marx, D.: Searching the k-change neighborhood for TSP is W[1]-hard. Oper. Res. Lett. 36(1), 31–36 (2008)

561. Marx, D., Schlotter, I.: Stable assignment with couples: parameterized complexity and local search. In: Chen, J., Fomin, F.V. (eds.) IWPEC 2009. LNCS, vol. 5917, pp. 300–311. Springer, Heidelberg (2009)
562. Marx, D., Schlotter, I.: Parameterized complexity and local search approaches for the stable marriage problem with ties. Algorithmica 58(1), 170–187 (2010)
563. Marx, D., Schlotter, I.: Stable assignment with couples: Parameterized complexity and local search. Discrete Optimization 8(1), 25–40 (2011)
564. Mashohor, S., Evans, J., Arslan, T.: Image registration of printed circuit boards using hybrid genetic algorithm. In: [117], pp. 2685–2690 (2006)
565. Mathias, K., Whitley, D.: Genetic Operators, the Fitness Landscape and the Traveling Salesman Problem. In: [551], pp. 219–228 (1992)
566. Mathias, K., Whitley, L., Stork, C., Kusuma, T.: Staged hybrid genetic search for seismic data imaging. In: IEEE Conference on Evolutionary Computation, vol. 1, pp. 356–361. IEEE Press, Los Alamitos (1994)
567. Mathieson, L.: The parameterized complexity of editing graphs for bounded degeneracy. Theor. Comput. Sci. 411(34-36), 3181–3187 (2010)
568. May, A., Johnson, M.: Protein-structure comparisons using a combination of a genetic algorithm, dynamic-programming and least-squares minimization. Protein Engineering 7(4), 475–485 (1994)
569. Maynard-Smith, J.: The Evolution of Sex. Cambridge University Press, Cambridge (1978)
570. Maynard-Smith, J., Száthmary, E.: The Major transitions in evolution. W.H. Freeman, New York (1995)
571. McBride, R.D., Yormark, J.S.: An Implicit Enumeration Algorithm for Quadratic Integer Programming. Management Science 26(3), 282–296 (1980)
572. Menczer, F., Parisi, D.: Evidence of hyperplanes in the genetic learning of neural networks. Biological Cybernetics 66, 283–289 (1992)
573. Mendes, A., Franca, P., Moscato, P.: Fitness landscapes for the total tardiness single machine scheduling problem. Neural Network World 2(2), 165–180 (2002)
574. Mendes, A., França, P.M., Moscato, P., Garcia, V.: Population studies for the gate matrix layout problem. In: Garijo, F.J., Riquelme, J.-C., Toro, M. (eds.) IBERAMIA 2002. LNCS (LNAI), vol. 2527, pp. 319–339. Springer, Heidelberg (2002)
575. Mendes, A., Muller, F., Franca, P., Moscato, P.: Comparing meta-heuristic approaches for parallel machine scheduling problems. Production Planning & Control 13(2), 143–154 (2002)
576. Mendes, A., Cotta, C., Garcia, V., França, P., Moscato, P.: Gene ordering in microarray data using parallel memetic algorithms. In: Skie, T., Yang, C.S. (eds.) 2005 International Conference on Parallel Processing Workshops, pp. 604–611. IEEE Press, Oslo (2005)
577. Mendes, A., Scott, R., Moscato, P.: Microarrays - identifying molecular portraits for prostate tumors with different gleason patterns. In: Trent, R. (ed.) Clinical Bioinformatics - Methods in Molecular Medicine, Methods in Molecular Medicine, vol. 141, pp. 131–151. Humana Press (2007)
578. Mendes, A.S., França, P.M., Moscato, P.: Fitness landscapes for the total tardiness single machine scheduling problem. Neural Network World 2(2), 165–180 (2002)
579. Mendoza, J.E., Castanier, B., Guéret, C., Medaglia, A.L., Velasco, N.: A memetic algorithm for the multi-compartment vehicle routing problem with stochastic demands. Computers & Operations Research 37(11), 1886–1898 (2010), metaheuristics for Logistics and Vehicle Routing

References

580. Guervós, J.J.M., Adamidis, P.A., Beyer, H.-G., Fernández-Villacañas, J.-L., Schwefel, H.-P. (eds.): PPSN 2002. LNCS, vol. 2439. Springer, Heidelberg (2002)
581. Merz, P.: Memetic algorithms for combinatorial optimization problems: Fitness landscapes and effective search strategies. PhD thesis, University of Siegen, Germany (2000)
582. Merz, P.: NK-Fitness Landscapes and Memetic Algorithms with Greedy Operators and k-opt Local Search. In: Krasnogor, N. (ed.) Proceedings of the Third International Workshop on Memetic Algorithms, WOMA III (2002)
583. Merz, P.: Advanced fitness landscape analysis and the performance of memetic algorithms. Evolutionary Computation 12(3), 303–326 (2004)
584. Merz, P., Freisleben, B.: Genetic Local Search for the TSP: New Results. In: Bäck, T., Michalewicz, Z., Yao, X. (eds.) Proceedings of the 1997 IEEE International Conference on Evolutionary Computation, pp. 159–164. IEEE Press, Piscataway (1997)
585. Merz, P., Freisleben, B.: Fitness landscapes and memetic algorithm design. In: [145], pp. 245–260 (1999)
586. Merz, P., Freisleben, B.: Genetic Algorithms for Binary Quadratic Programming. In: [42], pp. 417–424 (1999)
587. Merz, P., Freisleben, B.: Fitness Landscape Analysis and Memetic Algorithms for the Quadratic Assignment Problem. IEEE Transactions on Evolutionary Computation 4(4), 337–352 (2000)
588. Merz, P., Freisleben, B.: Memetic Algorithms for the Traveling Salesman Problem. Complex Systems 13(4), 297–345 (2001)
589. Merz, P., Freisleben, B.: Greedy and Local Search Heuristics for Unconstrained Binary Quadratic Programming. Journal of Heuristics 8(2), 197–213 (2002)
590. Merz, P., Huhse, J.: An Iterated Local Search Approach for Finding Provably Good Solutions for Very Large TSP Instances. In: [781], pp. 929–939 (2008)
591. Merz, P., Katayama, K.: Memetic Algorithms for the Unconstrained Binary Quadratic Programming Problem. Bio Systems 78(1-3), 99–118 (2004)
592. Merz, P., Zell, A.: Clustering gene expression profiles with memetic algorithms. In: [580], pp. 811–820 (2002)
593. Mezard, M., Virasoro, M.: The microstructure of ultrametricity. Journal de Physique 46(8), 1293–1307 (1985)
594. Mezard, M., Parisi, G., Virasoro, M.: Spin glass theory and beyond. World Scientific, Singapore (1987)
595. Michalewicz, Z.: A hierarchy of evolution programs: An experimental study. Evolutionary Computation 1(1), 51–76 (1993)
596. Michalewicz, Z.: Genetic Algorithms + Data Structures = Evolution Programs. Springer, Heidelberg (1996)
597. Michalewicz, Z., Schoenauer, M.: Evolutionary Algorithms for Constrained Parameter Optimization Problems. Evolutionary Computation 4(1), 1–32 (1996)
598. Michiels, W., Aarts, E., Korst, J.: Theoretical Aspects of Local Search. Springer, Heidelberg (2007)
599. Miettinen, K.: Nonlinear Multiobjective Optimization. International Series in Operations Research and Management Science, vol. 12. Kluwer, Dordrecht (1999)
600. Miller, G.F., Todd, P.M., Hegde, S.U.: Designing neural networks using genetic algorithms. In: [794], pp. 379–384 (1989)
601. Mininno, E., Neri, F.: A memetic differential evolution approach in noisy optimization. Journal of Memetic Computing 2, 111–135 (2010)
602. Mira, J., Álvarez, J.R. (eds.): IWINAC 2005. LNCS, vol. 3562. Springer, Heidelberg (2005)

603. Mitchell, T.J., Morris, M.D.: Bayesian design and analysis of computer experiments: Two examples. Statistica Sinica 2, 359–379 (1992)
604. Mladenović, N., Hansen, P.: Variable neighborhood search. Computers & OR 24(11), 1097–1100 (1997)
605. Molina, D., Herrera, F., Lozano, M.: Adaptive local search parameters for real-coded memetic algorithms. In: [116], pp. 888–895 (2005)
606. Molina, D., Lozano, M., Herrera, F.: Memetic algorithms for intense continuous local search methods. In: [69], pp. 58–71 (2008)
607. Molina, D., Lozano, M., Herrera, F.: Study of the influence of the local search method in memetic algorithms for large scale continuous optimization problems. In: Stützle, T. (ed.) LION 3. LNCS, vol. 5851, pp. 221–234. Springer, Heidelberg (2009)
608. Molina, D., Lozano, M., García-Martínez, C., Herrera, F.: Memetic algorithms for continuous optimization based on local search chains. Evolutionary Computation 18(1), 1–37 (2010)
609. Montes de Oca, M.A., Van den Enden, K., Stützle, T.: Incremental particle swarm-guided local search for continuous optimization. In: [69], pp. 72–86 (2008)
610. Morgan, B.W.: An Introduction to Bayesian Statistical Decision Processes. Prentice-Hall, Englewood Cliffs (1968)
611. Mori, H., Yoshida, T.: Probabilistic distribution network expansion planning with multi-objective memetic algorithm. In: IEEE Canada Electric Power Conference, pp. 1–6. IEEE Press, Los Alamitos (2008)
612. Morris, G., Goodsell, D., Halliday, R., Huey, R., Hart, W., Belew, R.: AJOlson Automated docking using a lamarckian genetic algorithm and an empirical binding free energy function. Journal of Computational Chemistry 19(14), 1639–1662 (1998)
613. Morrison, R., De Jong, K.: A test problem for nonstationary environments. In: [111], pp. 2047–2053 (1999)
614. Morrison, R., De Jong, K.: Measurement of population diversity. In: [137], pp. 31–41 (2002)
615. Moscato, P.: On evolution, search, optimization, genetic algorithms and martial arts: Toward memetic algorithms. Tech. Rep. 826, California Institute of Technology (1989)
616. Moscato, P.: An Introduction to Population Approaches for Optimization and Hierarchical Objective Functions: The Role of Tabu Search. Annals of Operations Research 41(1-4), 85–121 (1993)
617. Moscato, P.: Memetic algorithms: a short introduction. In: [145], pp. 219–234 (1999)
618. Moscato, P., Cotta, C.: A gentle introduction to memetic algorithms. In: [315], pp. 105–144 (2003)
619. Moscato, P., Cotta, C.: Memetic algorithms. In: González, T. (ed.) Handbook of Approximation Algorithms and Metaheuristics, ch. 22. Taylor & Francis, Abington (2006)
620. Moscato, P., Cotta, C.: A modern introduction to memetic algorithms. In: Gendrau, M., Potvin, J.Y. (eds.) Handbook of Metaheuristics, 2nd edn. International Series in Operations Research and Management Science, vol. 146, pp. 141–183. Springer, New York (2010)
621. Moscato, P., Norman, M.: A competitive and cooperative approach to complex combinatorial search. Tech. Rep. 790, Caltech Concurrent Computation Program (1989)
622. Moscato, P., Norman, M.G.: A "Memetic" Approach for the Traveling Salesman Problem Implementation of a Computational Ecology for Combinatorial Optimization on Message-passing Systems. In: Valero, M., Onate, E., Jane, M., Larriba, J.L., Suarez, B. (eds.) Parallel Computing and Transputer Applications, pp. 177–186. IOS Press, Amsterdam (1992)

623. Moscato, P., Norman, M.G.: On the performance of heuristics on finite and infinite fractal instances of the euclidean traveling salesman problem. INFORMS Journal on Computing 10(2), 121–132 (1998)
624. Moscato, P., Schaerf, A.: Local search techniques for scheduling problems. Notes of the tutorial given at the 13th European Conference on Artificial Intelligence, ECAI 1998 (1998)
625. Moscato, P., Tinetti, F.: Blending heuristics with a population-based approach: A memetic algorithm for the traveling salesman problem. Report 92-12, Universidad Nacional de La Plata, C.C. 75, La Plata, Argentina (1992)
626. Moscato, P., Cotta, C., Mendes, A.: Memetic algorithms. In: [692], pp. 53–85 (2004)
627. Moscato, P., Mendes, A., Cotta, C.: Scheduling and production & control. In: [692], pp. 655–680 (2004)
628. Moscato, P., Mendes, A., Linhares, A.: VLSI design: Gate matrix layout problem. In: [692], pp. 455–478 (2004)
629. Moscato, P., Berretta, R., Hourani, M., Mendes, A., Cotta, C.: Genes related with alzheimer's disease: A comparison of evolutionary search, statistical and integer programming approaches. In: Rothlauf, F., Branke, J., Cagnoni, S., Corne, D.W., Drechsler, R., Jin, Y., Machado, P., Marchiori, E., Romero, J., Smith, G.D., Squillero, G. (eds.) EvoWorkshops 2005. LNCS, vol. 3449, pp. 84–94. Springer, Heidelberg (2005)
630. Moscato, P., Berretta, R., Mendes, A.: A new memetic algorithm for ordering datasets: Applications in microarray analysis. In: Proceedings of MIC 2005 - The 6th Metaheuristics International Conference, Vienna, Austria, pp. 695–700 (2005)
631. Moscato, P., Mendes, A., Berretta, R.: Benchmarking a memetic algorithm for ordering microarray data. Biosystems 88(1-2), 56–75 (2007)
632. Moscato, P., Berretta, R., Cotta, C.: Memetic algorithms. In: Wiley Encyclopedia of Operations Research and Management Science. Wiley, Chichester (2011)
633. Moser, I., Chiong, R.: A hooke-jeeves based memetic algorithm for solving dynamic optimisation problems. In: Corchado, E., Wu, X., Oja, E., Herrero, Á., Baruque, B. (eds.) HAIS 2009. LNCS, vol. 5572, pp. 301–309. Springer, Heidelberg (2009)
634. Moser, I., Chiong, R.: Dynamic function optimisation with hybridised extremal dynamics. Journal of Memetic Computing 2, 137–148 (2010)
635. Moser, I., Hendtlass, T.: A simple and efficient multi-component algorithm for solving dynamic function optimisation problems. In: [118], pp. 252–259 (2007)
636. Müehlenbein, H., Gorges-Schleuter, M., Krämer, O.: Evolution algorithms in combinatorial optimization. Parallel Computing 7, 65–88 (1988)
637. Mühlenbein, H.: Parallel Genetic Algorithms, Population Genetics and Combinatorial Optimization. In: [794], pp. 416–421 (1989)
638. Mühlenbein, H.: Evolution in Time and Space – The Parallel Genetic Algorithm. In: Rawlins, G.J.E. (ed.) Foundations of Genetic Algorithms. Morgan Kaufmann, San Francisco (1991)
639. Mühlenbein, H., Schlierkamp-Voosen, D.: Predictive models for the breeder genetic algorithm, I: Continuous parameter optimization. Evolutionary Computation 1(1), 25–49 (1993)
640. Mühlenbein, H., Schomisch, M., Born, J.: The parallel genetic algorithm as function optimizer. In: [56], pp. 271–278 (1991)
641. Müller, A., Schneider, J., Schömer, E.: Packing a multidisperse system of hard disks in a circular environment. Physical Review E 79(021102) (2009)
642. Müller, A., Schneider, J., Schömer, E.: Ultrametricity property of energy landscapes of multidisperse packing problems. Physical Review E 79(031122) (2009)

643. Müller, C.L., Baumgartner, B., Sbalzarini, I.F.: Particle swarm CMA evolution strategy for the optimization of multi-funnel landscapes. In: [120], pp. 2685–2692 (2009)
644. Müller, H., Urner, R.: On a disparity between relative cliquewidth and relative nlc-width. Discrete Applied Mathematics 158(7), 828–840 (2010)
645. Murata, T., Ishibuchi, H., Tanaka, H.: Genetic algorithms for flowshop scheduling problems. Computers & Industrial Engineering 30(4), 1061–1071 (1996)
646. Myers, R.H., Montgomery, D.C.: Response Surface Methodology: Process and Product Optimization Using Designed Experiments. John Wiley and Sons, Chichester (1995)
647. Nagata, Y.: New EAX Crossover for Large TSP Instances. In: [783], pp. 372–381 (2006)
648. Nagata, Y., Kobayashi, S.: Edge assembly crossover: a high-power genetic algorithm for the travelling salesman problem. In: [33], pp. 450–457 (1997)
649. Nakano, R., Yamada, T.: Conventional genetic algorithm for job shop problems. In: [56], pp. 474–479 (1991)
650. Nannen, V., Eiben, A.: A method for parameter calibration and relevance estimation in evolutionary algorithms. In: [110], pp. 183–190 (2006)
651. Nannen, V., Eiben, A.E.: Relevance estimation and value calibration of evolutionary algorithm parameters. In: Veloso, M.M. (ed.) 20th International Joint Conference on Artificial Intelligence (IJCAI), Hyderabad, India, pp. 1034–1039. AAAI Press, Menlo Park (2007)
652. Nebro, A.J., Durillo, J.J.: A study of the parallelization of the multi-objective metaheuristic MOEA/D. In: Blum, C., Battiti, R. (eds.) LION 4. LNCS, vol. 6073, pp. 303–317. Springer, Heidelberg (2010)
653. Nelder, A., Mead, R.: A simplex method for function optimization. Computation Journal 7, 308–313 (1965)
654. Nemhauser, G., Wolsey, L.: Integer and Combinatorial Optimization. John Wiley & Sons, Chichester (1988)
655. Neri, F., Mininno, E.: Memetic compact differential evolution for cartesian robot control. IEEE Computational Intelligence Magazine 5(2), 54–65 (2010)
656. Neri, F., Kotilainen, N., Vapa, M.: An Adaptive Global-Local Memetic Algorithm to Discover Resources in P2P Networks. In: Giacobini, M. (ed.) EvoWorkshops 2007. LNCS, vol. 4448, pp. 61–70. Springer, Heidelberg (2007)
657. Neri, F., Tirronen, V., Kärkkäinen, T., Rossi, T.: Fitness diversity based adaptation in multimeme algorithms: A comparative study. In: [118], pp. 2374–2381 (2007)
658. Neri, F., Toivanen, J., Cascella, G.L., Ong, Y.S.: An adaptive multimeme algorithm for designing HIV multidrug therapies. IEEE/ACM Transactions on Computational Biology and Bioinformatics 4(2), 264–278 (2007)
659. Neri, F., Toivanen, J., Mäkinen, R.A.E.: An adaptive evolutionary algorithm with intelligent mutation local searchers for designing multidrug therapies for HIV. Applied Intelligence 27(3), 219–235 (2007)
660. Neri, F., Kotilainen, N., Vapa, M.: A memetic-neural approach to discover resources in P2P networks. In: Cotta, C., van Hemert, J. (eds.) Recent Advances in Evolutionary Computation for Combinatorial Optimization. Studies in Computational Intelligence, vol. 153, pp. 113–129. Springer, Heidelberg (2008)
661. Neumann, F., Sudholt, D., Witt, C.: Rigorous analyses for the combination of ant colony optimization and local search. In: Dorigo, M., Birattari, M., Blum, C., Clerc, M., Stützle, T., Winfield, A.F.T. (eds.) ANTS 2008. LNCS, vol. 5217, pp. 132–143. Springer, Heidelberg (2008)

References

662. Ngueveu, S.U., Prins, C., Calvo, R.W.: An effective memetic algorithm for the cumulative capacitated vehicle routing problem. Computers & Operations Research 37(11), 1877–1885 (2010)
663. Nguyen, H.D., Yoshihara, I., Yamamori, K., Yasunaga, M.: A New Three-Level Tree Data Structure for Representing TSP Tours in the Lin-Kernighan Heuristic. IEICE Transactions on Fundamentals of Electronics, Communications and Computer Sciences E90-A(10), 2187–2193 (2007)
664. Nguyen, H.D., Yoshihara, I., Yamamori, K., Yasunaga, M.: Implementation of an Effective Hybrid GA for Large-Scale Traveling Salesman Problems. IEEE Transactions on Systems, Man and Cybernetics, Part B 37(1), 92–99 (2007)
665. Nguyen, Q.H., Ong, Y.S., Krasnogor, N.: A study on the design issues of memetic algorithm. In: [118], pp. 2390–2397 (2007)
666. Niedermeier, R.: Invitation to Fixed Parameter Algorithms. Oxford Lecture Series in Mathematics and Its Applications. Oxford University Press, Oxford (2006)
667. Niedermeier, R., Rossmanith, P.: A general method to speed up fixed-parameter-tractable algorithms. Information Processing Letters 73, 125–129 (2000)
668. Nilsson, R., Björkegren, J., Tegnér, J.: On reliable discovery of molecular signatures. BMC Bioinformatics 10(38) (2009)
669. Nixon, K.C.: The parsimony ratchet, a new method for rapid parsimony analysis. Cladistics 15, 407–414 (1999)
670. Nobuhara, H., Han, C.W.: Evolutionary computation schemes based on max plus algebra and their application to image processing. In: International Symposium on Intelligent Signal Processing and Communications, pp. 538–541 (2006)
671. Noman, N., Iba, H.: Inferring gene regulatory networks using differential evolution with local search heuristics. IEEE/ACM Transactions on Computational Biology and Bioinformatics 4(4), 634–647 (2007)
672. Nonobe, K., Ibaraki, T.: A tabu search approach for the constraint satisfaction problem as a general problem solver. European Journal of Operational Research 106(2-3), 599–623 (1998)
673. Norman, M., Moscato, P.: A competitive-cooperative approach to complex combinatorial search. Tech. Rep. 790, California Institute of Technology (1989)
674. Norman, M., Moscato, P.: The euclidean traveling salesman problem and a space-filling curve. Chaos, Solitons and Fractals 6, 389–397 (1995)
675. Norman, M.G., Thanisch, P.: Models of machines and computation for mapping in multicomputers. ACM Comput. Surv. 25(3), 263–302 (1993)
676. Norman, M.G., Pelagatti, S., Thanisch, P.: On the complexity of scheduling with communication delay and contention. Parallel Processing Letters 5, 331–341 (1995)
677. Oakley, M., Barthel, D., Bykov, Y., Garibaldi, J., Burke, E., Krasnogor, N., Hirst, J.: Search strategies in structural bioinformatics. Current Protein and Peptide Science 9(3), 260–274 (2008)
678. Okushi, F.: Parallel cooperative propositional theorem proving. Annals of Mathematics and Artificial Intelligence 26(1-4), 59–85 (1999)
679. Ong, Y., Lum, K., Nair, P., Shi, D., Zhang, Z.: Global convergence of unconstrained and bound constrained surrogate-assisted evolutionary search in aerodynamic shape design. In: [114], pp. 1856–1863 (2003)
680. Ong, Y.S., Keane, A.J.: Meta-lamarckian learning in memetic algorithms. IEEE Transactions on Evolutionary Computation 8(2), 99–110 (2004)
681. Ong, Y.S., Nair, P.B., Keane, A.J.: Evolutionary optimization of computationally expensive problems via surrogate modeling. AIAA Journal 41(4), 687–696 (2003)

682. Ong, Y.S., Nair, P.B., Keane, A.J., Wong, K.W.: Surrogate-assisted evolutionary optimization frameworks for high-fidelity engineering design problems. In: Jin, Y. (ed.) Knowledge Incorporation in Evolutionary Computation, pp. 307–331. Springer, Berlin (2004)
683. Ong, Y.S., Lim, M.H., Zhu, N., Wong, K.W.: Classification of adaptive memetic algorithms: A comparative study. IEEE Transactions On Systems, Man and Cybernetics - Part B 36(1), 141–152 (2006)
684. Ong, Y.S., Nair, P.B., Lum, K.Y.: Max-min surrogate-assisted evolutionary algorithm for robust aerodynamic design. IEEE Transactions on Evolutionary Computation 10(4), 392–404 (2006)
685. Ong, Y.S., Zhou, Z., Lim, D.: Curse and blessing of uncertainty in evolutionary algorithm using approximation. In: [117], pp. 2928–2935 (2006)
686. Ong, Y.S., Krasnogor, N., Ishibuchi, H.: Special issue on memetic algorithms. IEEE Transactions on Systems Man and Cybernetics-part B 37(1) (2007)
687. Ong, Y.S., Lum, K.Y., Nair, P.B.: Hybrid evolutionary algorithm with Hermite radial basis function interpolants for computationally expensive adjoint solvers. Journal Computational Optimization and Applications 39(1), 97–119 (2008)
688. Ong, Y.S., Lim, M.H., Neri, F., Ishibuchi, H.: Special issue on emerging trends in soft computing–memetic algorithms. Journal of Soft Computing 13(8-9) (2009)
689. Ong, Y.S., Lim, M.H., Chen, X.: Memetic computation-past, present and future. IEEE Computational Intelligence Magazine 5(2), 24–31 (2010)
690. Ono, S., Hirotani, Y., Nakayama, S.: A memetic algorithm for robust optimal solution search–hybridization of multi-objective genetic algorithm and quasi-newton method. International Journal of Innovative Computing, Information and Control 5(12B), 5011–5019 (2009)
691. Ono, S., Yoshitake, Y., Nakayama, S.: Robust optimization using multi-objective particle swarm optimization. Artificial Life and Robotics 14(2) (2009)
692. Onwubolu, G., Babu, B. (eds.): New Optimization Techniques in Engineering. Studies in Fuzziness and Soft Computing, vol. 141. Springer, Berlin (2004)
693. Orantek, P.: Hybrid evolutionary algorithms in optimization of structures under dynamical loads. In: Burczyński, T., Osyczka, A. (eds.) IUTAM Symposium on Evolutionary Methods in Mechanics, Solid Mechanics and Its Applications, vol. 117, pp. 297–308. Springer, Heidelberg (2004)
694. Osman, I., Laporte, G.: Metaheuristics: A bibliography. Annals of Operations Research 65, 513–623 (1996)
695. Ozcan, E., Mohan, C.K.: Steady state memetic algorithm for partial shape matching. In: Porto, V.W., Waagen, D. (eds.) EP 1998. LNCS, vol. 1447, pp. 527–536. Springer, Heidelberg (1998)
696. Czyak, P., Jaszkiewicz, A.: Pareto simulated annealing - a metaheuristic technique for multiple-objective combinatorial optimisation. Journal of Multi-Criteria Decision Analysis 7, 34–47 (1998)
697. Paenke, I., Jin, Y., Branke, J.: Balancing population- and individual-level adaptation in changing environments. Adaptive Behavior – Animals, Animats, Software Agents, Robots, Adaptive Systems 17(2), 153–174 (2009)
698. Palacios, P., Pelta, D., Blanco, A.: Obtaining biclusters in microarrays with population-based heuristics. In: [779], pp. 115–126 (2006)
699. Palmers, P., McConaghy, T., Steyaert, M., Gielen, G.: Massively multi-topology sizing of analog integrated circuits. In: Conference on Design, Automation and Test in Europe, pp. 706–711. IEEE Press, Los Alamitos (2009)

700. Palubeckis, G.: Multistart Tabu Search Strategies for the Unconstrained Binary Quadratic Programming Problem. Annals of Operations Research 131, 259–282 (2004)
701. Papadimitriou, C.: Computational Complexity. Addison-Wesley, Reading (1994)
702. Papadimitriou, C., Yannakakis, M.: On limited nondeterminism and the complexity of the V-C dimension. In: Allender, J., et al. (eds.) 8th Annual Conference on Structure in Complexity Theory, IEEE Computer Society Press, pp. 12–18. IEEE Computer Society Press, San Diego (1993)
703. Papadimitriou, C., Yannakakis, M.: On limited nondeterminism and the complexity of the V-C dimension. Journal of Computer and System Sciences 53(2), 161–170 (1996)
704. Papadimitriou, C.H., Steiglitz, K.: Combinatorial optimization: algorithms and complexity. Prentice-Hall, Englewood Cliffs (1982)
705. Paquete, L., Chiarandini, M., Stützle, T.: Pareto local optimum sets in the biobjective traveling salesman problem: An experimental study. In: Gandibleux, X., Sevaux, M., Sörensen, K., Tkindt, V. (eds.) Meta-heuristics for Multiobjective Optimisation. Lecture Notes in Economics and Mathematical Systems, vol. 535, pp. 177–199. Springer, Heidelberg (2004)
706. Pardalos, P.M., Rodgers, G.P.: Computational Aspects of a Branch and Bound Algorithm for Unconstrained Quadratic Zero–One Programming. Computing 45, 131–144 (1990)
707. Pardalos, P.M., Rodgers, G.P.: A Branch and Bound Algorithm for the Maximum Clique Problem. Computers and Operations Research 19(5), 363–375 (1992)
708. Pardalos, P.M., Xue, J.: The Maximum Clique Problem. Journal of Global Optimization 4, 301–328 (1994)
709. Paredis, J.: The symbiotic evolution of solutions and their representations. In: Eshelman, L.J. (ed.) ICGA 1995, pp. 359–365. Morgan Kaufmann, Pittsburgh (1995)
710. Paredis, J.: Coevolutionary algorithms. In: Bäck, T., Fogel, D.B., Michalewicz, Z. (eds.) Evolutionary Computation 2 Advanced Algorithms and Operators, pp. 224–238. Taylor & Francis, Abington (2000)
711. Parga, N., Virasoro, M.: The ultrametric organization of memories in a neural network. Journal de Physique 47(11), 1857–1864 (1986)
712. Parisi, G.: Infinite number of order parameters for spin-glasses. Physical Review Letters (1979)
713. Park, Y.M., Park, J.B., Won, J.R.: A hybrid genetic algorithm/dynamic programming approach to optimal long-term generation expansion planning. International Journal of Electrical Power & Energy Systems 20(4), 295–303 (1998)
714. Parker, G., Blumenthal, H.: Varying sample sizes for the co-evolution of heterogeneous agents. In: [115], pp. 766–771 (2004)
715. Parthasarathy, P.V., Goldberg, D.E., Burns, S.A.: Tackling multimodal problems in hybrid genetic algorithms. In: [839], p. 775 (2001)
716. Pastorino, M.: Stochastic optimization methods applied to microwave imaging: A review. IEEE Transactions on Antennas and Propagation 55(3, Part 1), 538–548 (2007)
717. Paszkowicz, W.: Properties of a genetic algorithm extended by a random self-learning operator and asymmetric mutations: A convergence study for a task of powder-pattern indexing. Analytica Chimica Acta 566(1), 81–98 (2006)
718. Pawitan, Y.: In All Likelihood: Statistical Modelling and Inference Using Likelihood. Oxford Scientific Publishing (2001)
719. Perez, J.R., Basterrechea, J.: Comparison of different heuristic optimization methods for near-field antenna measurements. IEEE Transactions on Antennas and Propagation 55(3), 549–555 (2007)

720. Peter, M., Freisleben, B.: Memetic algorithms for the traveling salesman problem. Complex Systems 13(4), 297–345 (2001)
721. Pezzella, F., Morganti, G., Ciaschetti, G.: A genetic algorithm for the flexible job-shop scheduling problem. Computers & Operations Research 35(10), 3202–3212 (2008)
722. Phillips, A.T., Rosen, J.B.: A Quadratic Assignment Formulation for the Molecular Conformation Problem. Journal of Global Optimization 4, 229–241 (1994)
723. Pirkwieser, S., Raidl, G.R.: Finding consensus trees by evolutionary, variable neighborhood search, and hybrid algorithms. In: [784], pp. 323–330 (2008)
724. Poland, J., Knödler, K., Mitterer, A., Fleischhauer, T., Zuber-Goos, F., Zell, A.: Evolutionary search for smooth maps in motor control unit calibration. In: Steinhöfel, K. (ed.) SAGA 2001. LNCS, vol. 2264, pp. 107–116. Springer, Heidelberg (2001)
725. Poloni, C., Giurgevich, A., Onseti, L., Pediroda, V.: Hybridization of a multi-objective genetic algorithm, a neural network and a classical optimizer for a complex design problem in fluid dynamics. Computer Methods in Applied Mechanics and Engineering 186(2-4), 403–420 (2000)
726. Porumbel, D.C., Hao, J.K., Kuntz, P.: An evolutionary approach with diversity guarantee and well-informed grouping recombination for graph coloring. Computers and Operations Research 37(10), 1822–1832 (2010)
727. Potter, M., De Jong, K.: A cooperative coevolutionary approach to function optimisation. In: [185], pp. 248–257 (1994)
728. Powell, M.J.D.: An efficient method for finding the minimum of a function of several variables without calculating derivatives. The Computer Journal 7(2), 155–162 (1964)
729. Powell, M.J.D.: A fast algorithm for nonlinearly constrained optimization calculations. In: Watson, G. (ed.) Numerical Analysis, pp. 144–157. Springer, Heidelberg (1978)
730. Powell, M.J.D.: The NEWUOA software for unconstrained optimization. In: Pillo, G.D., Roma, M. (eds.) Large-Scale Nonlinear Optimization, pp. 255–297. Springer, Berlin (2006)
731. Press, W.H., Teukolsky, S.A., Vetterling, W.T., Flannery, B.P.: Numerical Recipes in C. The Art of Scientific Computing, 2nd edn. Cambridge University Press, New York (1992)
732. Price, K.V.: Mechanical engineering design optimization by differential evolution. In: [145], pp. 293–298 (1999)
733. Price, K.V., Storn, R., Lampinen, J.: Differential Evolution: A Practical Approach to Global Optimization. Springer, Heidelberg (2005)
734. Prins, C.: A simple and effective evolutionary algorithm for the vehicle routing problem. Computers & Operations Research 31(12), 1985–2002 (2004)
735. Prins, C.: Two memetic algorithms for heterogeneous fleet vehicle routing problems. Engineering Applications of Artificial Intelligence 22(6), 916–928 (2009), artificial Intelligence Techniques for Supply Chain Management
736. Prins, C., Prodhon, C., Calvo, R.W.: A memetic algorithm with population management (MA|PM) for the capacitated location-routing problem. In: Gottlieb, J., Raidl, G.R. (eds.) EvoCOP 2006. LNCS, vol. 3906, pp. 183–194. Springer, Heidelberg (2006)
737. Prodhom, C., Prins, C.: A memetic algorithm with population management (MA|PM) for the periodic location-routing problem. In: [69], pp. 43–57 (2008)
738. Păun, G.: Computing with membranes. Tech. Rep. TUCS Report 208, Turku Center for Computer Science (1998)
739. Păun, G.: Membrane Computing: An Introduction. Springer, Berlin (2002)
740. Puchinger, J., Raidl, G.: Cooperating memetic and branch-and-cut algorithms for solving the multidimensional knapsack problem. In: Proceedings of the 2005 Metaheuristics International Conference, Vienna, Austria, pp. 775–780 (2005)

References

741. Puchinger, J., Raidl, G.R.: Combining metaheuristics and exact algorithms in combinatorial optimization: a survey and classification. In: [602], pp. 41–53 (2005)
742. Puchinger, J., Raidl, G.R., Koller, G.: Solving a real-world glass cutting problem. In: Gottlieb, J., Raidl, G.R. (eds.) EvoCOP 2004. LNCS, vol. 3004, pp. 165–176. Springer, Heidelberg (2004)
743. Pudlák, P.: Complexity theory and genetics. In: Structure in Complexity Theory Conference, pp. 383–395 (1994)
744. Pudlák, P.: Complexity theory and genetics: The computational power of crossing over. Inf. Comput. 171(2), 201–223 (2001)
745. Qin, A.K., Suganthan, P.N.: Self-adaptive differential evolution algorithm for numerical optimization. In: [116], pp. 1785–1791 (2005)
746. Queiroz, L., Lyra, C.: Adaptive hybrid genetic algorithm for technical loss reduction in distribution networks under variable demands. IEEE Transactions on Power Systems 24(1), 445–453 (2009)
747. Quintero, A., Pierre, S.: A memetic algorithm for assigning cells to switches in cellular mobile networks. IEEE Communications Letters 6(11), 484–486 (2002)
748. Quintero, A., Pierre, S.: Sequential and multi-population memetic algorithms for assigning cells to switches in mobile networks. Computer Networks 43(3), 247–261 (2003)
749. Quintero, A., Pierre, S.: On the design of large-scale UMTS mobile networks using hybrid genetic algorithms. IEEE Transactions on Vehicular Technology 57(4), 2498–2508 (2008)
750. Radcliffe, N.: The algebra of genetic algorithms. Annals of Mathematics and Artificial Intelligence 10, 339–384 (1994)
751. Radcliffe, N., Surry, P.: Fitness Variance of Formae and Performance Prediction. In: Whitley, L., Vose, M. (eds.) Proceedings of the 3rd Workshop on Foundations of Genetic Algorithms, pp. 51–72. Morgan Kaufmann, San Francisco (1994)
752. Radcliffe, N., Surry, P.: Formal Memetic Algorithms. In: Fogarty, T.C. (ed.) AISB-WS 1994. LNCS, vol. 865, pp. 1–16. Springer, Heidelberg (1994)
753. Radcliffe, N.J.: Forma analysis and random respectful recombination. In: [56], pp. 222–229 (1991)
754. Radtke, P.V.W., Wong, T., Sabourin, R.: A multi-objective memetic algorithm for intelligent feature extraction. In: [135], pp. 767–781 (2005)
755. Rajesh, J., Gupta, K., Kusumakar, H.S., Jayaraman, V.K., Kulkarni, B.D.: Dynamic optimization of chemical processes using ant colony framework. Computers and Chemistry 25(6), 583–595 (2001)
756. Rammal, R., Toulouse, G., Virasoro, M.: Ultrametricity for physicists. Reviews of Modern Physics 58, 765–788 (1986)
757. Ratle, A.: Accelerating the convergence of evolutionary algorithms by fitness landscape approximations. In: [240], pp. 87–96 (1998)
758. Ray, T., Sarker, R.: Genetic algorithm for solving a gas lift optimization problem. Journal of Petroleum Science and Engineering 59(1-2), 84–96 (2007)
759. Ray, T., Singh, H.K., Isaacs, A., Smith, W.: Infeasibility driven evolutionary algorithm for constrained optimization. In: Mezura-Montes, E. (ed.) Constraint Handling in Evolutionary Optimization. Studies in Computational Intelligence, pp. 145–165. Springer, Heidelberg (2009)
760. Rechenberg, I.: Evolutionsstrategie – optimierung technischer Systeme nach Prinzipien der biologischen Evolution. PhD thesis, Technical University of Berlin (1971)

761. Rechenberg, I.: Evolutionsstrategie: Optimierung technischer Systeme nach Prinzipien der biologischen Evolution. Frommann-Holzboog Verlag, Stuttgart (1973)
762. Reeves, C.R., Rowe, J.E.: Genetic algorithms: principles and perspectives: a guide to GA theory. Kluwer Academic Publishers, Dordrecht (2003)
763. Régnier, S.: Sur quelques aspects mathématiques des problèmes de classification automatique. Mathématiques et Sciences Humaines, reprint of ICC Bulletin 4, 175–191 (1965)
764. Reinelt, G.: The Traveling Salesman: Computational Solutions for TSP Applications. LNCS, vol. 840. Springer, Heidelberg (1994)
765. Renderes, J.M., Flasse, S.P.: Hybrid methods using genetic algorithms for global optimization. IEEE Transactions on Systems, Man, and Cybernetics–Part B 26(2), 243–258 (1996)
766. Renders, J.M., Bersini, H.: Hybridizing genetic algorithms with hill-climbing methods for global optimization: Two possible ways. In: IEEE Conference on Evolutionary Computation, pp. 312–317. IEEE Press, Piscataway (1994)
767. Richer, J.-M., Goëffon, A., Hao, J.-K.: A memetic algorithm for phylogenetic reconstruction with maximum parsimony. In: Pizzuti, C., Ritchie, M.D., Giacobini, M. (eds.) EvoBIO 2009. LNCS, vol. 5483, pp. 164–175. Springer, Heidelberg (2009)
768. Riveros, C., et al.: A transcription factor map as revealed by a genome-wide gene expression analysis of whole-blood mRNA transcriptome in multiple sclerosis. PLoS ONE 5(12), e14,176 (2010)
769. Robic, T., Filipic, B.: DEMO: Differential evolution for multiobjective optimization. In: [135], pp. 520–533 (2005)
770. Robinson, D.: Comparison of labeled trees with valency three. Journal of Combinatorial Theory, Series B 11(2), 105–119 (1971)
771. Rodríguez, J.F., Renaud, J.E., Watson, L.T.: Trust region augmented Lagrangian methods for sequential response surface approximation and optimization. ASME Journal of Mechanical Design 120(1), 58–66 (1998)
772. Rodriguez-Tello, E., Hao, J.K., Torres-Jimenez, J.: An effective two-stage simulated annealing algorithm for the minimum linear arrangement problem. Computers and Operations Research 35(10), 3331–3346 (2008)
773. Romero-Campero, F., Cao, H., Camara, M., Krasnogor, N.: Structure and parameter estimation for cell systems biology models. In: [784], pp. 331–338 (2008)
774. Ronald, S.: Distance Functions for Order–Based Encodings. In: Proceedings of the 1997 IEEE International Conference on Evolutionary Computation, pp. 49–54. IEEE Press, Los Alamitos (1997)
775. Rosca, J.P.: Entropy-driven adaptive representation. In: Rosca, J.P. (ed.) Proceedings of the Workshop on Genetic Programming: from Theory to Real-World Applications, pp. 23–32 (1995)
776. Rosenbrock, H.H.: An automatic method for findong the greatest or least value of a function. The Computer Journal 3(3), 175–184 (1960)
777. Roshan, U., Moret, B.M.E., Warnow, T., Williams, T.L.: Rec-i-dcm3: A fast algorithmic technique for reconstructing large phylogenetic trees. In: IEEE Computer Society Bioinformatics Conference 2004, pp. 98–109. IEEE Press, Los Alamitos (2004)
778. Rosso, O., Mendes, A., Berretta, R., Rostas, J., Hunter, M., Moscato, P.: Distinguishing childhood absence epilepsy patients from controls by the analysis of their background brain electrical activity (ii): A combinatorial optimization approach for electrode selection. Journal of Neuroscience Methods 181(2), 257–267 (2009)

References

779. Rothlauf, F., Branke, J., Cagnoni, S., Costa, E., Cotta, C., Drechsler, R., Lutton, E., Machado, P., Moore, J.H., Romero, J., Smith, G.D., Squillero, G., Takagi, H. (eds.): EvoWorkshops 2006. LNCS, vol. 3907. Springer, Heidelberg (2006)
780. Rudolph, G.: How mutation and selection solve long-path problems in polynomial expected time. Evolutionary Computation 4(2), 195–205 (1997)
781. Rudolph, G., Jansen, T., Lucas, S., Poloni, C., Beume, N. (eds.): PPSN 2008. LNCS, vol. 5199. Springer, Heidelberg (2008)
782. Runarsson, T.P., Yao, X.: Stochastic ranking for constrained evolutionary optimization. IEEE Transactions on Evolutionary Compution 4, 284–294 (2000)
783. Runarsson, T.P., Beyer, H.-G., Burke, E.K., Merelo-Guervós, J.J., Whitley, L.D., Yao, X. (eds.): PPSN 2006. LNCS, vol. 4193. Springer, Heidelberg (2006)
784. Ryan, C., Keijzer, M. (eds.): Genetic and Evolutionary Computation Conference – GECCO 2008. ACM Press, Atlanta (2008)
785. Salcedo-Sanz, S., Yao, X.: A hybrid hopfield network-genetic algorithm approach for the terminal assignment problem. IEEE Transactions on Systems, Man, and Cybernetics, Part B: Cybernetics 34(6), 2343–2353 (2004)
786. Salehpour, A.A., Afzali-Kusha, A., Mohammadi, S.: Efficient clustering of wireless sensor networks based on memetic algorithm. In: International Conference on Innovations in Information Technology, pp. 450–454. IEEE Press, Los Alamitos (2008)
787. Salvatore, N., Cascella, G., Caponio, A., Stasi, S., Neri, F.: Optimization of DSKF-based algorithm for sensorless SFO-SM control of IMs using differential evolution. In: International Conference on Electrical Machines, paper ID 1225 (2008)
788. Sano, Y., Kita, H.: Optimization of noisy fitness functions by means of genetic algorithms and history of search. In: [796], pp. 571–580 (2000)
789. Santamaría, J., Cordón, O., Damas, S., García-Torres, J.M., Quirin, A.: Performance evaluation of memetic approaches in 3D reconstruction of forensic objects. Soft Computing 13(8-9), 883–904 (2009)
790. Santos, E., ESantos, J.: Effective computational reuse for energy evaluations in protein folding. International Journal of Artificial Intelligence Tools 15(5), 725–739 (2006)
791. Sarker, R., Newton, C.: Optimization Modelling: A Practical Approach. Taylor & Francis Group/CRC Press, USA (2008)
792. Sarker, R., Kamruzzaman, J., Newton, C.: Evolutionary optimization (evopt): A brief review and analysis. International Journal of Computational Intelligence and Applications 3(4), 311–330 (2003)
793. Schaffer, J., Morishima, A.: An adaptive crossover distribution mechanism for genetic algorithms. In: [333], pp. 36–40 (1987)
794. Schaffer, J.D. (ed.): Third International Conference on Genetic Algorithms. Morgan Kaufmann, San Mateo (1989)
795. Schiex, T., Fargier, H., Verfaillie, G.: Valued constraint satisfaction problems: hard and easy problems. In: 14th. International Join Conference on Artificial Intelligence, IJCAI 1995, Montreal, Canada, pp. 631–637 (1995)
796. Deb, K., Rudolph, G., Lutton, E., Merelo, J.J., Schoenauer, M., Schwefel, H.-P., Yao, X. (eds.): PPSN 2000. LNCS, vol. 1917. Springer, Heidelberg (2000)
797. Schuetze, O., Sanchez, G., Coello Coello, C.A.: A new memetic strategy for the numerical treatment of multi-objective optimization problems. In: [784], pp. 705–712 (2008)
798. Schwefel, H.P.: Kybernetische Evolution als Strategie der experimentellen Forschung in der Strömungstechnik. PhD thesis, Technical University of Berlin, Hermann Föttinger–Institute for Fluid Dynamics (1965)

799. Schwefel, H.P.: Numerical Optimisation of Computer Models. J. Wiley, Chichester (1981)
800. Schwefel, H.P.: Evolution strategies: A family of non-linear optimization techniques based on imitating some principles of natural evolution. Annals of Operations Research 1, 165–167 (1984)
801. Schwefel, H.P.: Evolution and Optimum Seeking. John Wiley & Sons, Inc., New York (1994)
802. Schwefel, H.-P., Männer, R. (eds.): PPSN 1990. LNCS, vol. 496. Springer, Heidelberg (1991)
803. Sefrioui, M., Périaux, J.: A hierarchical genetic algorithm using multiple models for optimization. In: [796], pp. 879–888 (2000)
804. Selman, B., Levesque, H., Mitchell, D.: A new method for solving hard satisfiability problems. In: National Conference on Artificial Intelligence, pp. 440–446. AAAI, Menlo Park (1992)
805. Serafini, P.: Simulated annealing for multiple objective optimization problems. In: Tenth International Conference on Multiple Criteria Decision Making, vol. 1, pp. 87–96 (1992)
806. Seront, G., Bersini, H.: A new GA-local search hybrid for continuous optimization based on multi-level single linkage clustering. In: [932], pp. 90–95 (2000)
807. Serpell, M., Smith, J.: Self-adaptation of mutation operator and probability for permutation representations in genetic algorithms. Evolutionary Computation 18(3), 491–514 (2009)
808. Sheskin, D.J.: Handbook of Parametric and Nonparametric Statistical Procedures, 4th edn. Chapman and Hall, Boca Raton (2007)
809. Shi, Y., Eberhart, R.: A modified particle swarm optimizer. In: IEEE World Congress on Computational Intelligence, pp. 69–73 (1998)
810. Shyr, W.-J.: The hybrid genetic algorithm for blind signal separation. In: King, I., Wang, J., Chan, L.-W., Wang, D. (eds.) ICONIP 2006. LNCS, vol. 4234, pp. 954–963. Springer, Heidelberg (2006)
811. Shyr, W.J.: Robust control design for aircraft controllers via memetic algorithms. International Journal of Innovative Computing, Information and Control 5(10A), 3133–3140 (2009)
812. Shyr, W.J., Wang, B.W., Yeh, Y.Y., Su, T.J.: Design of optimal PID controllers using memetic algorithm. In: American Control Conference, vol. 3, pp. 2130–2131 (2002)
813. Shyu, C., Sheneman, L., Foster, J.A.: Multiple sequence alignment with evolutionary computation. Genetic Programming and Evolvable Machines 5, 121–144 (2004)
814. Sinclair, M.: Minimum cost routing and wavelength allocation using a genetic-algorithm/heuristic hybrid approach. In: IEE Conference on Telecommunications (Conf. Publ. No. 451), pp. 67–71 (1998)
815. Sindhya, K., Deb, K., Miettinen, K.: A local search based evolutionary multi-objective optimization approach for fast and accurate convergence. In: [781], pp. 815–824 (2008)
816. Singh, H., Ray, T., Smith, W.: Performance of infeasibility empowered memetic algorithm for CEC 2010 constrained optimization problems. In: [121], pp. 1–8 (2010)
817. Singh, H.K., Isaacs, A., Ray, T., Smith, W.: Infeasibility Driven Evolutionary Algorithm (IDEA) for Engineering Design Optimization. In: 21st Australiasian Joint Conference on Artificial Intelligence AI 2008, pp. 104–115 (2008)
818. Sinha, A., Chen, Y., Goldberg, D.E.: Designing efficient genetic and evolutionary algorithm hybrids. In: [375], pp. 259–288 (2005)

References

819. Smit, V.N.S., Eiben, A.: Costs and benefits of tuning parameters of evolutionary algorithms. In: [781], pp. 528–538 (2008)
820. Smith, J.: Modelling GAs with self adaptive mutation rates. In: [839], pp. 599–606 (2001)
821. Smith, J.: Co-evolution of memetic algorithms: Initial investigations. In: [580], pp. 537–548 (2002)
822. Smith, J.: Co-evolving memetic algorithms: A learning approach to robust scalable optimisation. In: [114], pp. 498–505 (2003)
823. Smith, J.: Parameter perturbation mechanisms in binary coded gas with self-adaptive mutation. In: Rowe, J., Poli, R., De Jong, K., Cotta, C. (eds.) Foundations of Genetic Algorithms 7, pp. 329–346. Morgan Kauffman, San Francisco (2003)
824. Smith, J.: Protein structure prediction with co-evolving memetic algorithms. In: [114], pp. 2346–2353 (2003)
825. Smith, J.: The co-evolution of memetic algorithms for protein structure prediction. In: [375], pp. 105–128 (2005)
826. Smith, J., Fogarty, T.: Adaptively parameterised evolutionary systems: Self adaptive recombination and mutation in a genetic algorithm. In: [909], pp. 441–450 (1996)
827. Smith, J., Fogarty, T.: Recombination strategy adaptation via evolution of gene linkage. In: 1996 IEEE Conference on Evolutionary Computation, pp. 826–831 (1996)
828. Smith, J., Fogarty, T.: Self adaptation of mutation rates in a steady state genetic algorithm. In: 1996 IEEE Conference on Evolutionary Computation, pp. 318–323 (1996)
829. Smith, J., Fogarty, T.: Operator and parameter adaptation in genetic algorithms. Soft Computing 1(2), 81–87 (1997)
830. Smith, J.E.: Coevolving memetic algorithms: A review and progress report. IEEE Transactions on Systems, Man, and Cybernetics, Part B 37(1), 6–17 (2007)
831. Smith, J.E.: Credit assignment in adaptive memetic algorithms. In: Lipson, H. (ed.) GECCO 2007, pp. 1412–1419. ACM Press, London (2007)
832. Smith, R.E., Smith, J.E.: An Examination of Tunable, Random Search Landscapes. In: Foundations of Genetic Algorithms V, pp. 165–182. Morgan Kaufmann, San Francisco (1999)
833. Sneath, P., Sokal, R.: Numerical taxonomy. W.H. Freeman and Co, San Francisco (1973)
834. Solis, F.J., Wets, R.J.B.: Minimization by random search techniques. Mathematics of Operations Research 6(1), 19–30 (1981)
835. Song, W.: Multiobjective memetic algorithm and its application in robust airfoil shape optimization. In: Goh, C.K., Ong, K.Y. (eds.) Multi-Objective Memetic Algorithms. Studies in Computational Intelligence, vol. 171, pp. 389–402. Springer, Heidelberg (2009)
836. Sörensen, K., Sevaux, M.: MA|PM: memetic algorithms with population management. Computers and Operations Research 33(5), 1214–1225 (2006)
837. Sörensen, K., Sevaux, M.: A practical approach for robust and flexible vehicle routing using metaheuristics and Monte Carlo sampling. Journal of Mathematical Modelling and Algorithms 8(4), 387–407 (2009)
838. Spall, J.C.: Introduction to Stochastic Search and Optimization. Estimation, Simulation, and Control. John Wiley & Sons, Inc., Hoboken (2003)
839. Spector, L., et al. (eds.): Genetic and Evolutionary Computation Conference – GECCO 2001. Morgan Kaufmann, San Francisco (2001)
840. Speer, N., Merz, P., Spieth, C., Zell, A.: Clustering gene expression data with memetic algorithms based on minimum spanning trees. In: [114], pp. 1848–1855 (2003)

841. Speer, N., Spieth, C., Zell, A.: A memetic clustering algorithm for the functional partition of genes based on the gene ontology. In: IEEE Symposium on Computational Intelligence in Bioinformatics and Computational Biology, pp. 252–259. IEEE Press, Los Alamitos (2004)
842. Spieth, C., Streichert, F., Speer, N., Zell, A.: A memetic inference method for gene regulatory networks based on S-systems. In: [115], pp. 152–157 (2004)
843. Spieth, C., Streichert, F., Supper, J., Speer, N., Zell, A.: Feedback memetic algorithms for modeling gene regulatory networks. In: 2005 IEEE Symposium on Computational Intelligence in Bioinformatics and Computational Biology, pp. 61–67. IEEE Press, Los Alamitos (2005)
844. Srinivas, N., Deb, K.: Multiple objective optimization using nondominated sorting in genetic algorithms. Evolutionary Computation 2(2), 221–248 (1994)
845. Stadler, P.F.: Towards a Theory of Landscapes. In: Lopéz-Peña, R., Capovilla, R., García-Pelayo, R., Waelbroeck, H., Zertuche, F. (eds.) Complex Systems and Binary Networks. Lecture Notes in Physics, vol. 461, pp. 77–163. Springer, Berlin (1995)
846. Stadler, P.F.: Landscapes and their Correlation Functions. Joural of Mathematical Chemistry 20, 1–45 (1996)
847. Stadler, P.F., Schnabl, W.: The Landscape of the Travelling Salesman Problem. Physics Letters A 161, 337–344 (1992)
848. Steuer, R.: Multiple Criteria Optimization - Theory, Computation and Application. Wiley, Chichester (1986)
849. Stewart, G.W.: A modification of davidson's minimization method to accept difference approximations of derivatives. Journal of the ACM 14(1), 72–83 (1967)
850. Stone, C., Smith, J.: Strategy parameter variation in self-adaptation of mutation rates. In: [504], pp. 586–593 (2002)
851. Stone, C.J.: Optimal global rates of convergence for nonparametric regression. Annals of Statistics 10(4), 1040–1053 (1982)
852. Storch, T.: On the choice of the parent population size. Evolutionary Computation 16(4), 557–578 (2008)
853. Storn, R., Price, K.: Differential evolution - a simple and efficient adaptive scheme for global optimization over continuous spaces. Tech. Rep. TR-95-012, ICSI (1995)
854. Subudhi, B., Jena, D., Gupta, M.: Memetic differential evolution trained neural networks for nonlinear system identification. In: IEEE Region 10 and international Conference on Industrial and Information Systems, pp. 1–6. IEEE Press, Los Alamitos (2008)
855. Sudholt, D.: Local Search in Evolutionary Algorithms: The Impact of the Local Search Frequency. In: Asano, T. (ed.) ISAAC 2006. LNCS, vol. 4288, pp. 359–368. Springer, Heidelberg (2006)
856. Sudholt, D.: On the analysis of the (1+1) memetic algorithm. In: [110], pp. 493–500 (2006)
857. Sudholt, D.: Memetic algorithms with variable-depth search to overcome local optima. In: [784], pp. 787–794 (2008)
858. Sudholt, D.: The impact of parametrization in memetic evolutionary algorithms. Theoretical Computer Science 410(26), 2511–2528 (2009)
859. Sudholt, D.: Hybridizing evolutionary algorithms with variable-depth search to overcome local optima. Algorithmica 59(3), 343–368 (2011)
860. Suman, B.: Study of simulated annealing based algorithms for multiobjective optimization of a constrained problem. Computers & Chemical Engineering 28(9), 1849–1871 (2004)

References

861. Surry, P., Radcliffe, N.: Inoculation to initialise evolutionary search. In: Fogarty, T.C. (ed.) AISB-WS 1996. LNCS, vol. 1143, pp. 269–285. Springer, Heidelberg (1996)
862. Swofford, D.L., Olsen, G.J., Waddell, P.J., Hillis, D.M.: Phylogeny inference. In: Hillis, D.M., Moritz, C., Mable, B.K. (eds.) Molecular Systematics, ch. 11, pp. 407–514. Sinauer Associates, Inc. (1999)
863. Sylos Labini, M., Covitti, A., Delvecchio, G., Neri, F.: A quasi-genetic algorithm for searching the dangerous areas generated by a grounding system. COMPEL: International Journal for Computation and Mathematics in Electrical and Electronic Engineering 23(3), 724–732 (2004)
864. Syswerda, G.: A Study of Reproduction in Generational and Steady State Genetic Algorithms. In: Rawlings, G.J.E. (ed.) Foundations of Genetic Algorithms, pp. 94–101. Morgan Kaufmann, San Mateo (1991)
865. Szeider, S.: The parameterized complexity of k-flip local search for SAT and MAX SAT. In: Kullmann, O. (ed.) SAT 2009. LNCS, vol. 5584, pp. 276–283. Springer, Heidelberg (2009)
866. Szeider, S.: The parameterized complexity of k-flip local search for SAT and MAX SAT. Discrete Optimization 8(1), 139–145 (2011)
867. Szu, H., Hartley, R.: Fast simulated annealing. Physiscs Letters A 122, 157–162 (1987)
868. Tagawa, K., Matsuoka, M.: Optimum design of surface acoustic wave filters based on the taguchi's quality engineering with a memetic algorithm. In: [783], pp. 292–301 (2006)
869. Tagawa, K., Masuoka, M., Tsukamoto, M.: Robust optimum design of saw filters with the taguchi method and a memetic algorithm. In: [116], pp. 2146–2153 (2005)
870. Taguchi, G.: Robust Engineering. McGraw-Hill Book Company, New York (1990)
871. Tai, K., Wang, N., Yang, Y.: Hybrid GA multiobjective optimization for the design of compliant micro-actuators. In: IEEE International Conference on Systems, Man and Cybernetics, pp. 559–564 (2008)
872. Talbi, E.G.: A taxonomy of hybrid metaheuristics. Journal of Heuristics 8(5), 541–564 (2002)
873. Tang, J., Lim, M.H., Ong, Y.S.: Diversity-adaptive parallel memetic algorithm for solving large scale combinatorial optimization problems. Soft Computing-A Fusion of Foundations, Methodologies and Applications 11(9), 873–888 (2007)
874. Tang, M., Yao, X.: A memetic algorithm for VLSI floorplanning. IEEE Transactions on Systems, Man, and Cybernetics, Part B 37(1), 62–69 (2007)
875. Tao, W.H.: Fuzzy neural network control of truck backer-upper using hybrid genetic algorithms. In: International Conference on Information Acquisition, pp. 9–12. IEEE Press, Los Alamitos (2004)
876. Tenne, Y.: A model-assisted memetic algorithm for expensive optimization problems. In: Chiong, R. (ed.) Nature-Inspired Algorithms for Optimisation. Studies in Computational Intelligence, vol. 193, Springer, Heidelberg (2009)
877. Tenne, Y., Armfield, S.: A memetic algorithm assisted by an adaptive topology rbf network and variable local models for expensive optimization problems. In: Kosinski W (ed) Advances in Evolutionary Algorithms, IN-TECH (2008)
878. Tenne, Y., Armfield, S.W.: A memetic algorithm using a trust-region derivative-free optimization with quadratic modelling for optimization of expensive and noisy black-box functions. In: Yang, S., Ong, Y.S., Jin, Y. (eds.) Evolutionary Computation in Dynamic and Uncertain Environments. Studies in Computational Intelligence, vol. 51, pp. 389–415. Springer, Heidelberg (2007)

879. Tenne, Y., Armfield, S.W.: A versatile surrogate-assisted memetic algorithm for optimization of computationally expensive functions and its engineering applications. In: Yang, A., Shan, Y., Thu Bui, L. (eds.) Success in Evolutionary Computation. Studies in Computational Intelligence, vol. 92, pp. 43–72. Springer, Heidelberg (2008)
880. Tenne, Y., Armfield, S.W.: A framework for memetic optimization using variable global and local surrogate models. Journal of Soft Computing 13(8) (2009)
881. Tenne, Y., Goh, C.K.: Computational Intelligence in Expensive Optimization Problems. In: Evolutionary Learning and Optimization. Springer, Heidelberg (2010)
882. Thomas, R.: Sea Monster Tattoo. Polygon (1997)
883. Thomsen, R., Munkegade, N., Fogel, G.B., Krink, T., Group, E., Group, E.: A clustal alignment improver using evolutionary algorithms. In: [113], pp. 121–126 (2002)
884. Thorup, M.: Structured programs have small tree-width and good register allocation (extended abstract). In: Möhring, R.H. (ed.) WG 1997. LNCS, vol. 1335, pp. 318–332. Springer, Heidelberg (1997)
885. Tian, Z., Liu, Y., Yang, H., Wang, H.: A hybrid genetic algorithm with critical primary inputs sharing and minor primary inputs bits climbing for circuit maximum power estimation. In: International Conference on Natural Computation, vol. 4, pp. 183–187 (2007)
886. Tie-hua, J., Dong-lin, S., Ai-xin, C., Yan-jun, Z., Guo-yu, W.: Broadband matching network design for antennas using a hybrid genetic algorithm. In: International Symposium on Antennas, Propagation and EM Theory, pp. 90–93 (2008)
887. Tirronen, V., Neri, F.: Differential evolution with fitness diversity self-adaptation. In: Chiong, R. (ed.) Nature-Inspired Algorithms for Optimisation. Studies in Computational Intelligence, vol. 193, pp. 199–234. Springer, Heidelberg (2009)
888. Tirronen, V., Neri, F., Karkkainen, T., Majava, K., Rossi, T.: A memetic differential evolution in filter design for defect detection in paper production. In: Giacobini, M. (ed.) EvoWorkshops 2007. LNCS, vol. 4448, pp. 320–329. Springer, Heidelberg (2007)
889. Tirronen, V., Neri, F., Kärkkäinen, T., Majava, K., Rossi, T.: An enhanced memetic differential evolution in filter design for defect detection in paper production. Evolutionary Computation 16(4), 529–555 (2008)
890. Tirronen, V., Neri, F., Majava, K., Kärkkäinen, T.: The "natura non facit saltus" principle in memetic computing. In: [119], pp. 3881–3888 (2008)
891. Törn, A.: Cluster analysis using seed points and density-determined hyperspheres as an aid to global optimization. IEEE Transactions on Systems, Man, and Cybernetics 7(8), 610–616 (1977)
892. Trosset, M.W.: I know it when I see it: Toward a definition of direct search methods. SIAG/OPT Views-and-News 9, 7–10 (1997)
893. Tsai, K.Y., Wang, F.S.: Evolutionary optimization with data collocation for reverse engineering of biological networks. Bioinformatics 21(7), 1180–1188 (2005)
894. Tse, S.M., Liang, Y., Leung, K.S., Lee, K.H., Mok, T.: A memetic algorithm for multiple-drug cancer chemotherapy schedule optimization. IEEE Transactions on Systems, Man, and Cybernetics, Part B 37(1), 84–91 (2007)
895. Tsutsui, S., Ghosh, A.: Genetic algorithms with a robust solution scheme. IEEE Transactions on Evolutionary Computation 1(3), 201–208 (1997)
896. Ulder, N.L.J., Aarts, E.H.L., Bandelt, H.J., van Laarhoven, P.J.M., et al.: Genetic Local Search Algorithms for the Traveling Salesman Problems. In: [802], pp. 109–116 (1991)
897. Ulungu, E.L., Teghem, J., Fortemps, P., Tuyttens, D.: MOSA method: a tool for solving multiobjective combinatorial optimization problems. Journal of Multi-Criteria Decision Analysis 8, 221–236 (1999)

References

898. Umezawa, K., Yamazaki, K.: Tree-length equals branch-length. Discrete Mathematics 309(13), 4656–4660 (2009)
899. Vakil Baghmisheh, M., Alinia Ahandani, M., Talebi, M.: Frequency modulation sound parameter identification using novel hybrid evolutionary algorithms. In: International Symposium on Telecommunications, pp. 67–72. IEEE Press, Los Alamitos (2008)
900. Valenzuela, J., Smith, A.E.: A seeded memetic algorithm for large unit commitment problems. Journal of Heuristics 8, 173–195 (1999)
901. Vallejo, A., Cutiller, A.Z.D.V.D., Dalmau, J.: Implementation of traffic engineering in NGNs using hybrid genetic algorithms. In: International Conference on Systems and Networks Communications, pp. 262–267 (2008)
902. Vasquez, M., Hao, J.K.: A heuristic approach for antenna positioning in cellular networks. Journal of Heuristics 7(5), 443–472 (2001)
903. Vasquez, M., Hao, J.K.: A logic-constrained knapsack formulation and a tabu algorithm for the daily photograph scheduling of an earth observation satellite. Computational Optimization and Applications 20(2), 137–157 (2001)
904. Vavak, F., Jukes, K.A., Fogarty, T.C.: A genetic algorithm with variable range of local search for tracking changing environments. In: [909], pp. 376–385 (1996)
905. Vavak, F., Jukes, K.A., Fogarty, T.C.: Adaptive combustion balancing in multiple burner boiler using a genetic algorithm with variable range of local search. In: [33], pp. 719–726 (1997)
906. Vavak, F., Jukes, K.A., Fogarty, T.C.: Performance of a genetic algorithm with variable local search range relative to frequency of the environmental changes. In: Koza, J.R. (ed.) Third Annual Conference on Genetic Programming, pp. 602–608. Morgan Kaufmann, San Francisco (1998)
907. Vazirani, V.: Approximation Algorithms. Springer, Berlin (2001)
908. Venkatraman, S., Yen, G.: A generic framework for constrained optimization using genetic algorithms. IEEE Transactions on Evolutionary Computation 9(4), 424–435 (2005)
909. Ebeling, W., Rechenberg, I., Voigt, H.-M., Schwefel, H.-P. (eds.): PPSN 1996. LNCS, vol. 1141. Springer, Heidelberg (1996)
910. Volgenant, A., Jonker, R.: A branch and bound algorithm for the symmetric traveling salesman problem based on the 1-tree relaxation. European Journal of Operational Research 9, 83–88 (1982)
911. Volk, J., Herrmann, T., Wuthrich, K.: Automated sequence-specific protein nmr assignment using the memetic algorithm match. Journal of Biomolecular Nmr. 41(3), 127–138 (2008)
912. Walshaw, C.: A Multilevel Approach to the Travelling Salesman Problem. Operations Research 50(5), 862–877 (2002)
913. Wang, H., Wang, D., Yang, S.: A memetic algorithm with adaptive hill climbing strategy for dynamic optimization problems. Journal of Soft Computing 13, 763–780 (2009)
914. Wang, H., Qian, L., Dougherty, E.: Inference of gene regulatory networks using S-system: a unified approach. IET Systems Biology 4(2), 145–156 (2010)
915. Wang, H., Yang, S., Ip, W.H., Wang, D.: A particle swarm optimization based memetic algorithm for dynamic optimization problems. Natural Computing 3(9), 703–725 (2010)
916. Wang, J., Shan, H., Shasha, D., Piel, W.: Treerank: A similarity measure for nearest neighbor searching in phylogenetic databases. In: 15th International Conference on Scientific and Statistical Database Management, pp. 171–180. IEEE Press, Cambridge (2003)
917. Wang, N., Tai, K.: A hybrid genetic algorithm for multiobjective structural optimization. In: [118], pp. 2948–2955 (2007)

918. Wang, N., Yang, Y.: Target geometry matching problem for hybrid genetic algorithm used to design structures subjected to uncertainty. In: [120], pp. 1644–1651 (2009)
919. Wang, N., Yang, Y., Tai, K.: Hybrid genetic algorithm for designing structures subjected to uncertainty. In: IEEE International Conference on Systems, Man and Cybernetics, pp. 565–570. IEEE Press, Los Alamitos (2008)
920. Wang, N., Yang, Y., Tai, K.: Optimization of structures under load uncertainties based on hybrid genetic algorithm. In: [119], pp. 4039–4044 (2008)
921. Watanabe, I., Kurihara, I., Nakachi, Y.: A hybrid genetic algorithm for service restoration problems in power distribution systems. In: [117], pp. 3250–3257 (2006)
922. Waterman, M., Smith, T.: On the similarity of dendograms. Journal of Theoretical Biology 73, 789–800 (1978)
923. Watson, J.P., Howe, A.E., Whitley, L.D.: An analysis of iterated local search for job-shop scheduling. In: Ibaraki, T., Yoshitomi, Y. (eds.) Fifth Metaheuristics International Conference, MIC 2003 (2003)
924. Weaver, D., Workman, C., Stormo, G.: Modeling regulatory networks with weight matrices. In: Pacific Symposium on Biocomputing, vol. 4, pp. 112–123 (1999)
925. Wegener, I.: Complexity Theory—Exploring the Limits of Efficient Algorithms. Springer, Heidelberg (2005)
926. Wegener, I., Witt, C.: On the analysis of a simple evolutionary algorithm on quadratic pseudo-boolean functions. Journal of Discrete Algorithms 3(1), 61–78 (2005)
927. Weicker, K.: Performance measures for dynamic environments. In: [580], pp. 64–73 (2002)
928. Weinberger, E.D.: Correlated and Uncorrelated Fitness Landscapes and How to Tell the Difference. Biological Cybernetics 63, 325–336 (1990)
929. Weinberger, E.D.: NP Completeness of Kauffman's N-k Model, A Tuneable Rugged Fitness Landscape. Tech. Rep. 96-02-003, Santa Fe Institute, Santa Fe, New Mexico (1996)
930. Whitley, D.: Using reproductive evaluation to improve genetic search and heuristic discovery. In: [333], pp. 108–115 (1987)
931. Whitley, D., Starkweather, T., Fuquay, D.: Scheduling Problems and Traveling Salesman: The Genetic Edge Recombination Operator. In: [794], pp. 133–140 (1989)
932. Whitley, L.D., et al. (eds.): Genetic and Evolutionary Computation Conference – GECCO 2000. Morgan Kaufmann, Las Vegas (2000)
933. Wiegand, R., Liles, W., De Jong, K.: An empirical analysis of collaboration methods in cooperative coevolutionary algorithms. In: [839], pp. 1235–1245 (2001)
934. Wiegers, M.: The k-section of treewidth restricted graphs. In: Rovan, B. (ed.) MFCS 1990. LNCS, vol. 452, pp. 530–537. Springer, Heidelberg (1990)
935. Willett, P.: Genetic algorithms in molecular recognition and design. Trends in Biotechnology 13(12), 516–521 (1995)
936. Williams, H.P.: Model Building in Mathematical Programming, 4th edn. Wiley, Chichester (2000)
937. Williams, T., Smith, M.: The role of diverse populations in phylogenetic analysis. In: [110], pp. 287–294 (2006)
938. Witt, C.: Runtime analysis of the $(\mu+1)$ EA on simple pseudo-Boolean functions. Evolutionary Computation 14(1), 65–86 (2006)
939. Witt, C.: Population size versus runtime of a simple evolutionary algorithm. Theoretical Computer Science 403(1), 104–120 (2008)
940. Wolpert, D., Macready, W.: No free lunch theorems for optimization. IEEE Transactions on Evolutionary Computation 1(1), 67–82 (1997)

941. Wright, S.: The role of mutation, inbreeding, crossbreeding, and selection in evolution. In: Proceedigns of the Sixth International Congress on Genetics, vol. 1, pp. 356–366 (1932)
942. Wu, B., Chao, K.M., Tang, C.: Approximation and exact algorithms for constructing minimum ultrametric trees from distance matrices. Journal of Combinatorial Optimization 3(2), 199–211 (1999)
943. Wu, T., Li, D., Liu, X., Du, G., Han, R.: Model-adaptable MOSFET parameter extraction with a hybrid genetic algorithm. In: International Conference on Solid-State and Integrated Circuit Technology, pp. 1299–1302 (2006)
944. Xia, C., Xue, M., Chen, W., Xie, X.: Flux linkage characteristic measurement and parameter identification based on hybrid genetic algorithm for switched reluctance motors. In: IEEE Conference on Industrial Electronics and Applications, pp. 1619–1623. IEEE Press, Los Alamitos (2008)
945. Xu, C., Zou, X., Yuan, R., Wu, C.: Optimal coordination of protection relays using new hybrid evolutionary algorithm. In: [119], pp. 823–828 (2008)
946. Yamada, T.: Studies on metaheuristics for jobshop and flowshop scheduling problems. PhD thesis, Department of Applied Mathematics and Physics, Kyoto University, Kyoto, Japan (2003)
947. Yang, C.H., Cheng, Y.H., Chuang, L.Y., Chang, H.W.: Specific PCR product primer design using memetic algorithm. Biotechnol. Prog. 25(3), 745–753 (2009)
948. Yannakakis, M.: Computational complexity. In: Local Search in Combinatorial Optimization, pp. 19–55. Princeton University Press, Princeton (1997)
949. Yao, X., Liu, Y.: A new evolutionary systems for evolving artificial neural networks. IEEE Transactions on Neural Networks 8(3), 694–713 (1997)
950. Ying Xu, D.V. Olman Clustering gene expression data using a graph-theoretic approach: An application of minimum spanning tree. Bioinformatics 18(4), 526–535 (2002)
951. Yuan, B., Gallagher, M.: Combining meta-EAs and racing for difficult EA parameter tuning tasks. In: Lobo, F., Lima, C., Michalewicz, Z. (eds.) Parameter Setting in Evolutionary Algorithms, pp. 121–142. Springer, Heidelberg (2007)
952. Yuan, Y.: On the truncated conjugate gradient method. Mathematical Programming 87(3), 561–573 (2000)
953. Zhu, Z., Ong, M.Y.S.: Markov blanket-embedded genetic algorithm for gene selection. Pattern Recognition Archive 40(11), 3236–3248 (2007)
954. Zacharias, C.R., Lemes, M.R., Pino, A.D.: Combining genetic algorithm and simulated annealing: a molecular geometry optimization study. Journal of Molecular Structure: THEOCHEM 430, 29–39 (1998)
955. Zhang, Q., Li, H.: MOEA/D: A multi-objective evolutionary algorithm based on decomposition. IEEE Transactions on Evolutionary Computation 11(6), 712–731 (2007)
956. Zhang, Q., Sun, J., Tsang, E., Ford, J.: Hybrid estimation of distribution algorithm for global optimisation. Engineering Computations 21(1), 91–107 (2004)
957. Zhang, Q., Zhou, A., Jin, Y.: RM-MEDA: A regularity model based multiobjective estimation of distribution algorithm. IEEE Transactions on Evolutionary Computation 12(1), 41–63 (2008)
958. Zhang, Y.G., Huang, Y.M., Xie, L.M.: Robot inverse acceleration solution based on hybrid genetic algorithm. In: International Conference on Machine Learning and Cybernetics, vol. 4, pp. 2099–2103 (2008)
959. Zhao, X.C.: Advances on protein folding simulations based on the lattice hp models with natural computing. Applied Soft Computing 8(2), 1029–1040 (2008)

960. Zheng, H., Wong, A., Nahavandi, S.: Hybrid ant colony algorithm for texture classification. In: [114], pp. 2648–2652 (2003)
961. Zhou, A., Zhang, Q., Jin, Y.: Approximating the set of pareto optimal solutions in both the decision and objective spaces by an estimation of distribution algorithm. IEEE Transactions on Evolutionary Computation 13(5), 1167–1189 (2009)
962. Zhou, Z., Ong, Y.S., Lim, M.H., Lee, B.: Memetic algorithms using multi-surrogates for computationally expensive optimization problems. Journal of Soft Computing 11(10), 957–971 (2007)
963. Zhou, Z., Ong, Y.S., Nair, P.B., Keane, A.J., Lum, K.Y.: Combining global and local surrogate models to accelerate evolutionary optimization. IEEE Transactions on Systems, Man, and Cybernetics–Part C 37(1), 66–76 (2007)
964. Zhu, Z., Ong, Y.S.: Memetic algorithms for feature selection on microarray data. In: Liu, D., Fei, S., Hou, Z.-G., Zhang, H., Sun, C. (eds.) ISNN 2007. LNCS, vol. 4491, pp. 1327–1335. Springer, Heidelberg (2007)
965. Zhu, Z., Ong, Y.S., Dash, M.: Wrapper-filter feature selection algorithm using a memetic framework. IEEE Transactions on Systems, Man, and Cybernetics, Part B 37(1), 70–76 (2007)
966. Zhu, Z., Jia, S., Ji, Z.: Towards a memetic feature selection paradigm. Computational Intelligence Magazine 5, 41–53 (2010)
967. Zitzler, E., Künzli, S.: Indicator-based selection in multiobjective search. In: Yao, X., Burke, E.K., Lozano, J.A., Smith, J., Merelo-Guervós, J.J., Bullinaria, J.A., Rowe, J.E., Tiňo, P., Kabán, A., Schwefel, H.-P. (eds.) PPSN 2004. LNCS, vol. 3242, pp. 832–842. Springer, Heidelberg (2004)
968. Zitzler, E., Laumanns, M., Thiele, L.: SPEA2: Improving the strength pareto evolutionary algorithm for multiobjective optimization. In: EUROGEN 2001 Evolutionary Methods for Design, Optimisation and Control with Applications to Industrial Problems, pp. 12–21 (2001)
969. Zitzler, E., Thiele, L., Laumanns, M., Fonseca, C., Grunert da Fonseca, V.: Performance assesment of multiobjective optimizers: an analysis and review. IEEE Transactions on Evolutionary Computation 7(2), 117–132 (2003)
970. Zitzler, E., Knowles, J., Thiele, L.: Quality assessment of pareto set approximations. In: [84], pp. 373–404 (2008)
971. Zitzler, E., Deb, K., Thiele, L., Coello Coello, C.A., Corne, D.W. (eds.): EMO 2001. LNCS, vol. 1993. Springer, Heidelberg (2001)
972. Zou, L., Zhu, S., He, B.: Spatio-temporal EEG dipole estimation by means of a hybrid genetic algorithm. In: Annual International Conference of the IEEE Engineering in Medicine and Biology Society, vol. 2, pp. 4436–4439. IEEE Press, Los Alamitos (2004)

Author Index

Berretta, Regina 261

Caponio, Andrea 241
Cotta, Carlos 3, 29, 43, 121, 189, 261

de Oca, Marco A. Montes 29

Eiben, Ágoston E. 9

Gallardo, José E. 189

Hao, Jin-Kao 73

Ishibuchi, Hisao 201

Jaszkiewicz, Andrzej 201

Leiva, Antonio J. Fernández 189

Merz, Peter 95
Moscato, Pablo 261, 275

Neri, Ferrante 3, 29, 43, 121, 153, 241

Ray, Tapabrata 135

Sarker, Ruhul 135
Smith, James E. 9, 167
Sudholt, Dirk 55

Tenne, Yoel 219

Zhang, Qingfu 201

Subject Index

$(\mu+\lambda)$ MA 67

acronyms, list of XXIII
allele 13
alternative representation 140
ant colony optimization 268
anytime behaviour 19
applications 50
approximated objective functions 220
arity 15

backtracking 192
balance of global and local search 55–72
Baldwinian learning 228, 270
basin of attraction 123
beam search 196
beta distribution 162
binary quadratic programming 111–118
bioinformatics 261–271, 303
    cell models 270
    clustering 264–265
    conformational analysis 268
    consensus tree 267
    DNA sequencing 269
    feature selection 265–266
        filter vs wrapper methods 265
    gene ordering 264
    gene regulatory networks 270
    ligand docking 269
    microarray analysis 262–266
    molecular design 268–269
    molecular signature 263
    phylogenetic trees 197
    phylogeny 266–267
    maximum parsimony 89, 267
    ultrametric tree 266, 304
    polymerase chain reaction 269
    protein alignment 268
    protein structure analysis 267–268
    protein structure prediction 192, 267
        HP model 268
    sequence alignment 269
    sequence analysis 269–270
    shortest common supersequence 270
    systems biology 270–271
biomedicine 262
    drug therapy scheduling 262
    radiotherapy 262
    tomography 262
Boltzmann machine 279
branch and bound 51, 190, 195, 267
branch and cut 51, 109, 190, 195, 196
branchwidth 300
breeder genetic algorithm 156
brute force 124
bucket elimination 194
    mini-buckets 198, 199

candidate solution 13
CHC 126
    pseudocode 127
Checkers algorithm 232
child 15
chromosome 13
CMA-ES 41
co-evolution 51
coevolving MA 167
combinatorial local search 33

combinatorial optimization 96
complete techniques 51, 190
    approximation algorithms 190
    exact algorithms 190
complexity 5
    class 5
    intractability of local search 63–66
    polynomial hierarchy 5
    reduction 5
conjugate directions 39
constrained optimization 135–151
    benchmark 145
    weighted constraint satisfaction problem 194, 197
continuous optimization 121–134, 199
    local optimum 31
continuous space
    dense set 121
    optimization problem 122
control systems 255
conventional representation 138
covariance matrix adaptation evolution strategy, 126
crossover 16
crossover hill climbing, see recombination, hill climbing

Davidon-Fletcher-Powell method 40
decision space 4
decoding 13
dedication V, VII
differential evolution 129–131, 270
    mutation variants 130
    pseudocode 131
discrete optimization 73–94
Distance-Based information Preservation Crossover, 90
diversity 14, 83, 153–165
    χ measure 159
    ν measure 158
    ϕ measure 160
    ψ measure 159, 231
    τ_3 measure 161
    ξ measure 158, 257
    adaptive local search 155
    beta distribution 162
    crossover 155
    entropy 49
    exponential distribution 163

fitness diversity 157
    local search 155
    multi-search 156
    natura non facit saltus 162
    self-adaptation 155
    structured population 154
    truncation selection 154
domain decomposition 282
dominance 203
dynamic programming 190, 194
Dynasearch 193

encoding 13
engineering and design 241
    aerodynamic design 251, 253, 254
    antenna 247
    electrical and electronic engineering 247
        electic motors 248
        electromagnetism 250
        netwrok applications 250
        power systems 248
    electroenchephalogram 252
    filter design 251
    frequency modulation 246
    image processing 243
        forensic objects 244
        image registration 244
        tomography 244
    Internet applications 247
    radar design 246
    radio frequency assignment 246
    seismic analysis 251
    telecommunications 254
    thermal generator 251
engineering applications 241–260
environmental selection 17
epistasis 45
estimation of distribution 126
evaluation function 14
evaluation mechanism
    aggregation function 207
    dominance 206
evolution strategy 48, 124–125
    uncorrelated mutation 125
evolutionary algorithm 9–27
    Infeasibility Driven Evolutionary Algorithm, 142
    real coded 125

Subject Index

exploitation 18
exploration 18
exponential distribution 163
exponential time 62

fitness function 14
fitness landscape 45, 95–119, 171 208, 285
  basin of attraction 45
  distance 49
  distribution of local optima 50
  fitness distance correlation 101–103, 298
  multiobjective 208
  plateau 35
  random walk correlation 100
forma analysis 33, 82, 193, 296
Full Employment Theorem 46
fully polynomial time approximation scheme 299

gene 13
genetic algorithm 283
genotype space 13
genotypes 13
global optimization 123–131
gradient 122
graph coloring problems 87
GSAT 35

Hessian matrix 4
hill climbing 34–35
  crossover 133
  plateau 35
Hooke-Jeeves pattern search 230
Hopfield network 279
hybridization 189, 288
  collaborative models 192, 195–199
  exact techniques 189
  integrative models 192–195
  multilevel model 197
  taxonomy 191–192
  with backtracking 192
  with beam search 196, 198, 270
  with branch and bound 195
  with branch and cut 195, 196
  with evolution strategy 270
  with hill climbing 270
  with Hooke-Jeeves method 231

  with simulated annealing 231, 269, 284
hyperheuristics 51

ideal objective vector 204
IEMA 144
individual 13
infeasibility
  Infeasibility Driven Evolutionary Algorithm, 142
  Infeasibility Empowered Memetic Algorithm 144
initialization 17
innovative recombination 116
Ising bond 282
iterated local search 98

job-shop scheduling problem 147

Kauffman 101
knapsack problem 194, 196, 197
Kriging function 221

L-Systems 294
lagrangean relaxation 193
Lagrangian interpolation 221
Lamarckianism
  partial 194
Lin-Kernighan heuristic 49, 109, 298
linear programming 109
local branching 193
local optimum 123
local search 29–41, 48, 67, 78
  2-opt 50
  classification 31
  continuous domains 36–41
    classification 37
  depth 67
  frequency 69
  golden section search 270
  greedy 32
  iterated 98
  iterated local search 109
  Lin-Kernighan heuristic 49, 109, 298
  local optimum 31
  parameterized complexity 303
  parameters 50
  partial lamarckianism 50
  single vs multiple solution 32
  single-solution metaheuristics 33

Solis-Wets 269
  steepest ascent 32
  stochastic vs deterministic 31
locus 13

Markov blanket 265
mating selection 15
MAX-SAT problem 35, 195
maximum density still life problem 194, 199
maximum leaf spanning tree 301
maximum satisfiability 174
membrane computing 271
meme 175
memetic algorithm
  adaptive 51
  adaptive global-local 231
  co-evolution 51, 170–173
  combination with exact techniques 51
  combinatorial optimization 95
  complete 306
  continuous optimization 131
  design 49–50
  discrete local search 77
  discrete optimization 74
  fast adaptive 255, 257–258
  history 275–309
  Infeasibility Empowered Memetic Algorithm 144
  initial population 47
  local search 48
  meta-lamarckian learning 51
  metalamarckian 168
  multimeme 51, 168, 268
  multiobjective 50, 205–216
  need 44–46
  origins 275
  parallel 285
  Pareto archived evolution strategy 210
  performance measure 77
  philosophy 44, 98
  replacement 48
  reproduction 48
  restart 49
  self-adaptive 268
  self-generating 168
  template 46–49
  termination 47

memetic computing 51
metaheuristics 6–7
microarray analysis 262–266
minimum spanning tree 264
multi-layer perceptron network 231
multiobjective
  aggregation function 204
  archive of Pareto solutions 209
  evolutionary algorithm based on decomposition, 213
  fitness landscape 208
  genetic local search 210, 211
  ideal objective vector 204
  memetic algorithm 205–216
    engineering and design 252
  MOGLS 50
  MPAES 50
  nadir objective vector 204
  Pareto dominance 50, 203
  Pareto MA 50
  Pareto optimal solution 203
  Pareto optimal vector 203
  strength Pareto evolutionary algorithm 214
  Tchebycheff approach 205
  weighted sum approach 204
multiobjective optimization 201–217
mutation 15
  heavy 49

nadir objective vector 204
negative assortative mating 156
neighborhood 30
  binary 79
  combination 81
  design 78
  exploration 79
  integer 79
  permutation 33, 79
neighborhood generating function 171
neighborhood structure 45
neural network
  approximation 221
  Hopfield network 279
NEWUOA 40
NK-landscape 99–100
No Free Lunch Theorem 44, 58, 158, 208, 241
noisy problems, 229

Subject Index

explicit averaging   230
implicit averaging   230
NSGA-II   143, 211
number partitioning problem   298

objective function   14
offspring   15
operations research   50
optimization
   continuous   121–134
   discrete   73–94
   presence of uncertainties   219–237
optimization problem   3
optimum
   global   31
   local   31

P-systems   270
parameter tuning   25
parameterization   55–72
parameterized complexity   6, 190, 299–303, 305–309
   fixed-parameter tractable   6, 190, 303
   local search   303
   reduction rules   301
parent   15
parent selection   15
Pareto dominance   203
Pareto optimal solution   203
Pareto optimal solutions
   archive   209
Pareto optimal vector   203
particle swarm optimization   41, 127–129
   pseudocode   129
   variants   128
   velocity update, 128
path relinking   266
permanent magnet synchronous motor   255
permutation
   neighborhood   33
phenotype   13
phenotype space   13
polynomial local search   168
polynomial time   5, 62
polynomial time approximation scheme   109, 190
population   14

initialization   47
management   50
spatial structure   154, 284
population-based   12
Powell's algorithm   32, 39
predictive gradient   235
premature convergence   18, 48
prize-collecting Steiner tree problem   195
progressive neighborhood search   91

quadratic approximation   220
quadratic programming
   binary   111–118
   sequential   144

radial basis function   221
random walk   124
real coded evolutionary algorithms   125
Rec-I-DCM3   267
recombination   16
   BLX-α 126
   DPX   49
   dynastically optimal recombination   193
   hill climbing   133, 199
   optimal discrete   134
   parent centric   126
   PCX-α 126
replacement   17, 48
reproduction   48
representation   13
   alternative   140
   conventional   138
   non orthogonal   33
   orthogonal   33
restart   49
   heavy mutation   49
   random immigrant strategy   49
robust design   224

S-systems   270
saddle point   5
scatter search   266
self-adaptation   167–188
   meme coordination   176
   meme definition   176
   specific local search   171
self-adaptive MA   167
semantic combination operator   81
sequential quadratic programming   144

Shannon's entropy   49
shortest common supersequence
    problem   197
Simplex method   32
simplex method   38, 230
simulated annealing   35–36, 41, 231, 279
    cooling schedule   36
simultaneous perturbation stochastic
    approximation method   41
software agent   285
Solis and Wets' method   41
Steiner problem   90
stochastic global search   124
stop criterion   18
superpolynomial performance   66
survivor selection   17
symbols, list of   XXIII

tabu search   36, 88, 194, 268, 282
    aspiration criterion   36
    tabu list   36
    tenure   36
Tchebycheff approach   205
termination condition   18, 47
time complexity   61

time-dependency   232
    adaptive dual mapping   236
    triggered immigrants   236
trap function   176
travelling salesman problem   33, 49, 100, 105–111, 116, 193, 196, 279
    2-opt   50
    Lin-Kernighan heuristic   49, 109, 298
    very large instances   109
treewidth   299, 305
Trust region   40
Turing machine   5, 6

ultrametricity   286
uncertainties   219–237
unitation   174
Unweighted Pair Group Method with
    Arithmetic Mean, 91

Vapnik-Chervonenkis dimension   282
variable neighborhood search   168, 267
variation operators   15

Watson's Hierarchical-if-and-only-if   174

XHC   133

Printed by Publishers' Graphics LLC